# Frontiers in the History of Science

**Series Editor**

Vincenzo De Risi, Université Paris-Diderot – CNRS, PARIS CEDEX 13, Paris, France

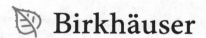

Frontiers in the History of Science is designed for publications of up-to-date research results encompassing all areas of history of science, primarily with a focus on the history of mathematics, physics, and their applications. Graduates and post-graduates as well as scientists will benefit from the selected and thoroughly peer-reviewed publications at the research frontiers of history of sciences and at interdisciplinary "frontiers": history of science crossing into neighboring fields such as history of epistemology, history of art, or history of culture. The series is curated by the Series Editor with the support of an international group of Associate Editors.

\* \* \*

**Series Editor:**
Vincenzo de Risi
Paris, France

**Associate Editors:**
Karine Chemla
Paris, France

Sven Dupré
Utrecht, The Netherlands

Moritz Epple
Frankfurt, Germany

Orna Harari
Tel Aviv, Israel

Dana Jalobeanu
Bucharest, Romania

Henrique Leitão
Lisboa, Portugal

David Marshal Miller
Ames, Iowa, USA

Aurélien Robert
Tours, France

Eric Schliesser
Amsterdam, The Netherlands

Michael Friedman

# Ramified Surfaces

## On Branch Curves and Algebraic Geometry in the 20th Century

  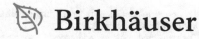

Michael Friedman
The Cohn Institute for the History and Philosophy of Science and Ideas, Humanities Faculty
Tel Aviv University
Tel Aviv, Israel

The author acknowledges the support of the Cluster of Excellence "Matters of Activity. Image Space Material" funded by the Deutsche Forschungsgemeinschaft (DFG, German Research Foundation) under Gemany's Excellence Strategy—EXC 2025—390648296.

ISSN 2662-2564          ISSN 2662-2572    (electronic)
Frontiers in the History of Science
ISBN 978-3-031-05719-9        ISBN 978-3-031-05720-5    (eBook)
https://doi.org/10.1007/978-3-031-05720-5

This book is published under the imprint Birkhäuser, www.birkhauser-science.com by the registered company Springer Nature Switzerland AG
The registered company address is: Gewerbestrasse 11, 6330 Cham, Switzerland

# Acknowledgements

I wish to thank and acknowledge the many scholars who have helped me in the process of writing this book. Vincenzo de Risi accompanied the conception and creation of this book from the beginning till its completion. Both Vincenzo de Risi and Moritz Epple gave extremely valuable comments and advice concerning the content and structure of the book, which helped to bring clarity, shape and form to it. I am also extremely grateful to Wolfgang Schäffner, who provided me with the space and time to pursue the research for this book at the Cluster of Excellence 'Matters of Activity', where a large part of it was written.

Moreover, the research on Boris Moishezon would not have been possible without the help of Mina Teicher; the numerous conversations with her led to thoughtful insights not only on his work and their mutual discoveries but also on the history of the research on branch curves in general, and I warmly thank her for that. I am also very grateful for the help given to me by the Tel Aviv University Archives, and especially from Ella Meirson and Gedalya Zhagov. Special thanks must also go to Donu Arapura, Ron Livne and Vitali Milman for the conversations on the work and life of Moishezon. Moreover, during my research on the Italian school of algebraic geometry I had various conversations with numerous scholars: Maria Dedò, Antonio Lanteri, Anatoly Libgober and Piera Manara helped me greatly understand the various mathematical configurations that were shaped not only during this period but also during other periods. I would also like to thank warmly Eliot Borenstein, Leo Corry, Christopher Hollings, François Lê, Klaus Volkert, Fernando Zalamea and the anonymous referees for important insights and advice. If I had forgotten anyone here, I deeply apologize.

Last but certainly not least, I thank Michael Lorber, who supported me constantly during the creation and writing of this book. Without his support, my own thought would have probably remained ramified and branched. Thank you.

# Contents

# Introduction

**1**

Monge, 1785: "The projection of a body's shadow on any surface is [...] the figure that the extensions of the rays of light tangent to the body's surface end on that surface. [...] In the following operations we will geometrically determine only the projections of the contours of the pure shadows, they are *the only ones* that it is necessary to have exactly in the drawings."[1]

At the end of the eighteenth century, the mathematician Gaspard Monge emphasized that to investigate surfaces properly, the "projections of the contours of the pure shadows" are the only curves necessary to draw accurately, and in a certain sense, whose properties are the only ones one should know exactly. An example of what should be drawn is given by Monge, as can be seen in Fig. 1.1, when the source of "light" is either a point or a spherical body. Monge's surfaces were real surfaces, that is, defined over the real numbers, and he also termed the "contour of the pure shadow" as "apparent contour."

Jumping to the twentieth century, when one takes these "surfaces" as complex and algebraic surfaces, embedded in a three-dimensional complex space, then the projection of this contour is called the 'branch curve.' Taking into consideration the fact that in the twenty-first century, Monge's requirements seem almost irrelevant, looking at the current research of complex algebraic surfaces, the question arises: What happened? How was this curve researched over decades, and how did its epistemic status change, especially during the twentieth century, in the then flourishing domain of algebraic geometry?

---

[1] "La projection de l'ombre d'un corps sur une surface quelconque est donc la figure que terminent sur cette surface les prolongements des rayons de lumière tangents à la surface du corps [...] Dans les opérations suivantes nous ne déterminerons géométriquement que les projections des contours des ombres pures, *ce sont les seules* qu'il soit nécessaire d'avoir exactement dans les dessins." (Monge 1847 [1785], p. 27, 29).

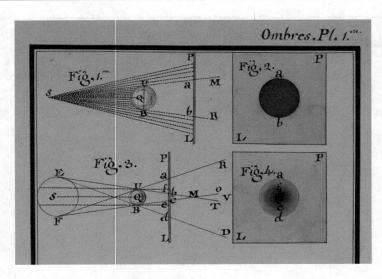

**Fig. 1.1** A part of the first plate from Monge's *Ombres*, from the 1780s. The circumference of the shape on Fig. 2 of the plate is an example of Monge's "contour of the pure shadow," today called the "branch curve." © Bibliothèque de l'Ecole polytechnique /Collections Ecole polytechnique/SABIX

This book will aim to answer these questions by presenting a certain cross section of the history of algebraic geometry during the twentieth century. I aim to show that the problem of how to define and characterize the branch curve has not only given rise to novel ways to consider algebraic surfaces and singular curves, but has also prompted research with new mathematical configurations. But in order to explicate what this curve is, which is the object of this book, let me take a step back, and instead of looking directly at complex algebraic surfaces, I will start by looking at complex algebraic curves. The next section— Sect. 1.1—will be somewhat technical, starting with the mathematical definition of branch points and branch curves, as I consider it essential to present, at the outset, a definition of the object of this book. Therefore, I ask the reader to bear with me while reading the next four pages.

## 1.1    On Branch Points and Branch Curves

In his 1851 dissertation, Bernhard Riemann (1826–1866) introduced the now well-known *Riemann surface*, defined as a covering of the complex (affine or projective) line for multi-valued analytical functions. In the following, I will present Riemann's results in *modern* mathematical language; below I will discuss Riemann's own formulations and present a more precise mathematical description (see Sect. 1.2.2).

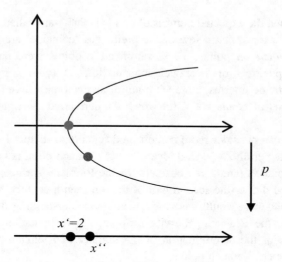

**Fig. 1.2** The real part of the curve $y^2 = x - 2$, the *ramification* point $(2,0)$ (blue point), and the preimages with respect to the projection $p$ of point $x''$, being close to the *branch* point $x' = 2$; these preimages are the two red points 'above' point $x''$ on the $x$-axis; one can imagine how these two red points 'coincide' into the blue point when $x''$ approaches $x'$. Hence one says that there are two *branches*, which 'come together' at the ramification point. The problem with this visualization is that it does not show how the complex part (or complex preimages) of the curve looks

Riemann considered complex algebraic functions of two variables as ramified coverings over a complex line. To consider this complex algebraic function—today considered as an algebraic curve—as a covering, means to look at an algebraic curve defined by $f(x, y) = 0$ of degree $n$, and to consider its projection $p$ to the $x$-axis:

$$p : \{(x, y) \in \mathbb{C}^2 : f(x, y) = 0\} \to \mathbb{C}, (x, y) \mapsto x.$$

A simple example would be to consider the function $y^2 = x - 2$, a function of degree 2; this function and its projection can be seen in Fig. 1.2. Note that for every $x' \neq 2$, the equation $(y')^2 = x' - 2$ has two solutions. Another way to formulate this is that with respect to the projection $p$, any point $x' \in \mathbb{C}$ (besides $x' = 2$) has two different preimages:[2] these are the points $(x', y_1)$ and $(x', y_2) \in \mathbb{C}^2$ when $(y_1)^2 = x' - 2$ and $(y_2)^2 = x' - 2$.

However, for $x' = 2$ the number of preimages is less than two—explicitly, there is only one preimage: $(2, 0)$, as can also be seen in Fig. 1.2. One might say that when considering the points $x'' \in \mathbb{C}$ which are close to $x' = 2$, the two preimages of $x''$ 'coincide' into one point when $x''$ approaches $x'$. These points, whose number of preimages is lower than the

---

[2] For $a \in \mathbb{C}$, the preimages $p^{-1}(a)$ of $a$ are all the points $(a, b)$ on the curve which are projected via $p$ to $a$. For example, $x' = 3$ has two preimages: $(3, 1)$ and $(3, -1)$, since $y' = \sqrt{3 - 2} = \sqrt{1} = \pm 1$.

expected one—when the expected number of points is $n$—are called *branch points*; the points on the curve for which a few of the preimages 'coincide' are called—in current terminology—*ramification* points. These ramification points were named by Riemann either "Windungspunkte"[3] or "Verzweigungspunkte,"[4] a terminology that I will also discuss below. In modern terms, a branch point of an algebraic curve $f(x, y) = 0$ can also be defined as the set of points $x \in \mathbb{C}$ for which the derivative $df/dx$ vanishes or does not exist.

Obviously one can consider more complicated functions—below I examine the curve $y^3 = x - 2$—but hopefully, a general idea of what a branch point is has been conveyed. Concerning how branch points can characterize the Riemann surface, several important results were proved during the second half of the nineteenth century, which will be also discussed below, and these results were fundamental to understanding algebraic curves and Riemann surfaces: the Riemann–Hurwitz formula, the computation of the moduli of Riemann surfaces, or the determination of the number of Riemann surfaces, branched along a given number of branch points.

$$* \; * \; *$$

This short, very partial and ahistorical overview of branch points implies that these points already played a central role in the research on coverings of the complex line $\mathbb{C}$. The question hence arises: What happens to the function and the epistemic status of these branch points when one increases the complex dimension by 1, that is, when one looks not at complex algebraic curves, but rather at complex algebraic *surfaces*, and considers their projection to the complex plane $\mathbb{C}^2$ and the resulting *branch curve*?[5] The answer is naively surprising: things get complicated.

But why? To answer this question, it might be worth noting that on the face of it, the definitions of a branch point and branch curve are surprisingly similar. Indeed, one may take a complex algebraic surface $S$ of degree $n$ defined by $f(x, y, z) = 0$,[6] and consider its projection $p$ to the complex plane $\mathbb{C}^2$:

$$p : \big\{ (x, y, z) \in \mathbb{C}^3 : f(x, y, z) = 0 \big\} \to \mathbb{C}^2, (x, y, z) \mapsto (x, y).$$

---

[3] In English: "turning points". The term was coined in 1851; see: Riemann (1851, p. 8). Note that the translation into English in 2004 translated "Windungspunkte" into "branch points" (Baker et al. 2004, p. 6).

[4] In English: "branch points". The term was coined in 1857; see e.g. Riemann (1857, p. 107). The different sheets of the function in the neighborhood of such a point are called "branches [Zweige]".

[5] A similar discussion can be presented regarding the complex projective plane. I will also discuss shortly below the situation when increasing the *real* dimension by 1, i.e. when one considers covers either of $\mathbb{R}^3$ or of the 3-sphere.

[6] The surface is considered as embedded in the 3-dimensional complex space. Similar definitions exist for algebraic surfaces embedded in an $m$-dimensional (projective) complex space.

The branch curve $B$ is defined as the set of all points on $\mathbb{C}^2$ for which some of the preimages on the algebraic surface $S$ 'coincide':

$$B = \big\{(x,y) \in \mathbb{C}^2 : (x,y) \text{ has less than } n \text{ preimages on } S\big\}.$$

Over the real numbers, when considering the surface as embedded in the three-dimensional *real* space, one may view such projections as a modern reformulation of Monge's proposal from 1785, to examine a "projection of a body's shadow" and its "contours." In more modern terms, given the surface $f(x,y,z) = 0$, one defines the branch curve $B$ as the set of all points $(x,y)$ in $\mathbb{C}^2$ for which the derivative *df/dz* vanishes or does not exist. Or, one can define the *ramification curve R* as the intersection of $\{f = 0\}$ and $\{df/dz = 0\}$, and then the *branch curve B* as the image of $R$ under $p$—as can be seen in Fig. 1.3a. Moreover, the curve $B$, as it turns out, is *singular*, having (generically) only nodes and cusps as singularities.[7] An example of a branch curve of a cubic surface is depicted in Fig. 1.3b.

So far, the definitions of branch point and branch curve are similar. Therefore, why do 'things get complicated'? Indeed, already at the end of the nineteenth century, with the research of Italian algebraic geometers Guido Castelnuovo and Federigo Enriques, it became evident, as I will elaborate below, that one cannot 'transfer' in an analogous way results from the research on algebraic curves to algebraic surfaces; or, more concretely, one cannot assume that results about branch points can be simply 'transferred' to branch curves.

So how is one to characterize complex algebraic surfaces? And more importantly, can the branch curve help with this inquiry in the same way that branch points illuminate essential aspects of Riemann surfaces and algebraic curves? In contrast to the situation with Riemann surfaces, where the central role of branch points was made clear by Riemann in the early years of his research, how branch curves were considered and what role they played have a more complicated history. Indeed, the definition of the branch curve given above was not the only one employed during the nineteenth and twentieth centuries; this implies—and this is one of the core claims of this book—that there was not only one, single 'branch curve' that was merely presented and researched in various ways over the decades.

This claim should be explicated: since *branch points* were considered as a way to characterize Riemann surfaces, one might have thought that *branch curves* could be used to characterize complex algebraic surfaces. However, as it turned out during the twentieth century, such a simple characterization was neither obvious nor immediate, partially because algebraic surfaces turned out to be more complicated objects than algebraic curves, and partially because branch curves—being curves with nodes and cusps—were discovered to be a special subset in the set of nodal–cuspidal plane curves. The present book aims to show that the problem of how to characterize this curve and its relation to the corresponding branched surface has given rise not only to novel ways of considering

---

[7] At a *node*, one has locally a curve of the form $xy = 0$. At a *cusp*, one has locally a curve of the form $y^2 = x^3$.

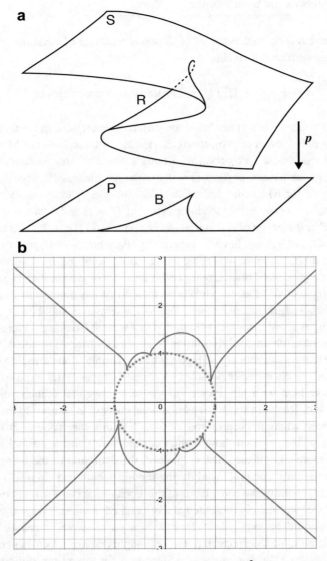

**Fig. 1.3** (**a**) A surface $S$, its projection to a complex plane $P = \mathbb{C}^2$ and the cuspidal branch (resp. ramification) curve $B$ (resp. $R$). (**b**) Given the projection $p$, the figure presents the real part of the branch curve (in red) of a smooth complex surface of degree 3: $f(z) = z^3 - 3az + b$, where $a = \{x^2 + y^2 - 1\}$, $b = \{y - 5(x^3 - 3x/4)\}$. The branch curve has six cusps, and these cusps lie on a conic $\{a = 0\}$ (in blue) (drawn by M.F. with https://www.desmos.com/calculator). This drawing (or any attempt to sketch this curve and the special position of the cusps) did not appear in any of the papers dealing with the subject during the twentieth century

algebraic surfaces and nodal–cuspidal curves, but has also prompted research with and within several mathematical domains that were not previously connected to the research on branch points and Riemann surfaces. In this sense, one can consider the research on branch curves as a cross-section through the history of algebraic geometry of the twentieth century, one that shows some of the main transitions and developments in this field. As will be explicated below, the object itself called 'branch curve' has changed through the decades, and it has been relocated and redefined within various mathematical settings, or more precisely, within various mathematical configurations. Hence, and this is essential to remember throughout this book, to talk about 'the' branch curve is highly misleading.

## 1.2   Dynamics of a Mathematical Object

> Enriques, 1949: "[...] it used to be said that while algebraic curves (already composed in a harmonic theory) are created by God, surfaces, instead, are the work of the devil. Now, on the contrary, it is clear that God chose to create for surfaces an order of more hidden harmonies, where a wonderful beauty shines forth [...]"[8]

As I have already implied, the research on algebraic surfaces turned out not to be analogous to the one on algebraic curves. The citation above from Enriques's *Le superficie algebriche* exemplifies this all too well, and we will see a few examples of this situation below. Moreover, the search for the "hidden harmonies" of surfaces was, one might say, in a constant process of being reformulated and reshaped. Concentrating on the branch curve as one of the objects used to detect those harmonies, I aim to show that while a plurality of approaches for investigating this curve was employed, those various approaches did not necessarily lead to the anticipated harmony, but rather forced shifts in context (or themselves underwent such shifts), sometimes overshadowing each other—or even prompting a marginalization of the original object of study, that is, of the branch curve itself. These shifts can be considered to have occurred in both the body of mathematical knowledge and the image of mathematical knowledge.

Here I employ the distinction introduced by Leo Corry between body and image of mathematical knowledge:[9] statements included in the body of knowledge are about the subject matter of the discipline involved, where these may be theories, conjectures, methods, problems, proofs, etc. Statements belonging to the image of knowledge function

---

[8] Enriques (1949, p. 464): "[...] si soleva dire che, mentre le curve algebriche (già composte in una teoria armonica) sono create da Dio, le superfici sono opera del Demonio. Ora si palesa invece che piacquea Dio di creare per le superficie un ordine di armonie più riposte ove rifulge una meravigliosa bellezza [...]"

[9] See: Corry (1989, 2004).

as "guiding principles or selectors,"[10] and answer questions about the discipline as such. These questions may be about authority, the correct and valid methods and proofs that can be used, methodology, and what, how and with whom one should investigate.[11] Though this distinction is essential for analytical purposes, Corry stresses not only that the "body and the image of mathematics appear as organically interconnected domains in the actual history of the discipline,"[12] but also that one should analyze "the subsequent transformations in both the body and the images of mathematics."[13] To examine these "subsequent transformations," I will frame the various research settings of the branch curve in terms of ephemeral local epistemic configurations—a term that I will explain further on. But first, to see how the branch curve reflects those shifts and transitions in the field of algebraic geometry, a field famous for being rewritten at least twice during the twentieth century—first by Oscar Zariski and André Weil and then by Alexander Grothendieck—it is instructive to look at several of the definitions given for the ramification and branch curves starting in the middle of the nineteenth century.

The following three definitions will be explicated throughout the various chapters of this book; the point of bringing them up now is not to demand or expect that the reader understand them mathematically, but rather to present the variety of mathematical settings in which the object called 'branch curve' was introduced and defined.

1. As we will see in Sect. 2.2, one of the common definitions of the branch curve was obtained by considering a surface $S$, defined by an equation $\{f = 0\}$, embedded in $\mathbb{C}^3$ (resp. $\mathbb{CP}^3$), and projecting it to the complex (projective) plane $\mathbb{C}^2$ (resp. $\mathbb{CP}^2$) from a point. The point was usually not on the surface, though one also considered projections of surfaces when the point of projection was on those surfaces. The surface itself could be singular, having either isolated singularities (such as nodes) or even a double curve. A concrete example of the calculation of the branch curve was given in Fig. 1.3b for a smooth cubic surface, by taking the projection of the intersection of $\{f = 0\}$ and $\{df/dz = 0\}$. During the 1930s, for example, Zariski took this method as the definition of the ramification curve.[14]

One indeed may think of this projection, and specifically, of the ramification curve, by considering drawing tangent lines to the surface, exiting from the given point. This method was made explicit by Monge, as we saw above. The ramification curve is then the

---

[10] Corry (1989, p. 411).

[11] Corry claims moreover that in "the particular case of mathematics, it brings to the fore a peculiar trait of this discipline, which will be of special interest for the discussion advanced in the present book: the possibility of formulating and proving metastatements about the discipline of mathematics, from within the body of mathematical knowledge." (Corry 2004, p. 4) Corry takes and extends this distinction from (Elkana 1981).

[12] Corry (2004, p. 5).

[13] Ibid.

[14] See: Zariski (1929, p. 306; 1935, p. 160).

collection of all tangent points.[15] Moreover, there are two special kinds of tangent lines: the first are tangent to the surface at two different points of it, hence corresponding to nodes of the branch curve; the second are also tangent to the ramification curve, hence corresponding to cusps of the branch curve (see Fig. 1.3a). Moreover, if the surface has isolated singularities, the branch curve has these singularities as well.

2. Other definitions slowly began to emerge during the 1950s, reflecting the algebraic rewriting of algebraic geometry. In 1958, Zariski gave a purely algebraic definition of a ramified point (with respect to a covering), using the machinery of ring and ideal theory, working with "an absolutely irreducible, $r$-dimensional normal algebraic variety" denoted by $V$, defined over "an arbitrary ground field" $k$. Taking $V^*$ as the normalization of $V$, Zariski then defined the following:

> Let $P^*$ be an arbitrary point of $V^*$ [...], and let $P$ be the corresponding point of $V$. We denote by $\mathfrak{o}$ the local ring of $P$ on $V/k$ and by $\mathfrak{m}$ the maximal ideal of $\mathfrak{o}$. Let $\mathfrak{o}^*$ and $\mathfrak{m}^*$ have a similar meaning for $P^*$ and $V^*/k^*$. It is well known that: (1) $\mathfrak{o}^*\mathfrak{m}$ is a primary ideal, with $\mathfrak{m}^*$ as associated prime ideal; (2) the residue field $k^*(P^*)$ $(= \mathfrak{o}^*/\mathfrak{m}^*)$ is a-finite algebraic extension of the field $k(P)$ $(= \mathfrak{o}/\mathfrak{m})$.

> *Definition*: The point $P^*$ is said to be *unramified* (with respect to $V$) if the following conditions are satisfied:

> > (a)  $\mathfrak{o}^*\mathfrak{m} = \mathfrak{m}^*$;
> > (b)  $k^*(P^*)$ is a separable extension of $k(P)$.

In the contrary case $P^*$ is said to be *ramified* (with respect to $V$).[16]

Any trace of the older definition of ramification (point, curve or variety) is hardly to be noticed at first glance.

3. In 1984, one finds the following definition in the book *Compact Complex Surfaces*, given now in another language, that of sheaf theory and canonical sections. The definition itself is as follows: for $p : X \to Y$ a covering, "let us assume that both $X$ and $Y$ are manifolds. Then the set of ramification points is the zero divisor $R$ of the canonical section in $\mathrm{Hom}(p^*(\mathcal{K}_Y), \mathcal{K}_X)$, i.e. $\mathcal{K}_X = p^*(\mathcal{K}_Y) \otimes O_X(R)$,"[17] where, for a variety $V$, $\mathcal{K}_V$ is defined by "$\mathcal{K}_V = \bigwedge^n \mathcal{T}_V^\vee$ is the canonical bundle of $V$."[18]

---

[15] Note that the tangent lines may also intersect the surface at other points. Note also that when the surface has a double curve, then this double curve does not count as a component of the ramification curve.

[16] Zariski (1958, p. 791; cursive by M.F.)

[17] Barth / Peters / Van de Ven (1984, p. 41).

[18] Ibid., p. 1.

There are and were other definitions as well,[19] along with variations of the above definitions; one of the common ways at the beginning of the twentieth century to construct surfaces with certain desired properties was to begin with a curve $f(x, y) = 0$, usually singular, and consider the double cover $z^2 = f(x, y)$ of the plane (also called "multiple plane"), ramified over this curve, or even the cyclic cover $z^n = f(x, y)$. Here one did not need to define the branch curve—it was already given by the construction, being the curve $\{f = 0\}$.

### 1.2.1    Ephemeral Epistemic Configurations and the Identity of the Mathematical Objects

What does this plurality of definitions imply? First, one should note that it implies, to follow one of the main insights of Lakatos,[20] that mathematical concepts and objects are always in a process of development and change, and no specific definition can capture the "essence" of the object (in our case, the branch curve). If such a mathematical object had a well-defined, static definition, one would be able to use it as a technical thing, to follow Hans-Jörg Rheinberger's differentiation between technical and epistemic things, the latter being objects of research.[21] If one focuses on mathematical research, it is instructive to note Moritz Epple's reflections on the various definitions of knots during the nineteenth and twentieth centuries. Concerning these definitions, Epple claims, it is clear that not only "[n]one of these definitions makes sense in mathematical practice without a technical framework,"[22] but also that "[t]he dynamics of the epistemic objects of mathematical

---

[19] See for example the definition of Grothendieck from the 1960s (Sect. 5.1). Here is the definition given in (Vakil 2017, p. 588): "Suppose $\pi : X \to Y$ is a morphism of schemes. The support of the quasicoherent sheaf $\Omega_\pi = \Omega_{X/Y}$ is called the *ramification locus*, and the image of its support, $\pi(Supp\ \Omega_{X/Y})$ is called the *branch locus*. If $\Omega_\pi = 0$, we say that $\pi$ is formally unramified, and if $\pi$ is also furthermore locally of finite type, we say $\pi$ is unramified."

[20] Lakatos (1976).

[21] According to Rheinberger epistemic objects possess an "irreducible vagueness." They are the object of research, and as they are in the process of being researched, their "vagueness is inevitable because, paradoxically, epistemic things embody what one does not yet know." (Rheinberger 1997, p. 28) These objects, their purpose, or the field of research that they open and the questions that they may propose are not yet defined or not yet canonically categorized. This is exactly what makes them into an epistemic object, being in a process of becoming 'well-defined' or 'stable' objects, while at the same time presenting an epistemic openness, in the sense that the questions resulting from the research on them are still open. In contrast, experimental conditions and technical objects "tend to be characteristically determined within the given standards of purity and precision. [...] they restrict and constrain" the scientific objects (ibid., p. 29). But while it seems that there is a clear distinction between the not yet defined epistemic object and the clearly defined technical one, Rheinberger stresses that the "difference between experimental conditions and epistemic things [...] is functional rather than structural." (Ibid., p. 30)

[22] Epple (2011, p. 485).

research [. . .] are secondary to the dynamics of the epistemic configurations as a whole."[23] What is meant here by "epistemic configuration"?

An epistemic mathematical configuration, to follow Epple,[24] is an array of epistemic mathematical objects, researched by a group of mathematicians in a certain, specific temporal and geographical setting, as well as of techniques developed to study those objects. An epistemic configuration is hence dependent on time and place; this, in turn, underlines two aspects: first, the *dynamic* character of the research done, dynamism being very much inherent to the epistemic objects themselves; second, the *local* nature of these configurations, which can be temporal, geographical or social. To stress the first aspect, the dynamical process does not necessarily entail the continuous development of mathematical knowledge: it may result in dead ends, unsolved problems or supplying wrong proofs, or in declarations about the image of the researched configuration that, for example, older research domains are obsolete. Following from this explanation of local epistemic configurations, it becomes clear that each such configuration should be considered an interweaving of local bodies and images of mathematical knowledge. While it may seem that epistemic configurations concentrate only on statements within the body of mathematical knowledge, each particular configuration should be considered with its accompanying image of knowledge. Moreover, every local configuration may have its own (sometimes or often implicit) criteria of what can and should count as a mathematical proof, which may lead to proofs that either have gaps or introduce fallible reasoning in some mathematical arguments.[25] Explicitly, these criteria may prompt the fallibility of either a proof or a justification, or a consideration of judgments of proofs as invalid—and indeed, we will see these phenomena in several of our configurations.

The emphasis on the dynamics of these configurations highlights that they are neither fixed nor static. This also indicates that the history unfolded here, which can be described as the history of various epistemic configurations of branch curves, should not be viewed as an 'internal' one; that is, the 'external' material, social and political conditions and events must also be taken into account when considering how those epistemic configurations transformed and emerged, since these conditions are sometimes the reason for their transformation. This is why, for some of the mathematicians discussed here, the unique social circumstances of the corresponding epistemic configurations are described, as they were very much relevant to the emergence and development of those mathematical configurations. As will become clear throughout the course of this book, several transformations between these configurations were also prompted by forced social and political changes: exile, captivity or emigration.

---

[23] Ibid., p. 488.

[24] See: (Epple 2004).

[25] On fallibilist account of mathematics, see the latest paper of Silvia de Toffoli (2021), who argues for "an alternative to the standard position in the philosophy of mathematics according to which justification requires a genuine proof and is therefore infallible." (ibid., p. 842)

If we return to the three definitions of branch curve given above, one could certainly claim that they are couched in different epistemic configurations. Taken separately, it may seem that none of those definitions of the branch curve has any connection with the other. But it is precisely the above three definitions that exemplify the shifts and transitions of the configurations: it is clear that without understanding "the dynamics of the epistemic configurations," that is, *not only* how frameworks of research on surfaces and curves in particular and algebraic geometry in general changed, *but also* how the image of algebraic geometry was transformed, one cannot comprehend the modifications occurring in the definitions of the branch curve, or the change in its status.

* * *

This discussion on the changes of mathematical objects and configurations is of course not new, as can be seen with Epple's analysis. This dynamic nature is expressed by Fernando Zalamea, who suggests the following, concerning how to think on the history of a mathematical object: "It is not that there exists 'one' fixed mathematical object that could be brought to life independently of the others [. . .]; instead [there] is the plural existence of webs incessantly evolving as they connect with new universes of mathematical interpretation."[26] Norbert Schappacher highlights this "plural existence of [mathematical] webs" when he describes the "explicit transformations of epistemic objects and techniques." Schappacher stresses the "notion of rewriting [. . .] [at a] microhistorical level,"[27] a notion which already underlines the processes of transformation of epistemic configurations. The emphasis on changes detected when focusing on microhistory is also noted by Catherine Goldstein: "the focus of recent history of mathematics has been much more on localised issues, short-term interests and ephemeral situations, on 'the era which produced' the mathematics in question; and moreover it has centred on diversity, differences and changes."[28] Ephemeral configuration, she continues, "implies links with social situations which have their own time scale and are most certainly not 'eternal truths'."[29] That being said, it must therefore be emphasized that two ephemeral configurations researching the 'same' object do not have to refer to each other, even if the second is subsequent to the first. Recognizing this, one may think of the interlacements of epistemic configurations as a braid of threads linking different times and contexts. The metaphor of the braid, as noted by Frédéric Brechenmacher, raises the notion of reconstructing the dynamics of knowledge through the multiplicity of its ephemeral configurations. This multiplicity amounts to establishing a mathematical object, concept, theorem, etc., historically by asking, again and again, about their identity.[30] To explicate: the focus on ephemeral configurations deconstructs the question, or rather the

---

[26]Zalamea (2012, p. 272–273).

[27]Schappacher (2011, p. 3260).

[28]Goldstein (2018, p. 489).

[29]Ibid., p. 490.

[30]Brechenmacher (2006, p. 1).

presupposition, of the identity of the mathematical object. Goldstein emphasizes the difficulty involved with such historiographic constructions in establishing the identity of a mathematical object, as they presuppose the category of the 'same' (as if in very different expressions one can recognize the same truths), not only at a given time, but also between periods.[31] The paradox is that the possibility of writing a history of a mathematical discovery is always based on the identification of certain identities between mathematical objects or domains: "writing the history of algebra presupposes the identification of what algebra is, or at least what could be part of this particular history."[32] Or, writing the history of 'the' branch curve presupposes the identification of what a 'branch curve' is and which methods, modes of reasoning and techniques one may be permitted to use. But it is precisely this identification that is a part of the image of the configuration, an image which is itself also ephemeral and subject to change.

There are numerous works on the history of mathematics that deconstruct this alleged retrospective identity of the mathematical object. Christian Gilain examines two versions of the fundamental theorem of algebra which are considered the same, and notes that the two "are fundamentally situated on distinct historical axes; their stories have neither the same origin, nor the same rhythm, nor the same duration."[33] Frédéric Brechenmacher analyses the dispute between Leopold Kronecker and Camille Jordan in 1874 about what we now see as the *same* reduction theorem for matrices.[34] Goldstein, in her work on a certain theorem of Fermat and its readers,[35] frames the question of identity by considering two proofs of this theorem, stressing that one should not consider mathematical knowledge as fixed and persistent. The presupposition of identity should be considered, according to Goldstein, as a problem, and not as a tautology; only in this way do the various epistemic configurations attest to the various practices involved.[36] Moreover, Goldstein stresses that seeing identity as tautology hinders understanding the process of mathematical creation.[37]

These studies hence pose the question of the identity of a mathematical object as it persists and is resituated in various ephemeral epistemic configurations. In the context of the current research, I claim that one can talk about ephemeral configurations of the research on branch curves, as most of the configurations studied here were in a process of transformation and re-evaluation, and were in a sense precarious, for a number of

---

[31] Goldstein (2010, p. 138–139).

[32] Ibid., p. 139.

[33] Gilain (1991, p. 121).

[34] Brechenmacher (2007).

[35] The theorem is that there is no right-angled triangle with rational sides whose area is a rational square.

[36] Goldstein (1995, p. 16).

[37] Ibid., p. 178. Other studies which pose the question of identity of a theorem or a concept at the centre are for example the study of Hourya Sinaceur on two versions of Sturm's theorem (Sinaceur 1991), or the recent research of François Lê (2020) on the different, though not unrelated terms: *genre*, *Geschlecht* and *connectivity* of an algebraic curve during the 1870s and the 1880s.

reasons that are unique to each particular configuration. Here, precarious configuration does not necessarily mean short-lived or marginal (though this may indeed be the case); it means much more, that each local configuration possesses its own temporality: "the time of mathematics, far from being a linear succession of events, possesses a dynamics."[38] Moreover, even when a relatively big community shares results, practices and objects associated to a certain domain—and hence "mathematics acquires its immanence only once when this process is completed"—this does not mean that the theory stops developing and that local configurations stop emerging or being transformed.[39]

Returning to the question of the identity of the mathematical object, as noted above, this question becomes secondary to the dynamic of the epistemic configuration. In other words, the formation of concepts is shaped and conditioned by reorganizations of ephemeral epistemic configurations. In the process of the consolidation of epistemic configurations, mathematical objects are being individuated and also consolidated as such (i.e., as objects) and as participating in the configuration. This process is essentially dynamic: epistemic configurations can be transformed and reorganized, or they can be extended or reduced when encountering new techniques, new notations or new images of knowledge, which in turn may lead to a reorganization of the very configuration in question.[40]

Understanding the dynamics of those configurations, in which the research on branch curves was situated, also implies that the goal of the current book is not to write a 'long-term' history of 'the' branch curve, as if there were a fixed, technical object, found beyond the various transformations of the local configurations.[41] One can think of the study presented here as considering the branch curve in a cross-section through the history of algebraic geometry of the twentieth century. Or rather, to employ a different, more geographical metaphor, the branch curve can be considered as found at the intersection of several research approaches; this also emphasizes the fact that there were various methods, existing at the same time, to investigate this curve. This can be seen when one notices that the branch curve sometimes functioned as an object of research, but at other times was used as a tool to research other objects.

To return to the metaphor of a geological cross-section, this cross-section does not imply capturing all layers in the history of algebraic geometry. This is also emphasized by concentrating on local ephemeral configurations, since locality does not necessarily imply that every aspect of a broader transformation in the field of algebraic geometry— to give one example, the algebraic rewriting project of algebraic geometry of Weil and Zariski during the 1940s and the 1950s—will be present in every local configuration examined here during this time period. Just as one may not take into account every geographical area when performing a cross-section, so too the branch curve is not present

---

[38] Ehrhardt (2012, p. 252).

[39] Ibid., p. 252–253.

[40] Cf. Feest / Sturm (2011, p. 294).

[41] See: Goldstein (1995, p. 179).

in every development of algebraic geometry. More to the point, not every study on surfaces in algebraic geometry saw branch curves as a way to characterize them. Moreover, this does not imply that every community of mathematicians dealing with branch curves was a leading community in algebraic geometry, just as a cross-section can also capture minor, smaller or thinner layers between the major ones. In this sense, unfolding the history of the research on branch curves also reveals a history of more minor traditions in algebraic geometry—or even marginal ones, and how they may (or may not) reflect (at the same point in time) or join (at a later point in time) the more major traditions. In this sense, the claim of this book is that the various branch curves and the way they were considered and researched during the twentieth century may serve as a touchstone for several of the changes and revolutions in algebraic geometry.

### 1.2.2  On Branch Points, Again: on Riemann's Terminology and How (Not) to Transfer Results

It is time to examine more closely the relations between branch points and branch curves and their corresponding configurations. Moreover, to delineate how the various epistemic configurations dealing with branch curves operated and were transformed, it is instructive to first return to Riemann's own research and terminology, and then to briefly examine whether the results and terms were successfully transferred to the research configurations dealing with branch curves.

(i) Riemann's formulation

The definition given in Sect. 1.1 is very much a modern one. From the description presented there, one might already have guessed that this was not the terminology used by Riemann.[42] As François Lê notes, "there is no algebraic curve in Riemann's memoir[s]. [...] Riemann only talked about algebraic equations, and never interpreted them as equations defining algebraic curves. Further, his proofs did not involve any other object related to algebraic curves, like tangents or cusps."[43] So how did Riemann describe coverings and branch points, and what did he term them?

Considering an algebraic function with two variables $x$ and $y$, in 1851, Riemann characterized a covering as follows: "[...] we permit $x, y$ to vary only over a finite region. The position of the point 0 is no longer considered as being in the plane $A$ [i.e., on the complex line], but in a surface $T$ spread out over the plane. We choose this wording since it

---

[42] For an extensive survey of Riemann's work and the responses to it, see for example (Gray 2015b, p. 153–194); see also (Bottazzini / Gray 2013, p. 259–341) for a similar discussion. On the development of the concept of manifold in Riemann's work and his concept of covering, see e.g., (Scholz 1999, p. 26–30). See also (Scholz 1980).

[43] Lê (2020, p. 78). Lê refers to Riemann's memoir from 1857 on Abelian functions: (Riemann 1857).

is inoffensive to speak of one surface lying on another, to leave open the possibility that the position of 0 can extend more than once over a given part of the plane [...]."[44] By the last sentence, Riemann meant that the degree $n$ of the covering may be greater than 1. Moreover, taking, for example, the multi-valued function $y = \sqrt[n]{(x - x')}$, one notes that the $n$ preimages of points in the neighborhood of $x'$ in fact come together when approaching $x'$. For example, the function $y = \sqrt[3]{(x - 2)}$ is of degree 3 and has a branch point at $x = 2$. The three preimages of points in the neighborhood of $x' = 2$ come together when approaching $x' = 2$.[45] The question that arises is how to describe the behavior of such a function, viewed as a covering, in the neighborhood of such a (preimage of such a) branch point.[46] An explicit description of this situation was given by Riemann in 1857.

Given a function $F(s, z) = 0$—when the degree of $s$ is $n$ and the degree of $z$ is $m$ and when $s$ is branched,[47] Riemann defined a simple "branch point" and a "branch point of multiplicity $\mu + 1$" as follows:

"A point of the surface $T$ at which only two branches are connected, so that one branch continues into the other and vice versa around this point, is called a *simple branch point* [*einfacher Verzweigungspunkt*].
    A point of the surface around which it winds [*windet*] $\mu + 1$ times can then be regarded as the equivalent of $\mu$ coincident (or infinitely near) simple branch points."[48]

Riemann specified that one can associate a permutation to each branch point that describes how the preimages of points (or more accurately, the sheets of the Riemann surface) near branch points "behave" and interchange. According to Riemann, above every point $z' \in \mathbb{C}$ (which is not an image of a branch point) there are $n$ preimages, i.e., $n$ values of $s$, being

---

[44] Riemann (1851, p. 7): "Für die folgenden Betrachtungen beschranken wir die Veränderlichkeit der Grössen $x$, $y$ auf ein endliches Gebiet, indem wir als Ort des Punktes 0 nicht mehr die Ebene A selbst, sondern eine über dieselbe ausgebreitete Fläche $T$ betrachten. Wir wählen diese Einkleidung, bei der es unanstössig sein wird, von auf einander liegenden Flächen zu reden, um die Möglichkeit offen zu lassen, dass der Ort des Punktes 0 über denselben Theil der Ebene sich mehrfach erstrecke [...]." Translation taken from: (Baker et al. 2004, p. 4).

[45] For example, for $x_0 = 3$, then $y_0 = \sqrt[3]{(3 - 2)} = \sqrt[3]{1}$, an equation which has (over the complex numbers) three solutions: $1, \frac{1}{2} - \frac{\sqrt{3}}{2}i$ and $-\frac{1}{2} + \frac{\sqrt{3}}{2}i$, and hence $p^{-1}(x_0)$ has three preimages.

[46] For Riemann, "branch point" often designated the preimage of the branch point on the surface (i.e. in modern terms, the *ramification* point).

[47] Baker et al. (2004, p. 101): "Let us now suppose that the irreducible equation $F(s^n, z^m) = 0$ has been given and that we have to determine the branching of the function $s$." (Riemann 1857, p. 110: "Es sei jetzt die irreductible Gleichung $F(s^n, z^m) = 0$ gegeben und die Art der Verzweigung der Function $s$ [...] zu bestimmen.")

[48] Ibid.: "Ein Punkt der Fläche $T$, in welchem nur zwei Zweige einer Function zusammenhängen, so dass sich um diesen Punkt der erste in den zweiten und dieser in jenen fortsetzt, heisse ein *einfacher Verzweigungspunkt*. Ein Punkt der Fläche, um welchen sie sich ($\mu + 1$) mal windet, kann dann angesehen werden als $\mu$ zusammengefallene (oder unendlich nahe) einfache Verzweigungspunkte." Translation taken from: (Baker et al. 2004, p. 101).

**Fig. 1.4** A model presenting a local neighborhood of a branch point of an algebraic curve $y = f(x)$; since the algebraic curve under consideration is embedded in the four-dimensional space $\mathbb{C}^2$, only a three-dimensional section of it can be presented. The model shows the three-dimensional section $\mathbb{C} \times Re\,(\mathbb{C})$; explicitly, what is modeled is a surface (over the real numbers) whose coordinates are $(Re(x), Im(x), Re(f(x)))$. The presented model is that of a "turning point" of the second order on a Riemann surface ("Riemann'sche Fläche [...] mit 1 Windungspunkt 2. Ordn. (dreiblättrig)"), built around the 1880s; described in Brill (1888, p. 48). © 2022 Model collection of the Mathematical Institute, Göttingen University

$s_1, \ldots, s_n$. If $z'$ is in the neighborhood of an image of a branch point of multiplicity $\mu + 1$, Riemann noted that $\mu + 1$ preimages $s_1, \ldots, s_{\mu + 1}$ coincide when $z'$ approaches the branch point. Moreover, when encircling the (image of the) branch point at the $z$-plane,[49] one obtains a cyclic permutation of the $\mu + 1$ preimages on the surface $T$.[50] Explicitly, Riemann stated that the following permutation results from this encircling: the preimages $s_1, s_2, \ldots, s_\mu, s_{\mu + 1}$ permute into $s_2, s_3, \ldots, s_{\mu + 1}, s_1$. Fig. 1.4 shows a plaster model from the 1880s, visualizing a three-dimensional section of a branch point of the second order on a Riemann surface.[51] In addition, Riemann specified that for a simple branch point, only $\frac{\partial F}{\partial s} = 0$, while $\frac{\partial F}{\partial z}$ and $\frac{\partial^2 F}{\partial s^2}$ do not vanish, thus distinguishing a simple branch point from a double branch point.[52]

---

[49] In Riemann's words: "nach einem positiven Umlaufe" (Riemann 1857, p. 110).

[50] Ibid. For example, as was noted above, the function $y = \sqrt[3]{(x - 2)}$ has a branch point at $x = 2$; if $x_0 = 3$, $p^{-1}(x_0)$ has three preimages. Therefore, to a small loop around $x = 2$, starting at $x_0$, corresponds a cyclic permutation of the three preimages of $x_0$ on the curve.

[51] For the various visualization techniques that Riemann used to illustrate branch points, see Friedman (2019a, p. 115–121). See also (ibid., p. 112–146) for how branch points were visualized by various mathematicians during the second half of the nineteenth century.

[52] Riemann (1857, p. 111).

This overview clarifies the important place of branch points in Riemann's research. But while this analytical investigation was (and still is) presented as one of the main ways to consider and calculate branch points, Riemann was also interested in how the branch points themselves determine the covering. To give only two examples: first, Riemann took an interest in the numerical relations between the different invariants of the curve, noting that $2(p-1) = w - 2n$, where $n$ is the degree of a covering $S \to \mathbb{CP}^1$, $w$ is the number of simple branch points, and $p$ is the genus of the surface $S$.[53] In fact, Riemann did not introduce the notion of 'genus' (or 'Geschlecht' in German), but rather referred to $p$ as what is derived from the order of 'connectivity.'[54] This formula was later generalized, and became known as the Riemann–Hurwitz formula for any covering of a Riemann surface. Second, Riemann considered a class of functions (or Riemann surfaces) $F(s, z) = 0$ containing all equations that can be rationally transformed into each other. In modern language, all of these surfaces have the same genus $p$.[55] For $p > 1$, Riemann showed that the class of these surfaces is characterized by $3p - 3$ complex parameters, which are the "number of moduli";[56] the computation was done by considering which values of the branch points could be determined and which could be freely deformed, giving the desired number of moduli.

Though not discovered by Riemann himself, one more result should also be mentioned: in 1891, Adolf Hurwitz asked what the number $N$ of degree $n$ Riemann surfaces, branched along a given number of branch points, is.[57] Assigning a permutation to each of the branch points, Hurwitz computed the number $N$; this was proven to be finite, and usually *greater* than 1.

(ii)  Riemann's terminology

As already noted above, Riemann coined two terms for designating branch points (or more accurately, the preimages of the branch points on the Riemann surface). In 1851, Riemann

---

[53] Ibid. p. 114.

[54] For example, a sphere has order of connectivity 1 and a torus 3. Riemann noted that "a surface spread over the whole complex plane could be seen either as a surface with a boundary situated at infinity or as a closed surface, provided the point $z = \infty$ is added to it. He then proved that the order of connectivity of such a closed surface is necessarily an odd number $2p + 1$." (Lê 2020, p. 77); on Riemann's notion of 'connectivity', see: ibid., p. 74–78; Popescu-Pampu (2016, p. 35–40). Note also (Lê 2020, p. 77): "[. . .] several sections of [Riemann's 1857 paper] were devoted to prove a formula expressing the number $p$ with the help of some of the characteristics of the considered algebraic function", although, as noted above, Riemann never interpreted algebraic equations as equations defining algebraic curves.

[55] In Riemann's terms, all these surfaces have order of "connectivity $2p + 1$" ["$(2p + 1)$-fach zusammenhängend"].

[56] Riemann (1857, p. 120): "die Anzahl der Moduln". For $p = 1$ Riemann proved that the number of moduli is 1.

[57] See: Hurwitz (1891, p. 2).

called these points on the covering "turning points" ("Windungspunkte"),[58] describing them, as we saw above, as follows: when another point on the complex line $\mathbb{C}$ encircles the (image of) such a point,[59] a permutation of the preimages of the moving point on the surface occurs.[60] In 1857, Riemann called this point a "branch point" ("Verzweigungspunkte"),[61] which is a point on the covering, whereas the different sheets of the function in the neighborhood of this point are called "branches"("Zweige"). Explicitly, the relation between the "branch point" and the "branches" of the surface are described as follows: "[. . .] the different prolongations of a given function in a given region [. . .] will be called *branches* of the original function and a point around which one branch continues into another a *branch point* of the function."[62] This change in terminology in the 1857 article from the 1851 article is neither neutral nor innocent, and the two terms: "Verzweigungspunkt" and "Windungspunkt" may point to different semantic fields and hence prompt different images in the mind of the reader. On the one hand, "Windungspunkt" is derived from the verb 'winden,' i.e., coil or wind up, and might have been employed by Riemann to refer to an image of how the various sheets behave (i.e., as they wind up) in the neighborhood of such a point. It is therefore also not surprising that when material models of such points were constructed (as can be seen in Fig. 1.4), these models were also termed models of "Windungspunkte."[63] On the other hand, the term "Verzweigungspunkt" is derived from 'verzweigen,' meaning branching, and this may prompt an image of sheets which branch out or ramify at a certain point—although, as seen above,[64] when Riemann described the behavior of points near a "Verzweigungspunkt," he used the verb 'winding,' which may indicate that the two images and semantic fields did not have a clear separation.

Mathematicians in the second half of the nineteenth century continued to use both terms;[65] however, toward the end of the nineteenth century, one finds that the term "Verzweigungspunkt" had slowly become more common, as Andrew Russell Forsyth noted in 1893 when presenting terms in various languages for "branch point": "[. . .] with

---

[58] Riemann (1851, p. 8).

[59] Instead of the complex line one can consider as well the projective complex line $\mathbb{CP}^1$.

[60] Cf. also Puiseux's very similar treatment from 1850 (Puiseux 1850).

[61] Riemann (1857, p. 107).

[62] Baker et al. (2004, p. 80–81); Riemann (1857, p. 90): "[. . .] sollen die verschiedenen Fortsetzungen *einer* Function für denselben Theil der z-Ebene *Zweige* dieser Function genannt werden und ein Punkt, um welchen sich ein Zweig einer Function in einen andern fortsetzt, eine *Verzweigungsstelle* dieser Function."

[63] See e.g. (Dyck 1892, p. 176).

[64] Riemann (1857, p. 110).

[65] See for example the various usages in (Friedman 2019a, p. 112–139); in addition, as noted above, when branch points were presented with material models, they were usually called "Windungspunkte". A study of classification of clusters of mathematicians and of the research settings in which each two of these terms was used is however outside the scope of this book.

German writers the title is *Verzweigungspunkt* and sometimes *Windungspunkt*. French writers use *point de ramification* and Italians *punto di giramento* and *punto di diramazione*."[66] As we will see, this terminological ambiguity will reappear in the research on branch curves, albeit with other terms.

(iii)  An impossible transfer of results and terms?

Having surveyed Riemann's research on branch points and the terms that he employed more accurately, the question arises as to whether analogous results and terms were found and coined within the configurations dealing with branch curves. The answer is rather negative; to see this, let us review a few examples, which will also be examined in this book.

The fact that this transfer was anything but a simple analogous transfer of results was already clear in the 1890s: while it was known that a complex algebraic curve is rational if its (geometric) genus is 0,[67] it became clear to Castelnuovo and Enriques that even when both the geometric genus and the arithmetic genus of a complex algebraic surface are 0, this is not enough to characterize rational surfaces.[68] With respect to the branch curve, two concrete examples of the failure of this transfer should be mentioned.

First, for the question paralleling Hurwitz's aforementioned query: how many complex algebraic surfaces are there, ramified over a given branch curve? To recall, Hurwitz showed that for a given number of branch points, the number $N$ of degree $n$ Riemann surfaces branched over them is usually greater than 1. The above question concerning the number of ramified algebraic surfaces suggests a similar inquiry. Nevertheless, the answer to this question became remarkable at the end of the twentieth century: if the degree of the algebraic surface is greater than 4, then it was proven in 1999 that there is only *one* surface;[69] moreover, Oscar Chisini had proposed this conjecture—but with a wrong proof—already in 1944. To emphasize: the analogous result of Hurwitz's calculations for the number of branched Riemann surfaces (given the branch points) was not at all analogous when dealing with the same question concerning the number of branched algebraic complex surfaces (given a branch curve). Moreover, Hurwitz showed in 1891 that Riemann surfaces are classified by the permutations associated to the branch points,

---

[66] Forsyth (1893, p. 15).

[67] In modern formulation: A *rational curve* is a curve which is birationally equivalent to a line. Two algebraic varieties *X, Y* are *birationally equivalent* if there is a rational map (i.e. not necessarily a polynomial one, but composed of rational functions) from $X$ to $Y$ and another rational map from $Y$ to $X$, when both maps are inverse to one another (outside lower-dimensional subsets, where these maps may not be defined).

[68] See Sect. 3.1.2. For a short explanation what are these genera mathematically, see the appendix to Chap. 3 (Sect. 3.4).

[69] See: Kulikov (1999, 2008). The proof was first given in 1999 to surfaces of degree greater than 11, and in 2008 it was completed to all surfaces of degree greater than 4.

fulfilling certain conditions.[70] This led Enriques to investigate, at the beginning of the twentieth century, the parallel situation for algebraic surfaces and branch curves, by examining how to associate permutations to the branch curve; but his investigation eventually led Zariski to an investigation of the *fundamental group*: the group of loops in $\mathbb{C}^2$ encircling the branch curve. This group turned out to have a more intricate structure than expected, with no equivalent in the parallel configuration for Riemann surfaces and their branch points.

The second example to the failure of this transfer appears, as we will see in Sect. 3.1.3, when Enriques attempted in 1912 to prove a formula for the moduli of algebraic surfaces. Presenting his results as analogous to those of Riemann,[71] Enriques was inspecting the numerical invariants of the branch curve; but in the course of his proof for his formula, he made an assumption which turned out to be—years later, in 1974—incorrect.

These examples and others will be thoroughly explored in the following chapters. But it is not only research configurations and results on branch points that could not have been transferred and transformed analogously to branch curves. The terms used for "branch curve" themselves were also unstable. As we will see in Chaps. 2 and 3, during the nineteenth century, this curve had several names ("contour apparent," "curve of contact," "Verzweigungskurve"); this is not surprising since before 1851, one could not have transferred any terminology from the (non-existing) research on Riemann's surfaces; and although the term "Verzweigungskurve" slowly became the 'official' name to term this curve in the last decade of the nineteenth century, this acceptance was not evident: other terms were also employed, such as the "Uebergangscurve," which had no equivalent for Riemann surfaces.

### 1.2.3   On Branch Curves, Again: Plurality of Notations

Section 1.2.2 showed the lack of clear, analogous transfer of results or terms from the research on branch points to branch curves. Moreover, the various definitions of branch curves given in Sect. 1.2, as well as the discussion in Sect. 1.2.1, emphasize the variety of local research configurations, all dealing with the 'same' branch curve.

This plurality can also be seen in the notations and graphical means of representing this curve. That being said, I do not aim to suggest a bijective correlation between the various definitions and notations or textual means. Hence, along with the historical investigation of the dynamics of the configurations in which branch curves are to be found, this book also

---

[70] If there are $w$ branch points, and if $S_1$, $S_2$,..., $S_w$ are the associated permutations, then those conditions are that the composition of all of the permutations is the identity permutation: $S_1 S_2 \ldots S_w = 1$ and that the group generated by the $S_i$'s is transitive, which is equivalent to saying that the Riemann surface constructed would be connected.

[71] Recall that Riemann showed for algebraic curves of genus $p > 1$, that the class of these surfaces is characterized by $3p - 3$ complex parameters.

aims to explore the graphematic representations and the spaces of inscription that were employed to represent, and at the same time investigate the branch curve.[72] Fig. 1.5 presents several of the ways used to represent a branch curve of a smooth surface in $\mathbb{CP}^3$.

What does this plurality of inscriptions and methods of representation indicate? First of all, as we will see, the reasoning used by several of the involved players was not only symbolical, but also more broadly graphic and diagrammatic. The various ways to consider the branch curve gave rise to various notations and graphic inscriptions, but were also conditioned by them. To reiterate: these various notations and graphic means were not ways of representing an 'identical' object, functioning merely as a mnemonic device for a mere representation. To be found concretely in either a two-dimensional plane (written on a piece of paper), or in a three-dimensional space (in the form of a three-dimensional model), those various representations were used to explore the object they represent, and also eventually mobilized either new theorems, or new conceptual activities. These different spaces of script again raise the question of identity of the mathematical object and stress the different epistemic settings, which will be discussed throughout the book. Moreover, several notations considerably increased the possibilities of manipulation, making it possible not only to store information and process it, but also to create new epistemic configurations; in other words, these novel methods of inscription were not only representations, but also prompted discovery and invention.

Figure 1.5 shows these aspects clearly. I will discuss these various notations and inscriptions thoroughly in the book, but I would like to examine this figure briefly here. Notation can be considered as epistemic, since it can be manipulated; this is to be seen with what is called the 'characteristic braid' associated to a singular curve, used by Oscar Chisini and later also by his students (Modesto Dedò, for example), presented in Fig. 1.5b. This figure presents two modes of representation, introduced by Dedò, of this 'braid' associated to the branch curve of a smooth cubic surface in $\mathbb{CP}^3$. Boris Moishezon, on the other hand, while using another notation for the branch curve, representing it as a factorization of elements in a group (see Fig. 1.5c), was more cautious than Chisini and Dedò with the manipulation of notation, in the sense that he was working less with diagrams and more with "symbolical" expressions, to cite Moishezon's term.[73] Oscar Zariski preferred to describe the curve that Dedò presented in Fig. 1.5b with words and equations—see Fig. 1.5a. Those three different ways of representing and notating the branch curve emphasize that notation is "not only a medium for the representation of the objects [. . .] but also [. . .] an instrument through which those very objects can be *generated* and *explored*."[74] That is, notation is what takes part in the organization and reorganization of

---

[72] See: (Lenoir 1998) on the role of inscription practices and their materiality within science and how "inscription devices actually constitute the signifying scene" in science in general and in mathematics in particular (ibid., p. 12).

[73] Moishezon (1981, p. 117)

[74] Krämer (2014, p. 3).

**a**

Let

(14)
$$F(x, y, z) = 0$$

be the equation of the cubic surface. The branch curve $f$ of the function $z$ is a sextic with 6 cusps on a conic, the equation of which has the following form:

(15)
$$f(x, y) = \phi_2{}^3(x, y) + \phi_3{}^2(x, y) = 0,$$

where $\phi_2 = 0$ and $\phi_3 = 0$ represent a conic and a cubic curve respectively.

**b**

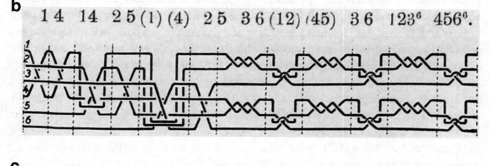

**c**
$$\Delta^2 = [\Pi_{j=h-2}^{0} (\Pi_{k=0}^{n-2} z_{j,k;j+1,k})$$

$$\cdot \Pi_{j=0}^{n-1} [z_{j,0;j,1}^{3} \cdot \Pi_{k=2}^{n-2} ((\Pi_{k'=0}^{k-2} z_{j,k',j,k}^{2}) \cdot z_{j,k-1;j,k}^{3})]]^{n-1}$$

$$\cdot \Pi_{j=n-2}^{0} (\Pi_{k=0}^{n-2} z_{j,k;j+1,k}).$$

**Fig. 1.5** (a) Zariski's description of the branch curve of a smooth cubic surface in $\mathbb{CP}^3$ (Zariski 1929, p. 320). (b) Dedò's two ways of notating the same branch curve: with a braid diagram (below) and with a numerical notation of the same braid (Dedò 1950, p. 257). (c) Moishezon's braid monodromy factorization of the branch curve of a smooth surface of degree $n$ in $\mathbb{CP}^3$ (Moishezon 1981, p. 161)

the epistemic configuration. However, while notation may be "advantageous for mathe-matical creativity,"[75] it is essential to emphasize that this exploration can terminate in a dead end; a few of the various textual practices exemplify this, as we will see throughout this book.

## 1.2.4   Transformations Between Epistemic Configurations

Given the plurality of the epistemic configurations and their notations, the various chapters in this book also aim to elucidate the transformations between these configurations. Transformations between one epistemic configuration and another can occur in various ways. One way was already seen in Sect. 1.2.2: the attempts to transfer analogous results and methods. Other ways include a rereading of a problem or a proof;[76] a new solution to a problem; a rewriting of an entire mathematical domain with new tools and with a new language, etc. ... These transformations are not without consequences, and may prompt "proof" of wrong results or unfounded assumptions concerning the *target* configuration. But other consequences may also result for the *transformed* configuration: the marginali-zation of former practices or the conscious, deliberate ignoring or silencing of other, older or concurrent traditions of research.

"Ignorance," as Robert N. Proctor notes, may be "a product of inattention, and since we cannot study all things, some by necessity—almost all, in fact—must be left out."[77] This, however, might be far from evident, especially when it comes to scientific research, mathematics in particular. While at this stage, unfolding these ignored and marginalized spaces and configurations would be impossible (but they will be made explicit throughout this book), one example can already be given. As already hinted above, the field of algebraic geometry went through two rewriting projects. On this subject, Schappacher notes that there "are different ways to tell the story of a rewriting."[78] One of these ways is to note that not every group of mathematicians chose to participate; Chisini and his students (among them Dedò) decided, in the 1950s, not to use algebraic tools or practices, which were already common in algebraic geometry due to the first rewriting project by Weil and Zariski. As a consequence, the results of this group of mathematicians were ignored, only to be partially revealed and revised later on, when their results were sometimes disproven, decades later. In that way, one can speak about an epistemic configuration made ephemeral and even precarious.

---

[75] de Cruz / de Smedt (2013, p. 12). See also (Giardino 2018; Krämer / Totzke 2012)

[76] See: Goldstein (1995), who unfolds the history of these rewritings and rereadings of one of Fermat's theorems over a period of more than 300 years.

[77] Proctor (2008, p. 7).

[78] Schappacher (2011, p. 3285).

These transformations—seen, for example, in the form of various rewritings and rereadings—show that it is precisely those rereadings and re-embeddings of the results of other (sometimes former) mathematicians that are never neutral or innocent with respect to other later epistemic configurations. Quite the contrary: these later configurations may not only reorganize the concurrent epistemic configuration, but also prompt the marginalization and ephemerality of the former one.

## 1.3 An Overview: Historical Literature, Structure and Argument

The branch curve—if one can even talk about *the* 'branch curve' as a single, well-defined object that has retained its definitions over decades—has not gone completely unnoticed in the historical research, but the accounts given on how this curve was investigated are fairly concise and somewhat too terse.[79] In Jean Dieudonné's *History of Algebraic Geometry* one finds the following remark: "Ever since the invention of Riemann surfaces, the concept of *ramified covering* of an algebraic variety had been familiar (in a more or less precise way) to algebraic geometers, and the Italian school had used it not only for curves in $\mathbb{CP}^2$, but also for surfaces in $\mathbb{CP}^3$."[80] Although Dieudonné does mention the fundamental group, which was researched intensively during the 1930s for the case of branch curves (Dieudonné mentioned in general the fundamental group of the complement of the branch variety), an explicit discussion of the history of the research on the branch curve goes unmentioned. In 1995, one finds a somewhat more detailed account:

> Enriques "considered the representation of [algebraic complex] surfaces as *multiple planes*, that is, as multiple covers of the plane, branched along appropriate curves endowed with cusps and nodes. [. . .] [T]he characterization of curves with nodes and cusps that are branch curves for some multiple plane is due to Enriques (1923), a characterization translated into topological terms by Zariski (1935), while Chisini (1934–1939) tried to construct—with interesting considerations of degenerations—branch curves for vast classes of multiple planes, and thus of surfaces, with given invariants; in this he was followed by students in his school and in more recent times by [. . .] B.C. Moishezon [during the 1980s]."[81]

An even more terse description, also pointing to other mathematical investigations, is to be found 10 years later, in 2005:

> "It was known early on, by algebraic geometers (see Zariski 1929) that, for proposed applications to the study of algebraic surfaces, the most important knot groups [i.e., fundamental groups of the complement] of complex plane curves were those of cusp curves. Starting

---

[79] This is in contrast to how branch points of Riemann surfaces were studied historically, see for example (Epple 1995).

[80] Dieudonné (1985, p. 143).

[81] Brigaglia / Ciliberto (1995, p. 102).

in 1933 and continuing through the 1950s, Oscar Chisini and his school published a series of
papers around that subject [. . .] The tools they brought to bear were what would later be called
'braid monodromy' and 'the arithmetic of braids' [by] [. . .] Moishezon and Teicher."[82]

Other accounts recall a similar story.[83] While the two accounts from 1995 and 2005 note
the main players: Federigo Enriques (1871–1946), Oscar Zariski (1899–1986), Oscar
Chisini (1889–1967), Boris Moishezon (1937–1993) and Mina Teicher (born in 1950),
to name a few—the terseness of these accounts almost prevents paying attention to the
different epistemic configurations, and the various notations, mathematical techniques and
traditions and associated images of algebraic geometry which were involved in the course
of the investigation. That is, those descriptions ignore not only the different epistemic
configurations, but also their transformations and the fact that there is no one constant
"branch curve," identical to itself, in all of those configurations. Moreover, the question
remains as to whether these historical accounts are indeed complete and faithfully reflect all
of the traditions and transformations that the research on branch curves went through. For
example, several questions follow: Can one indeed consider Moishezon's methods as
following those of Chisini? In which ways was Chisini followed by his students? Was
Zariski's early research on the branch curve relevant for his later rewriting of algebraic
geometry? Or were Enriques and Zariski really the first to deal with these curves? These
questions already warrant a more refined historical account.

### 1.3.1   Omitted Traditions

As I mentioned above, taking the branch curve as a cross-section also implies that this
research does not intersect all of the layers in the history of algebraic geometry. In the
context of the present book, I decided not to discuss, in depth, two of the main rewriting
projects to which algebraic geometry was subjected during the twentieth century: the first
with the language of commutative algebra (with Zariski and Weil), and the second with the
language of schemes and sheaves (with Grothendieck). While the phenomenon of ramification in rings, ideals and especially in what was later called Dedekind domains was
already defined and researched with Dedekind and Weber during the 1880s (see below),
and was also discussed by Zariski in the context of commutative algebra (resp.
Grothendieck in the context of schemes, see: *SGA I* or *EGA IV*), branch curves were not
considered an explicit object of research. Moreover, the history of the term 'ramification'
(resp. 'Verzweigung') and how it has spread and been used in those mathematical domains

---

[82] Rudolph (2005, p. 404)

[83] See: Libgober (2011, 2014), or (Conte / Ciliberto 2004, under the section "Superfici algebriche").

deserves more thorough attention, but this would take us outside the scope of this book.[84] Therefore, I deliberately chose to delineate a different history of algebraic geometry that sometimes focuses on more minor schools and epistemic configurations which are less hegemonic, appearing in the form of micro-schools or local communities, for example, and were less accepted or even ignored, eventually becoming either obsolete or rather forgotten.

I would like, however, to mention briefly two other mathematical configurations that are not discussed in the book, though the research on them and their development parallel how branch curves were conceptualized. The first question, which I do not discuss in detail, was already implicitly hinted at above: if Riemann considered ramified coverings over the complex (projective) line, that is, as the coverings of a real two-dimensional manifold, then what happens when one increases the *real* dimension by 1, that is, considering three-dimensional real manifolds as real coverings of the 3-sphere? These manifolds are then branched over a *real* curve, which is a knot or a link. James W. Alexander noted this in 1920: "In the 3-dimensional case [...] the branch system may be [...] a set of simple, non-intersecting, closed curves such that only two sheets come together at a curve. The curves may, however, be knotted and linking."[85] Indeed, these coverings were researched at the beginning of the twentieth century, with Poul Heegaard, Wilhelm Wirtinger, Alexander, G. B. Briggs and Heinrich Tietze, among others, mostly in the context of knot theory. As Alexander and Briggs noted in 1927, regarding a lecture that Alexander had delivered in 1920, "[he] pointed out that if the space of a knotted curve be covered by an *n*-sheeted 'Riemann 3-spread' (the three-dimensional analogue of a Riemann surface) with a *branch curve* of order $n − 1$ covering the knot itself, then, the topological invariants of the covering spread will also be invariants of the knot."[86] Though the term "branch curve" was also used by those mathematicians, it is not the (complex) branch curve of our study, but rather a knot or a link (i.e., a real curve). Hence a study of those 'branch curves' would take us into the epistemic configurations of knot theory, which I will not discuss,[87]

---

[84] Hence, for example, in the context of sheaves, I will not deal with how unramified morphisms were define via the cotangent sheaf $\Omega_{X/Y}$ for a morphism $f : X \to Y$. See: *EGA IV*, corollary 17.4.2, in: (Grothendieck 1967, p. 65) or (Vakil 2017, p. 588).

[85] Alexander (1920, p. 372). Cf. also: Alexander (1928, p. 303): one can consider an "*n*-sheeted Riemann spread $S^n$ (the 3-dimensional generalization of an *n*-sheeted Riemann surface) with the knot as *branch curve* (generalized branch point)."

[86] Alexander / Briggs (1927, p. 562). As Epple notes "the main thrust of this talk was the statement that all topological invariants of a Riemann space were also invariants of the system of its branch curves [...]." (2004, p. 157).

[87] There exists already extensive literature on how knots and links were considered as the branch curves of 3-dimensional real ramified covers. See: (Epple 1999, p. 330–354; Epple 2004; Stillwell 2012).

even though the two rather different research settings on those branch curves did intersect at a certain point—as we will see with Zariski.[88] Moreover, this emphasizes a comment noted above: the definition and characteristic of a mathematical object is only secondary to the epistemic configuration; as Poincaré noted in 1908, "mathematics is the art of giving the same name to different things."[89] Though Poincaré meant that the same mathematical structure may be found to clarify the same relations between various, different mathematical objects, one may also offer an opposite conclusion: the same name is given to two 'different things,' when these two different things are embedded in two different epistemic settings, hence having no essential 'structure' in common.

While the above research on branch curves arose from changing the settings from real two-dimensional to real three-dimensional coverings, the question that might arise is what occurs when investigating the phenomenon of ramification from an algebraic point of view? More explicitly, what happens when one deals with the factorization of prime ideals in a field extension, hence viewing this field as a covering?[90] The origin of this point of view is to be found with Dedekind and Weber in 1882, who clearly define the "ramification ideal" ("Verzweigungideal").[91] Although they were influenced by Riemann's work, there were obviously essential differences, as they were attempting a "rewriting of Riemannian function theory [as] [. . .] an arithmetical rewriting."[92] Though they were referring directly to ramified Riemann surfaces and to ramification points in their work,[93] the question remains whether one may find an equivalence to branch or ramification curves; this is another mathematical configuration I will not discuss here.[94]

---

[88] See Sect. 3.2.3.

[89] Poincaré (1952 [1908], p. 34).

[90] In a modern language, if $p$ is a prime ideal in a ring of integers $O_K$ of an algebraic number field $K$, then given a finite field extension $L/K$, one considers the ideal $pO_L$ of $O_L$. This ideal factorizes into a factorization of distinct prime ideals of $O_L$: $p_1^{e_1} \cdot \cdots \cdot p_k^{e_k}$; one says that $p$ is *ramified* in $L$ if there is an $i$ such that $e_i > 1$.

[91] Dedekind / Weber (1882, p. 226).

[92] Haffner (2014, p. 175). See also Haffner (2014) for an extensive discussion on Dedekind's work; for a discussion on Dedkind's and Weber's conception of the ramification ideal, and the similarity to branch points, see (ibid., p. 153–154, 162–166, 175–176).

[93] Dedekind / Weber (1882, p. 243).

[94] Note that another mathematical configuration, which I do not follow here but which is very much related to Dedekind and Weber's configuration, is Heinrich W. E. Jung's arithmetic theory of algebraic functions in two variables, developed during the first half of the twentieth century. Jung's research also dealt with the branch curve (or more concretely, with "branch places" ["Verzweigungsstellen"]), but from an algebraic-arithemtic point of view. For a summary of his results, see his book: Jung (1951).

## 1.3.2   Structure of the Book: The Twentieth Century

In this book, I decided to concentrate on the events and the research that took place mainly during the twentieth century. The reason for this is that it is only during this century— starting in fact from the last decade of the nineteenth century—that branch curves (of complex algebraic surfaces) in various epistemic configurations were taken into consideration within the emerging research field of algebraic surfaces. The emphasis on those configurations should be taken into account when reading the various chapters: there is no one 'branch curve' which remained fixed throughout the nineteenth and twentieth centuries; rather, the different episodes recounted in the chapters of this book highlight the dynamic character of this object, which was constantly being resituated in various configurations. I will therefore begin each chapter with a survey of the different configurations appearing in that chapter.

This is not to suggest that during the nineteenth century, branch curves were not considered at all, as Chap. 2 will show. Monge, as we saw in the opening of this introduction, was already calling for future research dedicated to these curves at the end of the eighteenth century. Chapter 2 will therefore delineate the works of several nineteenth century mathematicians who did deal with branch curves (Monge, Étienne Bobillier, George Salmon, Heegaard and Wirtinger) in different configurations, although this research is characterized by a certain discontinuity between these configurations, as well as shifts to other domains of research (such as knot theory); this might not come as a surprise, as the research on branch points of Riemann surfaces was only developed during the second half of the nineteenth century.

Chapter 3 moves to the beginning of the twentieth century and presents two of the protagonists of the book: Federigo Enriques and Oscar Zariski, each viewing the branch curve in a different way. For Enriques (see Sect. 3.1), one can and should investigate this curve with a plurality of methods, a fact which underlines Enriques' epistemological sensitivity to various possible configurations of mathematical investigations. Indeed, the research of Enriques on and with the branch curve during the first three decades of the twentieth century was characterized by this plurality. For example, Enriques showed how one could use the branch curve to construct a double cover (or what was called a "double plane"), or even a multiple cover, hence classifying surfaces with this curve. From this, several conclusions can be drawn about the properties of the branch curve concerning the moduli space of surfaces. To give another example, Enriques also carried out research on the necessary and sufficient conditions for a curve to be a branch curve, research which also emphasizes how this curve operated as an epistemic object. This line of research prompted a future investigation of the algebraic group $\pi_1(\mathbb{C}^2 - B)$: the group of all loops of the complement to the branch curve $B$, called the "fundamental group"—though Enriques was not investigating it explicitly.

In contrast to Enriques' approach, it was Zariski (see Sects. 3.2.1 and 3.2.3) who concentrated explicitly, from 1927 and 1937, on the algebraic structure of $\pi_1(\mathbb{C}^2 - B)$, as well as on how the topology of a cover depends on the position of the singularities of its associated branch curve. The position of the singular points of the branch curve was a subject that Beniamino Segre also dealt with (see Sect. 3.2.2). Those works, situated within Zariski's studies on nodal–cuspidal curves, arise from the background of Zariski's work on topology during the same period. As is already clear from the above review: the three mathematicians mentioned here (Enriques, Zariski, Segre) all worked in different, though not unconnected, epistemic configurations. Chapter 3 ends with the reflections of various mathematicians during the 1950s on the work done during those first decades of the twentieth century, reflections which (re)shaped the image of classical algebraic geometry.

Chapter 4 concentrates on a more 'minor' protagonist within the research of branch curves: Oscar Chisini, as well as on two of his students, Modesto Dedò and Cesarina Tibiletti. The research of this rather small Italian mathematical community between the 1930s and 1950s on the "characteristic braid" associated to a singular plane curve, as well on branch curves, can be considered a 'minor' tradition in the history of algebraic geometry. This is partially because it was ignored by the mathematical community outside of Italy, and partially because the Italian mathematical community was to a certain extent isolated during those years. Chisini and his students introduced not only novel diagrams to represent the branch curve, but also a new notation for braids (see Fig. 1.5b). At the same time, they were representative of the decline of the classical approach to algebraic geometry.

In Chap. 5, after briefly sketching the developments in algebraic geometry during the 1960s and the beginning of the 1970s concerning the research on coverings and on the branch curve (Sect. 5.1), I concentrate on the contribution of Boris Moishezon to the research on this curve via the introduction of what Moishzon termed "braid monodromy factorization." Moishezon, also together with Mina Teicher, while disproving the 'watershed conjecture' of Bogomolov as well as a 'theorem' by Chisini from 1954, on the one hand deepened the understanding of the structure of the fundamental group of the complement of the branch curve using advanced techniques of group theory. On the other, both Moishezon and Teicher posited the above-mentioned factorization at the center of their research.

The concluding Chap. 6 returns to the question of the identity of the mathematical object. It attempts to analyze how taking the various configurations of research on branch curves into consideration—as a cross-section of (the history of) algebraic geometry—not only shows the opening of new spaces of representation, but also, and in a parallel fashion, delineates a history of ignorance and marginalization. As such, the history of how the branch curve was researched tells us not only a story of the emergence and consolidation of mathematical configurations, practices and communities, but also of the forgetting and ignoring of mathematical knowledge. It is the aim of this book to shed light on traditions that may have been bypassed or ignored in the conventional history of algebraic geometry.

# Prologue: Separate Beginnings During the Nineteenth Century

**2**

As stated in the introduction, it was only in the last decade of the nineteenth century that the branch curve began to receive considerable attention as a mathematical object that deserves to be researched in its own right and in various epistemic configurations. The question remains, however: What happened earlier in the nineteenth century, before this period? In this chapter, I would like to briefly sketch several mathematical configurations which dealt with ramification and branch curves during that century. These configurations appeared separately and within different contexts. For the most part, they did not prompt—indefinitely or at least till the twentieth century—new investigations concerning branch curves. This does not imply, however, that these research trajectories did not function epistemically in any respect. Nevertheless, in most cases, this did not lead to research of the curve itself. I will therefore survey, here, several examples, concentrating on Gaspard Monge and the rise of descriptive geometry in Sect. 2.1; other mathematicians, such as Étienne Bobillier or George Salmon (see Sect. 2.2), or Poul Heegaard's research and Wilhelm Wirtinger's subsequent shift to knot theory (see Sect. 2.3) will be treated more briefly.

As we will see in the following sections, Monge's specific configuration of treating projections of surfaces and their ramification curves (or, in Monge's terms, "contour apparent" or in English, "apparent contour") was embedded in his general conception and development of descriptive geometry, and how spatial curves and surfaces can be represented by, and reconstructed from two-dimensional figures. Monge's treatment can be characterized as a practical investigation of shadows, via projections from a source of light. This practical configuration was restructured by Bobillier, who also coined a term for another surface: the "surface polaire," whose intersection with the researched surface results in the "apparent contour." Salmon, who was probably unaware of this particular research of Bobillier's but used the coined terminology, was more interested, in the late 1840s till the 1860s, in the numerical properties associated with this curve; at the same time, any trace of Monge's practical research is not to be found in Salmon's investigation.

M. Friedman, *Ramified Surfaces*, Frontiers in the History of Science,
https://doi.org/10.1007/978-3-031-05720-5_2

Whereas Salmon gave no term for the branch or the ramification curve, this situation changed at the end of the nineteenth century. In the configuration of Wirtinger and Heegaard, the branch curve (or the branch manifold) was named as such: "Verzweigungsmannigfaltigkeit," and though first researching the branch curve, they afterwards—and especially Wirtinger—later focused only on its singular points and their local neighborhoods. These various research configurations show not only their plurality, but also their epistemic character: numerous directions were proposed and questions asked in those settings, and though several of those questions were researched or answered, others were left open or were not sufficiently developed.

## 2.1    The Beginning of the Nineteenth Century: Monge and the "Contour Apparent"

Gaspard Monge (1746–1818) is sometimes today considered not only as the founder of descriptive geometry, but also as one of the driving forces behind the foundation of the *École polytechnique*, one of the most famous engineering schools in France at the turn of the eighteenth century.[1] Descriptive geometry was considered as the study of projections of three-dimensional objects to two-dimensional planes, as well as the attempts to reconstruct these three-dimensional objects. As Joël Sakarovitch notes, Monge did not present at any moment descriptive geometry as a new science that he himself had founded; on the contrary, he emphasized that this geometry existed a long time before the nineteenth century and even before the modern treatments of algebra, as was practiced by carpenters and fitters.[2] Nevertheless, Monge's treatment of branch curves was nonetheless novel. To stress: the curves he investigated were not called "branch curves" (or the equivalent term in French) but, as the motto at the beginning of the introduction shows, the projection of the

---

[1] For a recent survey of Monge's conception of descriptive geometry, and how descriptive geometry spread in France see: (Barbin 2019a, b). Both papers contain numerous references concerning Monge and the beginning of descriptive geometry. Concerning the foundation of the *École Polytechnique*, I turn to Brechenmacher's recent account: "In 1793, the schools of instruction and teaching were disorganized by the war [. . .]. In 1794, Jacques-Élie Lamblardie, director of the École des ponts, who lost a great number of his pupils, thought of creating a preparatory school for bridges and roads, and then for all engineers. Monge was enthusiastic about this idea and convinced several members of the *Comité de Salut Public* [. . .]. Under support of figures such as the chemist François Fourcroy, a decree of March 11, 1794 created the Central school of public works, which would be renamed *École polytechnique* one year later, on September 1, 1795. Its mission was to provide its students with a well-rounded scientific education with a strong emphasis on mathematics, physics, and chemistry. The *Comité de Salut Public* entrusted Monge, Lazare Carnot, and several other scholars with enlisting, by means of a competitive recruitment process, the best minds of their time, and teaching them science for the benefit of the French Republic." (Brechenmacher 2022, p. 70–71) On the history of the creation of Polytechnique, see for example: (Langins 1987; Dhombres / Dhombres 1989; Belhoste 2003).

[2] Sakarovitch (1998, p. 212).

"contour apparent". This was the term Monge employed. This curve was therefore considered in Monge's epistemic configuration as a curve from which one may derive several properties of the associated projected surface.

How was descriptive geometry described by Monge? Monge emphasized the geometric visualization of mathematical and physical problems and defined the aim of descriptive geometry as follows:

> "Descriptive geometry has two main objectives: the first is to give the methods to represent on a drawing sheet that has only two dimensions, namely, length and width, all the bodies of nature that have three, length, width and depth, provided however that these bodies can be strictly defined.
>      The second object is to give the way to recognize, according to an exact description, the shapes of the bodies and to deduce from them all the truths that result from them and from their respective shapes and positions."[3]

Taking this statement into consideration, when introducing any new, possibly abstract, concept, Monge described either the concrete situations or the possible applications in which the concept appears.[4] This was done through drawing and modeling: indeed, Monge was one of the first mathematicians in the *École polytechnique* who built and collected three-dimensional material models of maps, models of architecture or fortifications, mechanical devices and also of geometrical objects and mathematical surfaces,[5] a tradition which was further developed by his student Théodore Olivier;[6] see Fig. 2.1. When discussing descriptive geometry, Monge emphasized its connections to practical concerns (such as the needs of the army) as well as the methods of stone carvers. He also stressed the importance of the intimate connections between analysis, analytical geometry and descriptive geometry: "There is no construction in descriptive geometry which cannot be

---

[3] See the first lesson that Monge gave in *L'École normale de l'an III* in 1795: "La géométrie descriptive a deux objets: le premier, de donner les méthodes pour représenter sur une feuille de dessin qui n'a que deux dimensions, savoir, longueur et largeur, tous les corps de la nature qui en ont trois, longueur, largeur et profondeur, pourvu néanmoins que ces corps puissent être définis rigoureusement. Le second objet est de donner la manière de reconnaître, d'après une description exacte, les formes des corps et d'en déduire toutes les vérités qui résultent et de leur forme et de leurs positions respectives." (Dhombres 1992, p. 308)

[4] Sakarovitch (1998, p. 214).

[5] On Monge's "cabinet des modèles", see (Brechenmacher 2022, p. 78–81). See also (Sattelmacher 2021, p. 73–114). In the recent years there is a renewed interest in the history of material mathematical models. See for example: (Cimorelli 2014; Seidl / Loose / Bierende 2018; Sattelmacher 2021; Friedman / Krauthausen 2022).

[6] Felix Klein (1926, p. 78) described Olivier's models as the origin of the mathematical models of the nineteenth century. See also: Sakarovitch (1998, p. 325–331) regarding the models of Olivier. Concerning the role of the *Conservatoire national des arts et métiers*, established in 1794, in the dissemination of the tradition of mathematical material models in France, see (Brenchenmacher 2022, p. 91–95).

**Fig. 2.1** Théodore Olivier,
*Intersection of Two Cones of
Unequal Diameter* (red and
green), ca. 1830–1845,
fruitwood, brass, thread, lead
weights, Union College
Permanent Collection, 1868.18
© Courtesy of the Union
College Permanent Collection

translated into analysis; and if the problems have no more than three unknowns, then every analytic operation can be regarded as the script for a play [*spectacle*] in geometry."[7]

The "spectacle of geometry" points out that Monge chose the most visual graphic representation for the introduced concepts. However, this was not necessarily a representation of the object itself. To provide the most relevant example for this book, given a surface and a projection point, and recalling that—using modern terminology—the ramification curve can be determined via drawing tangent lines to the surface exiting from the projection point, in the determination of the plane tangent to this surface neither this surface nor any tangent line or plane appear explicitly, but rather what Monge called the "apparent contour"[8] (i.e. in modern terms—the ramification curve).

So how did Monge research ramification and branch curves? In the following I would like to sketch more precisely the specific terms and the concrete mathematical configuration with which Monge described this curve in three works published around the 1780s: "Des Ombres", "Mémoire sur la théorie des déblais et remblais", and "Mémoire sur les propriétés de plusieurs genres de surfaces courbes et particulièrement sur celles des surfaces

---

[7]In: Dhombres (1992, p. 317): "Il n'y a aucune construction de Géométrie descriptive qui ne puisse être traduite en Analyse; et lorsque les questions ne comportent pas plus de trois inconnues, chaque opération analytique peut être regardée comme l'écriture d'un spectacle en Géométrie."

[8]See: Sakarovitch (1998, p. 216).

développables avec une application à la théorie générale des ombres et des pénombres". This theme was present in Monge's writings not only during the 1780s, but was also taught by Monge during the 1790s during his lessons on descriptive geometry at the *École normale de l'an Ill*, where he devoted an entire lesson to it.[9] Additionally, in 1807 Monge defined the ramification curve more precisely, indicating how the "apparent contour" should be calculated.

In Monge's teaching, the projections of three-dimensional bodies to the plane were usually considered in the context of tracing the shadows of a body, when the source of light is either another body or a point, which can be also located infinitely far away. In "Des Ombres",[10] Monge first looked at the case of a luminous source, which is found at a finite distance and assumed to be spherical, and introduced the notions of penumbra and degradation of hues, giving also an estimate of the size of the penumbra. After these general investigations, he noted that tracing the shadow is done via rays (tangents) going out from the luminous point (in the case the luminous body is a point), saying: "The projection of a body's shadow on any surface is therefore the figure that the extensions of the rays of light tangent to the body's surface end on that surface. [. . .] In the following operations we will geometrically determine only the projections of the contours of the pure shadows, they are *the only ones* that it is necessary to have exactly in the drawings."[11] (see also Fig. 2.2). The projection of the contour of the pure shadow is—in modern terminology—the branch curve of the projected body.

It is essential to again emphasize that Monge did not use this modern terminology, but rather referred to "projection of contours" or to "apparent contours", hence situating this contour under the framework of descriptive geometry, and more specifically, within the mathematical research of shadows and (light) projections. The term 'branch curve' as such did not yet exist.

In his 1781 manuscript "Mémoire sur la théorie des déblais et remblais" a shorter discussion is given, treating specifically the case of a projection from a point, and naming the curve separating the shadowed part from the luminous one as "contour apparent": "If we suppose that an eye reduced to a single point is placed in any way in relation to a curved surface, I call the *apparent contour* of this surface the one composed of the extreme points of this surface that the eye can see."[12] However, the most elaborate discussion on the apparent contour appeared in the manuscript "Mémoire sur les propriétés de plusieurs

---

[9]See: Dhombres (1992, p. 421–435).

[10]The manuscript was published in 1785, according to Olivier.

[11]"La projection de l'ombre d'un corps sur une surface quelconque est donc la figure que terminent sur cette surface les prolongements des rayons de lumière tangents à la surface du corps [. . .] Dans les opérations suivantes nous ne déterminerons géométriquement que les projections des contours des ombres pures, *ce sont les seules* qu'il soit nécessaire d'avoir exactement dans les dessins." (Monge 1847 [1785], p. 27, 29).

[12]Monge (1781, p. 694): "Si l'on suppose qu'un œil réduit à un point unique, soit placé d'une manière quelconque, par rapport à une surface courbe, j'appelle ligne de contour apparent de cette surface, celle qui est composée des points extrêmes de cette surface que l'œil peut apercevoir." (cursive by M.F.)

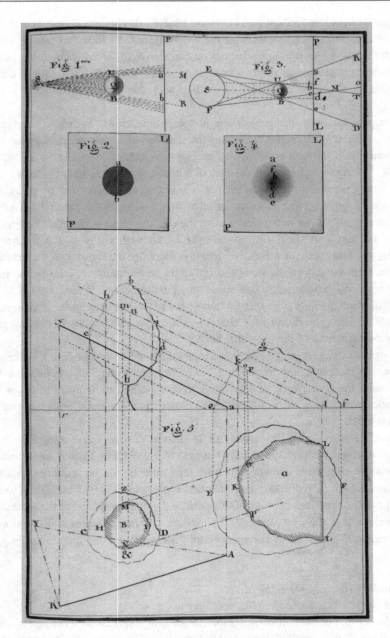

**Fig. 2.2** From Monge's "Petit traité des ombres à l'usage de l'école du génie" (notes taken by a student of Monge): Fig. 1 and 2 describe the illumination of a sphere from a point outside of it. © SABIX/Bibliotheque de l'Ecole polytechnique

genres de surfaces courbes", presented to the French academy in 1775 but published in 1781. After discussing several results of Euler's research on developable surfaces, Monge noted that he had already deduced "the determination of the shadow and penumbra of any figure body, illuminated by a luminous body of any figure";[13] he then presented the advantage of descriptive geometry over analysis: "Analysis can only derive a very great advantage from its application to this kind of Geometry; because I give the solution to several analytical problems, which we would perhaps have a lot of difficulty solving, without geometric considerations."[14]

Discussing apparent contours, a mathematical treatment was nevertheless given by Monge with the tools of analysis. He first noted the lack of serious treatments on the subject: "I know of no Treatise of Perspective where this question [on the theory of shadows] is resolved so generally. [...] these [other treatises] have said nothing about the line of contact of the body with the circumscribed conical surface, which is the most important line [curve], since it is that of the apparent contour of the body."[15] Here one may ask why Monge chose the term "circumscribed conical surface." The reason may be due to the fact that Monge chose as examples mainly surfaces of degree $m = 2$, hence the cone of light would be also of degree 2. This may be a more reasonable reason to name this surface as "conical"—i.e. as a cone—as this "circumscribed surface" is obtained by fixing one end of a line segment at the point of projection and moving the other end along the "apparent contour"; it is hence a cone over this "contour".

Immediately after his remark on the "most important" curve, Monge presented the following problem, dealing with an analytic investigation of finding the equation of the enveloping surface, which is both tangent to the luminous body as well as to the opaque body: "Problem V. Find the equations of the surfaces that envelop the Shade and Penumbra of any opaque body, illuminated by any luminous body, both bodies being given figures and positions in space."[16] To see how Monge approached this problem analytically, a closer examination of Monge's solution is necessary.

While the problem is formulated only in terms of shadows and illuminations, Monge's solution turns to analytical geometry and calculus. Denoting by $(x, y, z)$ the coordinates of a tangent plane to both bodies, $(x', y', z')$ the coordinates of the luminous body, and by

---

[13]Monge (1780 [1775], p. 383): "[...] la détermination de l'ombre & de la pénombre d'un corps de figure quelconque, éclairé par un corps lumineux de figure quelconque."

[14]Ibid.: "L'analyse ne peut que retirer un très-grand avantage de son application à ce genre de Géométrie; car je donne la solution de plusieurs problèmes d'analyse, qu'on aurait peut-être beaucoup de peine à résoudre, sans les considérations géométriques."

[15]Ibid., p. 406: "Je ne connais aucun Traité de Perspective où cette question soit résolue aussi généralement [...] ils n'ont rien dit de la ligne de contact du corps avec la surface conique circonscrite, ligne qui est la plus importante, puisqu'elle est celle du contour apparent du corps." Earlier in this paragraph Monge refers also to the case when the source of light is a point.

[16]Ibid., p. 407: "Problème V. Trouver les équations des surfaces qui enveloppent l'Ombre & la Pénombre d'un corps opaque quelconque, éclairé par un corps lumineux quelconque, les deux corps étant donnés de figures & de postions dans l'espace."

$(x'', y'', z'')$ the coordinates of the opaque body, Monge noted that the general equation of a tangent plane to the luminous body is $z = p'(x - x') + q'(y - y') + K'$ (where $p' = dz'/dx'$, $q' = dz'/dy'$), whereas in a similar fashion, the general equation of a tangent plane to the opaque body is $z = p''(x - x'') + q''(y - y'') + K''$. Demanding both tangent planes to coincide is equivalent, so Monge, to demanding that $p' = p''$, $q' = q''$ and $K' - p'x' - q'y' = K'' - p''x'' - q''y''$; eliminating variables, Monge noted that the general equation of the tangent plane to both bodies is of the form "(E): $z = Ax + By + C$", when $A$, $B$ and $C$ are functions dependent only on $x'$.[17] The question that interested Monge was what happened to this tangent plane when one changed $x'$—now taken as the point of contact of the tangent plane with the luminous body—infinitesimally: "if one conceives that $x'$ becomes $x' + dx'$, and that $\partial$ is the character of the space of differential, equation (E) will become (F): $z = (A + \partial A) x + (B + \partial B) y + C + \partial C$."[18] Monge then underlined that the two planes intersect at a line $EF$, whose projection to the plane, denoted as $ef$, is given by equating the equations (E) and (F), obtaining equation (G): $x\frac{\partial A}{dx'} + y\frac{\partial B}{dy'} + \frac{\partial C}{dz'} = 0$. Monge, also referring to Fig. 2.3, then concluded that the points on the line $ef$ are always outside the shadowed area; they hence determine the surface that envelope it; solving now equations (G) and (E) together, one can eliminate the variable $x'$ and obtain the equation of the enveloping surface of the shadow.

The only explicit example that Monge gave for this procedure is for the case that both the luminous body and the opaque body are spheres; in that case, the surface enveloping the shadow is a cone.[19] However, the next problem ("Probleme VI") deals briefly with the way of finding the contours of the shadow and the penumbra: "Given figures and positions in space of two bodies, one opaque and the other luminous, find the equations of the contours of the shade and the penumbra of the opaque body, projected over any given surface [...]."[20] Here Monge instructed finding the equation of the projection of the apparent contour. The solution provided is descriptive: given the equation of the enveloping surface and the opaque body, one should eliminate the variable $z$, and hence the obtained equations of the variables $x$ and $y$ are the ones of the demanded curves. A more condensed treatment was provided in 1807 in Monge's book *Application de l'analyse à la géométrie*: "Given the coordinates of the vertex, find the equation of the individual conical surface circumscribed [i.e. the cone above the apparent contour] to a given curved surface."[21] Instructing again to find the equation of the tangent plane, tangent both to the surface and to the "conical surface

---

[17] Ibid., p. 408–409.

[18] Ibid., p. 411: "[...] si l'on conçoit que $x'$ devienne $x' + dx'$, & que $\partial$ soit lé caractère de cette espèce de différentielle, l'équation (E) deviendra (F) $z = (A + \partial A) x + (B + \partial B) y + C + \partial C$."

[19] Ibid., p. 412–414.

[20] Ibid., p. 414: "Étant donnés de figures & de positions dans l'espace deux corps, l'un opaque & l'autre lumineux, trouver les équations des contours de l'Ombre & de la Pénombre du corps opaque, projetés fur une surface quelconque aussi donnée de figure & de position."

[21] Monge (1807, p. 13): "Étant données les coordonnées du sommet, trouver l'équation de la surface conique individuelle circonscrite à une surface courbe donnée."

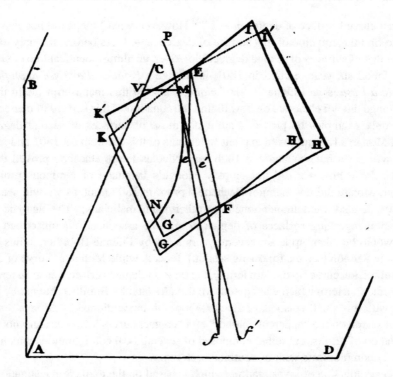

**Fig. 2.3** Figure 5 from Monge's "Mémoire sur les propriétés de plusieurs genres de surfaces courbes" (1780); the line *ef* is the one enveloping the shadowed area

circumscribing" it, Monge did not give the precise equation of this conical surface, but noted the following:

> "When the summit of the cone is a luminous point, the curve of contact of which we have just found the equations, is that which, on the given surface, separates the illuminated part from the dark part. When the top of the cone is the site of an eye, this contact curve is the line of the apparent contour of the surface given to the eye placed at the top of the cone."[22]

By that he related the analytical treatment to the practical research of shadows in descriptive geometry. He then remarked, as if in passing, how "it is shown with the same facility that, if the surface is of degree *m*, its contact curve with a circumscribed conical surface is

---

[22]Ibid., p. 14: "Lorsque le sommet du cône est un point lumineux, la courbe de contact dont nous venons de trouver les équations, est celle qui, sur la surface donnée, sépare la partie éclairée de la partie obscure. Lorsque le sommet du cône est le lieu d'un œil, cette courbe de contact est la ligne du contour apparent de la surface donnée pour l'œil placé au sommet du cône."

on another curved surface of degree $m - 1$."[23] However, what Monge did not elaborate on is how to find the equation of this surface of degree $m - 1$. As before, the only examples given are that of surfaces of second degree. Moreover, an almost identical line of reasoning is to be found six years earlier, in 1801, in Monge's *Feuilles d'analyse appliquée à la géométrie à l'usage de l'Ecole Polytechnique*;[24] also in this manuscript (as in the 1807 book) Monge did not elaborate on how to find the equation of this surface of degree $m - 1$, and the only examples he provided are again those of surfaces of second degree. But between Monge's two unproved explicit statements of this theorem (in 1801 and in 1807), a proof was nevertheless found. In 1806 Jean Nicolas Pierre Hachette proved the theorem.[25] Hachette however did not employ Monge's language of luminous projections. Moreover, Monge did not mention Hachette's proof in 1807 (and, as we will see, during the 1820s, it also went unmentioned by other mathematicians). The analytic line of investigation regarding surfaces of degree $m$ and the associated circumscribed conical surface would be taken up again several years later, by Étienne Bobillier, Julius Plücker and George Salmon (see the following section). Indeed, while Monge's "curve of contact" is the ramification curve (in modern terms), the $(m - 1)$-degree surface was to be termed the "polar surface", a term which was coined two decades later by Bobillier, but not by Monge. Already with this small episode, of Hachette's proof, unmentioned by Monge, one notes that the research on the "apparent contour" and "contact curves" was certainly not a linear accumulation of results, but rather composed of several local configurations which did not necessarily communicate with or continue each other.

Moreover, this line of investigation (which is based on the explicit investigation of the equations), which was implied in 1801 (and in 1807), and even before, during the 1780s, is somewhat ignored in the later manuscripts on descriptive geometry, although discussions within descriptive geometry on the "apparent contour" still appear. For example, in the sixth edition of Monge's *Géométrie descriptive* from 1838, edited by Barnabé Brisson (1777–1828), there is no analytic description, but rather the following description is given by Brisson himself in an appendix: "The apparent contour [. . .] [is] a curve to be determined on the surface of the body, using its particular character, which is to separate the part of the body that is visible from the part that is not, from an eye whose position is given."[26] A similar account is provided in 1855, when Jules de la Gournerie writes: "In practice, to obtain an apparent contour, a number of lines were drawn on the surface, and

---

[23] Ibid., p. 14–15: "On démontre avec la même facilité que, si la surface est du degré $m$, sa courbe de contact avec une surface conique circonscrite est sur une autre surface courbe du degré $m - 1$."

[24] Monge (1801, no. 5).

[25] Hachette in fact gave two proofs: one of his own and the other by Cauchy (Hachette 1806, p. 190–191), showing that the resulting surfaces of degree $m - 1$ are identical.

[26] "[. . .] le contour apparent [. . .] [est] une courbe qu'il faut déterminer sur la surface du corps, à l'aide de son caractère particulier, qui est de séparer la partie du corps qui est visible de celle qui ne l'est pas, par rapport à un œil dont la position est donnée" (Monge 1838, p. 164). This edition is identical to the 1820 edition of Monge's *Géométrie descriptive*, also edited by Brisson. As (Dhombres 1992, p. 333,

one enveloped their perspective in a curve. [. . .] Monge, looking for a general method to determine the shadow line of a body, imagined cutting it by planes parallel to the rays of light and leading to sections of tangents parallel to these rays";[27] but de La Gournerie did not mention the analytic calculation that Monge conducted. Gino Loria in 1908, in *Vorlesungen über Geschichte der Mathematik* also similarly described Monge's method, noting that the latter had already discovered it in 1775.[28] Strangely enough, any remark that other, more advanced methods, to determine this curve, were already found before 1908, is absent. These three examples exemplify the history of rereading Monge's treatment of the "apparent contour": each such rereading is neither neutral not an attempt to present Monge's investigations at the beginning of the nineteenth century in their own context, but rather re-situates those investigations in a new, local configuration.[29]

Before turning to the research on the "apparent contour" done during the 1820s, the 1840s and the 1860s by Bobillier and Salmon, one additional question should be raised when considering the examples given by Monge and his own drawings: why did Monge not consider explicitly more complicated surfaces, which would give rise to more complicated (projections of) apparent contours of the shadowed area, which would have nodes or cusps? Monge's research in his 1785 *Mémoire sur les développées* certainly shows that he was very much aware of cuspidal curves and the role they play in the research of surfaces, and especially of the "cuspidal edges" ["arête de rebroussement"] of developable surfaces.[30] But this research, though also continuing during the 1790s, was not about the projection of surfaces. A possible reason why Monge's research of projections did not account for cusps or nodes could be found in Monge's terminology: he always termed the illuminated body the opaque body. But in order that the image of an apparent contour (under projection) would have singular points such as nodes (or respectively cusps), the body should be transparent, or half-transparent, and the tangent lines exiting from the luminous point should be bi-tangent to the illuminated surface (or respectively tangent also to the apparent contour of this surface). These cases are not even mentioned in Monge's treatments; it was Salmon who was one of the first mathematicians to consider these singularities, but from a completely different point of view. One may claim that the parallels between the visual 'spectacle' of geometry and calculations of analysis

---

footnote 99) comments, the additions of Brisson (done after Monge's death) are somehow confusing and misleading.

[27] de la Gournerie (1855, p. 21, 25): "Dans la pratique, pour obtenir un contour apparent, on concevait sur la surface un certain nombre de lignes, et on enveloppait leur perspective d'une courbe. [. . .] Monge, cherchant une méthode générale pour déterminer la ligne d'ombre d'un corps, imagina de le couper par des plans parallèles aux rayons de lumière et de mener aux sections des tangentes parallèles à ces rayons."

[28] Loria (1908, p. 633–634).

[29] Cf. here also Goldstein (1995) on the history of rereading of one of Fermat's theorems.

[30] This research was done in the framework of finding the evolute of a space curve. See: (Delcourt 2011, p. 240–247).

encountered their limits when dealing with the projection of the "apparent contour": in order to obtain such a projection with cusps one would need to consider a surface of degree 3 (see Fig. 1.3b); to obtain nodes, one would need to consider a surface of degree 4—but both surfaces hardly appear in Monge's treatments on projections of surfaces.

## 2.2    1820s–1860s: Étienne Bobillier and George Salmon

The second example of mathematicians treating ramification curves revolves around the works of Étienne Bobillier (1798–1840) and George Salmon (1819–1904), Bobillier working in the 1820s and Salmon in the 1840s till the 1860s. I will start with Bobillier's research, who was working also on descriptive geometry, and was influenced by Monge's ideas.[31]

Before discussing Bobillier's work, one should note that in 1826 the mathematician François Vallès proved the following theorem regarding the "lines of contact", or in Monge's terminology, the "curve of contact": "The line of contact of a surface of any order with the circumscribed conical surface, always belongs to a surface of a lower order."[32] While Vallès did use Monge's terminology of "lines of contact", he did not mention Monge's or Hachette's results, did not introduce any terminology regarding this surface of "lower order", and also did not state explicitly that if the surface is of degree $m$, then the other surface is of degree $m - 1$. It was however Bobillier, who, following Vallès's methods, definitively introduced, in a series of articles between 1827 and 1828, the explicit nomenclature for this surface of degree $m - 1$: "surface polaire". Bobillier was a pupil of Siméon Denis Poisson, and at the age of 19 he was accepted into the École Polytechnique and studied there for a year; he had to cancel his studies due to financial reasons and in 1818 became a teacher at the École des Arts et Métiers in Châlons-sur-Marne. Besides his work on the theory of curves, surfaces and duality, he is best known for his book *Cours de géométrie*, first published in 1832, which ran through 10 editions, most of them published after Bobillier's death.[33]

Given a surface whose equation is $M(x, y, z) = 0$, in 1827 Bobillier called an equation of a certain surface "the polar equation of a curved surface [$M$] which cuts [. . .] along its lines of contact with the circumscribed conical surface which would have its summit at the origin."[34] Monge's "apparent contour" is now presented as the intersection of the "polar"

---

[31] A more detailed account of Bobillier's work on ramification curve, especially in the context of duality, see (Friedman 2022).

[32] Vallès (1826, p. 315): "La ligne de contact d'une surface d'un ordre quelconque avec la surface conique circonscrite, appartient toujours à une surface d'un ordre inférieur."

[33] For an extensive research on Bobillier's life and work, see (Haubrichs dos Santos 2015).

[34] Bobillier (1827–1828a, p. 92): "L'équation [...] est l'équation polaire d'une surface courbe qui coupe [...] suivant ses lignes de contact avec la surface conique circonscrite qui aurait son sommet à l'origine."

with the surface. While Monge also considered the "apparent contour" as an intersection of two surfaces, Bobilier individuated the second surface (by terming its equation explicitly), hence reshaping Monge's configuration for the investigation of projections. Bobillier's "polar equation" given at first is one possible equation of a polar surface, since he gave several possible equations; he did prove the theorem of Vallès, but now stated explicitly that if the original surface is of degree $m$, then the "lines of contact" are on surface of degree $m - 1$.[35] At this point Bobillier also introduced another equation, of what later would be termed the "polar surface", given by $a\frac{dM}{dx} + b\frac{dM}{dy} + \frac{dM}{dz} = 0$ (when working in $\mathbb{C}^3$).[36] The explicit term "polar surface" appeared in another paper of Bobillier later that year;[37] to emphasize, Bobillier was the one who coined the term.

That Bobillier's work resituated Monge's results in another mathematical configuration may be also noticed in that fact that he mentioned neither the works of Monge from 1801 and 1807 on the subject, nor the work of Hachette. Moreover, in a footnote added by Joseph Diaz Gergonne to Bobillier's 1827 paper, Gergonne noted this cynically: "M. Poncelet [in an article from 1828] observes, with much reason [...] that it is by mistake that M. Bobillier and we have attributed this theorem to M. Vallès since it is formally stated, although without proof, in Monge's [1807] book *L'Application de l'analyse à la géométrie* [...]. This proves that neither of us should compete with the memory of M. Poncelet."[38] But somewhat ironically, also Poncelet mentioned in this account neither the 1801 work of Monge nor Hachette's proof.[39]

---

[35] Ibid., p. 95.

[36] Compare Monge's equation (G) from 1775 in the previous section.

[37] Bobillier (1827–1828b, p. 258). In (Bobillier 1827–1828b), Gergonne reorganized Bobillier's results in two columns (ibid., p. 258), emphasizing the dual nature of Bobillier's theorems, thus causing Bobillier's work to take part in the argument between Poncelet and Gergonne concerning polar reciprocity vs. duality as the "right" principle to perform research in geometry. This reorganization was taking place although Bobillier did think of himself as being situated between Poncelet and Gergonne, as he used also the theory of "reciprocal polar" (ibid., p. 256), though these methods were later marginalized in a later article of Bobillier (See: Haubrichs dos Santos 2015, p. 194. See also: ibid., p. 157–159 on Gergonne's new vocabulary of geometry, using duality as a syntactic principle; regarding the relationships between Bobillier, Gergonne and Poncelet, see: ibid., p. 162–163).

[38] This note can be considered as a part of Gergonne's dispute with Jean-Victor Poncelet. Bobillier (1828–1829a, p. 140–141): "M Poncelet observe, avec beaucoup de raison [...] que c'est par erreur que M Bobillier et nous avons attribué ce théorème à M Vallès attendu qu il se trouve formellement énoncé, bien que sans démonstration, dans l'Application de l'analyse à la géométrie de Monge [...]. Cela prouve que nous ne devons ni l'un ni l'autre lutter de mémoire avec M Poncelet." For the dispute between Poncelet and Gergonne, see the footnote above. Poncelet's article, which Gergonne refers, is (Poncelet 1828).

[39] Poncelet not only mentions Monge's result only from 1807, but also notes that several of Bobillier's calculations are "prolix", which could have been spared by using the methods of Poncelet (Poncelet 1828, p. 301).

\* \* \*

Bobillier stopped working on this subject after 1829, and practically no one continued his line of research. More than a decade later, George Salmon, starting from the 1840s, also began investigating branch and ramification curves, without any knowledge of Bobillier's work. Though Salmon was influenced from Poncelet and Plücker, who followed, in one way or another, Monge's approach to descriptive geometry, one may notice in Salmon's treatment of ramification curves his claim, that he researched alone and for the first time this subject. To emphasize: here neither the term 'branch curve' nor the term 'ramification curve' were employed by Salmon—or the terms 'branch' or 'ramification point' for that matter. This is not at all surprising, as Riemann coined these terms only in the 1850s. We will see below how Salmon did term these curves.

Salmon, an Irish mathematician and theologian, is known today especially for his joint research, together with Cayley, on the 27 lines of the cubic surface.[40] His name and research, as Gow mentions, "would scarcely attract any attention among mathematicians so many years after his death if his reputation was based only on his research papers. [. . .] [His] fame lies in the influence exerted by four textbooks he wrote. These were: *A Treatise on Conic Sections*; *A Treatise on the Higher Plane Curves*; *Lessons Introductory to the Modern Higher Algebra*; *A Treatise on the Analytic Geometry of Three Dimensions*."[41]

How did these influential books treat surfaces and their branch curves? I surveyed in a previous work the work of Salmon on—in modern terms—ramification and branch points,[42] and I will concentrate here briefly only on the case of ramification and branch curves of complex surfaces. To stress: for Salmon, ramification points of complex projective curves are not called "ramification points" but rather are presented as the intersection points of the original curve with another curve, called the (first) "polar curve", whose equation is

$$x_1 \frac{dU}{dx} + y_1 \frac{dU}{dy} + z_1 \frac{dU}{dz} = 0.$$

when the point $(x_1 : y_1 : z_1)$ is not on the curve and $U = 0$ is the equation of the curve.[43] Here Salmon did employ the terminology "polar curve" and "first polar" often.[44] Note that if the degree of the curve is $m$, then the degree of the polar curve is $m - 1$. Formulated in modern

---

[40] Salmon however was not involved in the preparation of the various material models of the cubic surface.

[41] Gow (1997, p. 38). See also Gow's paper for a summary of Salmon's mathematical work and (Flood 2006, p. 208–209).

[42] Friedman (2019a, p. 148–153).

[43] Recall that the projective (complex) plane $\mathbb{CP}^2$ has three coordinates $x$, $y$, $z$ and hence a projective plane curve is expressed with three variables.

[44] Salmon (1852, p. 59ff).

language, the ramification points can be obtained by drawing all of the tangents exiting from the point $(x_1 : y_1 : z_1)$. Also, the result of Salmon was not new: Poncelet already noted in 1818 that the number of tangents from a given point to a curve of degree $m$ is $m(m-1)$, when considering the case when the point is at infinity, when in this case one deals with a "primitive figure".[45]

Salmon termed the intersection points of the curve with its polar "contact of tangents"; he emphasized that the tangents "can be drawn",[46] noting that "[t]he degree of the polar reciprocal of any curve is equal to the number of tangents which can be drawn from any point to that curve", and that "if $S$ be a conic section, two, and only two, tangents, real or imaginary, can be drawn to it from any point."[47] Though it is certainly not clear how one can *draw* imaginary tangents, this reflects the insight of Jemma Lorenat regarding the usage of "virtual figures" within projective geometry,[48] to be found especially in the constructions of Poncelet, Plücker and other geometers, and hence, as an implicit result, in the work of Salmon, as he was well acquainted with the results of these mathematicians. A figure hence would be "described and manipulated with enough unambiguous detail for the reader to construct their own illustration."[49] As Lorenat adds, regarding Poncelet, his "treatment of imaginary and infinite points admitted objects into pure geometry that no longer had an immediate figurative correspondence, and his definitions of the ideal challenged the intuitive nature of geometry."[50] Poncelet's method regarding "virtual figures" had obviously echoed the visualization tradition of Monge, and Salmon's approach, that one "can draw" even imaginary tangent lines, can be seen as a part of this visual tradition. In this sense, the (tangents to these) points could be virtually drawn.

The question arises, how do ramification and branch curves enter the picture, even if they are not termed as such? In a paper from 1847, after discussing the formula for the number of tangent lines to a curve of degree $m$ exiting from a given point, Salmon noted: "As I am not aware that the corresponding question as to reciprocal surfaces has been

---

[45] In: Poncelet (1817–18, p. 214); see also: ibid., p. 213–215 (question VII). This indicates that already Poncelet might have thought on drawing all of the tangents; however, there is no consideration of ramification points as 'special points', or an attempt to look at the corresponding situation regarding surfaces. To recall, Poncelet indicated, in his 1822 *Traité des propriétés projectives des figures* concerning the described figure—that "one never loses it from view [. . .] one never reaches consequences that cannot be painted in the imagination or in sight, by sensible objects." (Poncelet 1822, p. xxi: "[. . .] la figure est décrite, jamais on ne la perd de vue, [. . .] et jamais on ne tire de conséquences qui ne puissent se peindre, à l'imagination ou à la vue, par des objets sensibles."; Translation taken from: Lorenat (2015, p. 60).

[46] Salmon (1852, p. 62).

[47] Salmon (1855, p. 254).

[48] Lorenat borrows this concept from Dominique Tournès; according to Tournès, a "virtual diagram [is one] that one must have in mind, but that is no longer physically drawn on the paper, or at least which is left to the reader to draw." (2012, p. 272). See also (Lorenat 2014).

[49] Lorenat (2015, p. 65).

[50] Ibid., p. 115.

before investigated, I propose in the present paper to enquire [into this] [. . .]."[51] Obviously, what is missing here is any reference to Bobillier's research, of which Salmon was indeed unaware. Salmon then added: "The degree of the reciprocal surface is plainly the same as the number of tangent planes which can be drawn to the surface through a given line: now we know that all the points of contact of tangent lines passing through a given point lie on a surface of the $n - 1^{st}$ degree, which we call the $(n - 1)^{st}$ polar surface of that point."[52] Salmon noted that "the point of contact of tangent lines"[53] (in modern terms—the ramification curve) is the intersection of $U = 0$ and $dU/dz = 0$, when $U$ is the equation of the surface,[54] and proved the following two theorems: "(1) Every cone touching a surface of the $m^{th}$ degree must in general have $\frac{m(m-1)^2(m-2)}{2}$ double sides, real or imaginary. (2) Of these double sides $m(m - 1)(m - 2)$ are cuspidal lines, and consequently there are only $\frac{m(m-1)(m-2)(m-3)}{2}$ ordinary double lines."[55] This numerical investigation of the number of special tangents to the ramification curve points already to a future investigation (of Salmon) of the branch curve (though not termed as such), as will become clearer in 1862. However, to emphasize now, an "ordinary double side" (or "line") of a cone is a line on the cone (exiting from the vertex of it), which is tangent *twice* to the surface, when each tangent point is a simple one; this results, as we will see, in a node of the branch curve. If a line on the cone is tangent to the surface, such that it is also tangent to the ramification curve, then this line is also a "double side" and is called a "cuspidal line", resulting in a cusp of the branch curve. This numerical investigation of special tangents can be seen as a restructuring of the epistemic configuration which was set by Bobillier and beforehand by Monge, a restructuring which I will discuss later. Nevertheless, it should be stressed that the terms themselves "cusp" and "node" were not used in the 1847 paper as associated to the "cuspidal line" resp. to the "double tangent". In 1849 Salmon published another paper, where he explicitly wrote the number of the cusps and the nodes of the branch curve, as well as its degree—but as with the ramification curve, the branch curve is not termed as such, but rather described as the intersection of the cone with "any plane".[56] Here however, the nodes—termed "double points"—and the "cusps" of this cut are named as such.[57]

In 1862, Salmon returned to investigate this topic, when he published his book *A Treatise on the Analytic Geometry of Three Dimensions*, considering complex projective surfaces in the complex projective three-dimensional space $\mathbb{CP}^3$. His treatment of projection of surfaces followed a similar line of interpretation as his treatment of complex curves: given a smooth surface $U$ in $\mathbb{CP}^3$, choosing a point $O = (x' : y' : z' : w')$ not lying on the

---

[51] Salmon 1847, p. 65.

[52] Ibid., p. 66.

[53] Ibid.

[54] Ibid.

[55] Ibid.

[56] Salmon (1849, p. 188–189).

[57] Ibid., p. 189.

surface, and examining the tangent lines to $U$ passing through $O$. Salmon called this collection of all lines tangent to the surface the "tangent cone", and called to:

> "consider the case of tangents drawn through a point not on the surface. [...] we see that the points of contact of all tangent lines (or of all tangent planes) which *can be drawn* through $x'y'z'w'$, lie on the first polar [denoted by $\Delta U$], which is of the degree $(n-1)$: viz.
>
> $$x'\frac{dU}{dx} + y'\frac{dU}{dy} + z'\frac{dU}{dz} + w'\frac{dU}{dw} = 0.$$
>
> And since the points of contact lie also on the given surface, their locus is the curve of the degree $n(n-1)$, which is the intersection of the surface with the polar."[58]

As was already noted above, Salmon did not mention the previous research that had been done concerning ramification curves, though he certainly used the terminology of Bobillier, which might indicate that the *terms* themselves (which Bobillier coined) were well known. As Salmon was autodidact, it may very well be he encountered these terms when reading the works of Poncelet, which he also mentioned in his works.[59]

* * *

In contrast to Monge, not a single *actual* drawing of the "curve of the degree $n(n-1)$" appeared in Salmon's book, but rather a virtual drawing is implied, in the form of tangent lines or a "tangent cone" which "can be drawn". This virtual drawing stands in contrast to the actual drawing of this cone by Monge.[60] Here Salmon implicitly considered projections of the surface from a point.

Considering the fact that Salmon did draw curves in general,[61] one might wonder why he did not draw any branch curves, being the projection of the ramification curve on a complex plane. This question is justified given that Salmon did take the branch curve into account in his investigations, as I will show later. However, the lack of explicit reference to visual means in the 1862 book was somehow corrected in the translation of Salmon's book into German, done by Wilhelm Fiedler, a translation whose second edition was published in 1874.[62] The translation mentions material models of several surfaces in an appendix,

---

[58] Salmon (1862, p. 190).

[59] Gow (1997, p. 45–46, 54).

[60] Salmon did not draw a single sketch of any complex surface in the 1862 book, but rather only partial images of concrete situations (e.g. tangent planes or tangent lines, for example; see Salmon (1862, p. 274 or p. 296)). This might be also due to constraints on printing techniques during this period, but also in accordance to how Monge and his followers were using concrete images.

[61] See: Friedman (2019a, p. 149).

[62] Felix Klein (1926, p. 165) notes that while Salmon's works lack systematic development and resemble a walk in a garden with "no systematic representation", Fiedler's translations of them corrected the situation. On Fiedler's work, see: (Volkert 2018, 2019).

models mostly made in Germany.[63] Fiedler did point explicitly to such models of cubic surfaces: "Of the general surface of third-degree with 27 real lines, the editor owns a rod model since 1865 [...]. Since 1869 the plaster model, constructed by [Christian] Wiener, has been disseminated; it was created due to the suggestion of Clebsch. [...] Recently, the surfaces of third-degree with four nodes [...] have been modeled. The principle of rod models is convenient for many other cases."[64] Moreover, in the text itself, when Fiedler discussed the possible number of nodes that a surface of degree 3 can have,[65] he also discussed ways of gaining "intuition [Anschauung]" to the various algebraic forms of nodal surface of degree 3.[66] This paragraph did not appear in Salmon's original text, and was certainly Fiedler's addition. Giving references to Klein's work on the subject from 1873,[67] it is clear that Salmon did not refer to this work in his 1862 book. Klein refered in this paper both to the various material models of the cubic surface, as well as to how to approach them intuitively, describing the "concrete *Anschauung* of the shape [Gestalt]" of the cubic surface,[68] and that the material model enables this concrete *Anschauung*,[69] also referring constantly to figures added to the paper. Hence one may suggest that Fiedler's additional remarks situate the translation in another configuration, i.e. in the then flourishing German tradition of material model construction of mathematical curves and surfaces,[70] as well as in the German discourse of the late nineteenth century regarding the role of "Anschauung" within mathematics.[71] This also indicates that the material, three-dimensional models were operating at an opposite epistemological pole to the "virtual images", as these models were concrete and material, in contrast to the virtual images. Notwithstanding this opposition, Fiedler neither added in the book any figure nor did he edit or remark about the treatment of branch curves and their lack of illustrations.[72]

---

[63] Salmon (1874, p. 622–623, 663, 667).

[64] Ibid., p. 663, footnote 118, added by Fiedler: "Von der allgemeinen Fläche dritten Grades mit 27 reellen Grades besitzt der Herausgeber ein Stabmodell seit 1865 [...]. Seit 1869 ist ein von Wiener construiertes Gyps modell mehrfach verbreitet, das auf Anregung von Clebsch entstanden ist. [...] Neuerdings sind auch die Flächen dritter Ordnung mit vier Knotenpunkten [...] modelliert worden. Das Princip der Stabmodelle ist auf sehr viele andere Fälle bequem anwendbar."

[65] Fiedler terms nodes of plane curves "Doppelpunkte" and nodes of surfaces "Knotenpunkte" (Ibid., p. 355)

[66] Ibid.

[67] In: ibid., p. 663, footnote 121; the reference is to: (Klein 1873).

[68] Ibid., p. 554.

[69] Ibid., p. 574.

[70] On Fiedler's models and his conception of modeling in mathematics, see: (Volkert 2018, 2019).

[71] The literature on the acceptance and development of the concept of "Anschauung" during the nineteenth century is vast. See for example: (Volkert 1986; Friedman 2000, 2012; Bouriau et al. 2016).

[72] One may ask why there was no model of the cubic surface with a ramification curve drawn on it made during the nineteenth century. To recall, in 1849 Arthur Cayley and Salmon proved that every

One should additionally note that Fiedler himself also added remarks concerning the ramification curve. More precisely, in the second German edition of Salmon's book *Analytische Geometrie des Raums*[73] a detailed appendix with references and bibliography was added as well as remarks concerning what happens to the number of cuspidal and bi-tangent lines when the projection point lies on the surface.[74] However, neither Bobillier's work on the polar surface was mentioned, nor the terms "branch curve", "ramification curve", i.e. in German "Verzweigungskurve" were employed.[75] This, however, one might argue, was precisely the place where Bobillier's work on this surface should have been discussed. Indeed, Fiedler did know of Bobillier's work on polars, as he referred in 1873 to Bobillier's article "Théorèmes sur les polaires successives".[76]

* * *

Returning to Salmon's 1862 book, after defining the ramification curve as the "locus" of all the points on the "curve of [. . .] degree $n(n - 1)$",[77] he continued to investigate the "cuspidal lines" and the "double lines"—the two types of special tangents to it, as he did in the 1847 paper, though in 1862 he gave a more elaborate description. It is important to emphasize that these two types of tangents were not discussed in Bobillier's works, which does suggest the novelty of Salmon's mathematical configuration. The first are tangent lines called "inflexional tangents", which are not only tangent to the surface, but in addition

---

non-singular complex cubic surface, embedded in the projective complex space $\mathbb{CP}^3$ contains exactly 27 lines. Several mathematicians were in fact occupying themselves with the task of finding a specific equation of a cubic surface, such that all of the equations of the 27 lines would be real equations. The idea of building a material model was explicitly expressed in 1861 by James Joseph Sylvester (1861, p. 979): "I propose to build a system of 27 straight lines out of strings or brass." But without getting into the history of the construction of models of cubic surfaces (Lê 2015), smooth or singular,—or of surfaces of higher degree—by inspecting the models of surfaces themselves, one cannot find any drawings of ramification curves on them. One may argue that in order to visualize the ramification curve on the (real part of the) surface, one had to model also the projection to the plane. For curves there were actually models, which showed how spatial curves are projected onto plane curves (see for example the models from 1879 of Christian Wiener, in: (Dyck 1892, p. 298)). For surfaces, however, no such projection was considered when modeling them. This actually might be due to technical difficulties: to produce a surface with a possibility of projecting it to a plane, such that one would see its branch or ramification curve, the model of the surface would have to be made out of a semi-transparent material—which was, when considering the materials from which the models were produced at the end of the nineteenth century (usually strings, plaster or wood), quite expensive and technically complicated.

[73] The first edition of the German translation was published in 1865; the appendix section ("Zusätze") does not contain any references concerning polar surfaces (See: Salmon 1865, p. 553–562).

[74] Salmon (1874, p. 619).

[75] That these terms are not mentioned is not surprising: "Verzweigungskurve" was used in the literature only starting the end of the nineteenth century (see the short discussion on terminology below).

[76] See: Salmon (1873, p. 455). Fiedler refers to (Bobillier 1828–1829b). I thank Klaus Volkert for pointing this out.

[77] Salmon (1862, p. 190).

are also tangent to the ramification curve itself. Salmon proved that these points lie on the intersection of the ramification curve (itself defined and termed as the intersection of the surface and its "first polar") and the *second* polar of the surface, i.e. the surface $\Delta(\Delta U)$ or $\Delta^2 U$, repeating his calculation from 1847: "Through a point not on the surface can in general be drawn $n(n - 1)(n - 2)$ inflexional tangents" (being the intersection of the surface, the first and the second polar). The second type of special tangents to the surface are lines which are tangent to the surface at two different points: "Through a point not on the surface can in general be drawn $\frac{1}{2}n(n - 1)(n - 2)(n - 3)$ double tangents to it."[78]

Salmon also noted that "[w]e have proved then that the tangent cone which is of the degree $n(n - 1)$ has $n(n - 1)(n - 2)$ cuspidal edges, and $\frac{1}{2}n(n - 1)(n - 2)(n - 3)$ double edges; that is to say, any plane meets the cone in a section having such a number of cusps and such a number of double points."[79] Here the branch curve is defined and termed by Salmon as the curve obtained by a "plane [that] meets the cone in a section". This curve, which hardly stands at the center of Salmon's investigation, therefore has $n(n - 1)(n - 2)$ cusps and $\frac{1}{2}n(n - 1)(n - 2)(n - 3)$ nodes, as already discussed in 1847. Salmon did not even consider the option that the tangent cone may have more "complicated" edges (or equivalently, that the branch curve has singularities of higher order). Moreover, one may wonder why Salmon did not take notice of the special properties of the branch curve of a cubic surface, a surface he knew well. Regarding this surface, Salmon only briefly indicated that "[t]he tangent cone whose vertex is any point, and which envelops such a surface is, in general, of the sixth degree, having six cuspidal edges and no ordinary double edge."[80]

Salmon however did use the characteristics of the branch curve of a cubic surface to show that the number of "double points" (i.e. nodes, in modern formulation) of a singular cubic surface cannot be arbitrary. In Salmon's own language:

"since every double point on the surface adds a double edge to the tangent cone, there cannot be more double points than will make up the total number of double edges of the tangent cone to the maximum number which such a cone can have. Thus a curve of the sixth degree having six cusps can have only four other double points; therefore since the tangent cone to a cubic is of the sixth order, having six cuspidal edges, the surface can at most have four double points."[81]

Salmon noted that a "curve of the sixth degree having six cusps"—being the branch curve of a cubic surface—can have at most four nodes; to prove this he might have used

---

[78] Ibid., p. 191.

[79] Ibid., p. 192.

[80] Ibid., p. 376. However, as we will see later (see Sect. 3.2.1, and also Fig. 1.3b), when considering the branch curve of the (smooth) cubic surface, its six cusps lie on a conic. This is a non-trivial theorem, which Salmon might have had the tools to discover.

[81] Ibid., p. 379.

*numerical restrictions* stemming from examining the branch curve to impose restrictions on the singularities of the surface itself.[82]

While Salmon's numerical investigation was part of his method of work, for the discussed surface (a cubic surface with 4 nodes) no possible visualization was mentioned, also not of the branch or ramification curve. Indeed, whereas Salmon was not interested in explicit, actual visualization of curves on surfaces,[83] let alone ramification curves or their projection, other attempts, however, were made to employ visual tools to study the branch curve. But the employment of visual tools was done in another setting: as we will see, Wilhelm Wirtinger was the first to explicitly consider *visually* a neighborhood of a singular point of the branch curve.

## 2.3   1890s–1900s: Wirtinger's and Heegaard's Turn Towards Knot Theory

Wilhelm Wirtinger (1865–1945), an Austrian mathematician, is known for his work in complex analysis and knot theory, and especially for his presentation of the fundamental group of the complement of a knot in $\mathbb{R}^3$.[84] He began in the 1890s a research on branch curves of complex surfaces, which did not mature into a full-blown theory, but nevertheless prompted a re-contextualization of the research of singular complex plane curves by means of knots associated to their singular points; this re-contextualization was also prompted by the doctoral thesis of Poul Heegaard (1871–1948) on algebraic surfaces. Here the issue of terminology must be stressed: Wirtinger was probably the first one among the mathematicians discussed here who employed the Riemannian term "Verzweigung" to describe "branch curves" (and more generally "branch manifolds"). This usage of terminology is not surprising: Riemann's term ("Verzweigungspunkt") was in the 1890s well accepted, and the term "Verzweigung" was used as well in number theory starting the 1880s (by Weber and Dedeking; see Sect. 1.3.1).[85]

In the 1890s, Wirtinger considered projections of complex surfaces. Since his work on the subject has already been discussed extensively,[86] I will only survey his research briefly, concentrating on his study of branch curves. Wirtinger dealt with these curves in a letter

---

[82] To explain: otherwise, i.e. when the number of nodes would be greater than 4, then the number of the inflection points would be negative. The formula of the number of inflection points $3m(m-2) - 6n - 8c$ (when $m$ is the degree of the curve, $n$, $c$ are the number of nodes and cusps respectively) was discovered by Plücker and hence was known by Salmon (though this argument did not appear explicitly in the above citation).

[83] Except perhaps the visualization of lines of surfaces, as in the case of the cubic surface.

[84] I will discuss this group (in the context of branch curves) in more detail in Sect. 3.2.1.

[85] On the other hand, as we will see in Sect. 3.1.2, during the 1870s one also employed the term "Uebergangscurve" to designate the ramification curve.

[86] Epple (1995); Friedman (2019a, p. 153–158).

written to Felix Klein on 22 December 1895 and in a lecture held in 1905. In 1895, his aim "is to show that an arbitrary system of $n - 1$ dimensional algebraic varieties with an associated branching scheme can *always* be understood as a system of branch manifolds of a function of $n$ variables",[87] indicating however the possibility that not every $n - 1$ dimensional manifold can be a branch manifold of an $n$-dimensional function. This is since one has to distinguish between two cases: when the branch manifold ["Verzweigungsmannigfaltigkeit"] is smooth and when the branch manifold is singular. The example that Wirtinger gave for a singular branch curve is a cubic surface in $\mathbb{C}^3 : \{(x, y, z) \in \mathbb{C}^3 : z^3 + 3zx + 2y = 0\}$. He however did not specify explicitly the branch curve, being the curve $\{x^3 + y^2 = 0\}$ under a projection $(x, y, z) \to (x, y)$, which has a cusp at $(0, 0)$. Nevertheless, he noted that the Galois group[88] is not cyclic, and that it is the whole symmetric group on three letters. The main question that arises was formulated by Wirtinger as follows: "Can this group be arbitrarily pre-determined, or is it bound to conditions, so that associated functions exist?"[89] Hence, Wirtinger asked what the possible singular points (or subvarieties) may be that the branch manifold might have under a generic projection.

In 1901, in his article on algebraic functions and their integrals for the *Enzyklopädie der mathematischen Wissenschaften*, Wirtinger noted, "one has not yet succeeded in determining a given algebraic variety by a finite number of data in a similar way as is possible with the different forms of a Riemann surface."[90] Four years later, in a 1905 lecture entitled "Über die Verzweigungen bei Funktionen von zwei Veränderlichen" ["On the branching of functions with two variables"] a shift within the configuration Wirtinger worked in occured in the way he examined branch curves. Although the lecture itself is not available, Moritz Epple has reconstructed it and showed that Wirtinger, inspired by Heegaard's 1898 dissertation, decided to consider only the local neighborhood of singular points of branch curves. What was the nature of Heegaard's influence? It is this work that I would like to examine in detail, since it has hardly been dealt with in the literature.

* * *

Poul Heegaard (1871–1948), a Danish mathematician, is well known today for his research in topology. In his thesis he introduced what is today known as Heegaard splitting

---

[87] Epple (1995, p. 398): "Mein Ziel ist dabei zu erweisen, dass man ein beliebiges System algebraischer Gebilde von $n - 1$ Dimensionen mit zugehörigem Verzweigungsschema *immer* als System von Verzweigungsmannigfaltigkeiten einer Function von $n$ Variablen auffassen kann."

[88] The Galois group is the group generated by permutations that can be induced by encircling the neighborhood of the singular point of the branch curve, examining the resulting permutations of the preimages.

[89] Ibid.: "Kann man diese Gruppe willkürlich vorgeben, oder ist sie an Bedingungen gebunden, damit zugehörige Functionen existieren?"

[90] Translation taken from: (ibid., p. 383).

of a 3-manifold, followed by the Heegaard diagrams. Though not dealing with 3-manifolds (i.e. manifolds whose dimension is 3), I would like to briefly examine his thesis.

The thesis of Heegaard, titled *Forstudier til en topologisk Teori for de algebraiske Fladers Sammenhang* [*Preliminary studies for a topological theory in the context of algebraic surfaces*], deals with four-dimensional real manifolds, and especially with complex algebraic surfaces, being a specific kind of these four-dimensional manifolds.[91] Heegaard noted that the task of visualization of these manifolds is more difficult in comparison to the case of complex algebraic curves:

> "Were one to build a theory for functions of two independent variables, it would therefore be natural to seek a similar presentation [to the one of the theory of functions of one complex variable]. [. . .] [But] investigations for two independent variables are much more difficult than for one. The multitude of possibilities produces phenomena for which no analogues exist for one variable. Thus, Picard notes 'On voit, par ce qui précède, les différences profondes qui séparent la théorie des fonctions algébriques d'une variable de la théorie des fonctions algébriques de deux variables indépendantes. L'analogie qui souvent est un guide excellent, peut devenir ici trompeuse.'"[92]

Although the analogy with the one complex variable case might be deceitful ["trompeuse"], Heegaard did list the research directions that were taken, which attempted to make the analogy between functions of one and two complex variables. He noted the "elementary algebraic investigations" by adjoint polynomials by Max Noether and Alfred Clebsch, linear systems of curves of Guido Castelnuovo and Federigo Enriques and "[s]urface transformations [. . .] by enumerative geometry by Zeuthen"; he highlighted also the "investigations through transcendental functions" and the "topological investigations".[93] However, these analogies—especially regarding the topological approach—are, according to Heegaard, not complete, and further research must be done; Heegaard criticized Poincaré's and Picard's writings on the subject, and concludes the introduction as follows:

> "For the algebraic surfaces there had to be created something which would correspond to the Riemann surfaces for the algebraic curves. A topological criterion for the bijective correspondences among such had to be established. [. . .] the material that existed beforehand for such an investigation was either insufficient or full of mistakes. The contents of the following pages originate in my attempts to correct these mistakes and to fill these gaps."[94]

---

[91] See: Heegaard (1898). The thesis was translated to French in 1916, titled "Sur l'Analysis situs". The motivation for the translation is first and foremost, as mentioned in the introduction, to promote the dissemination of Poincaré's *Analysis situs* (Heegaard 1916, p. 161). The English translation of the 1898 thesis is taken from preprint of a translation into English done in 2007 by Hans J. Munkholm.

[92] Heegaard (1898, p. 1).

[93] Ibid., p. 4–5.

[94] Ibid., p. 6.

One of the main tasks standing at the center of the thesis is that of the visualization of four-dimensional manifolds. The first chapter begins with comparing ramification points (on a Riemann surface) and the ramification curve (on an algebraic surface):

> "one [...] constructs a Riemann surface, [which] is essentially characterized by the number of leaves and the number of branching points on the corresponding [...] [Riemann] surface [...]. [One can] carry out an analogous investigation of the connectivity of an algebraic surface $z = f(x, y)$ [...]. By multiple covering [of] this collection [of points on the surface] and introducing suitable 'branching surfaces' ['Forgreningsflader'] and connecting these by 3-dimensional creations through which the different 'layers' are connected to one another, one can create a 4-dimensional manifold in which the connectivity of the algebraic surfaces can be studied."[95]

How this "creat[ion of] a 4-dimensional manifold" can be done is explained in the thesis. Heegaard noted that in the investigation "everything depends on a clear visualization of the elements" and that the chosen ways to describe mathematically the four-dimensional manifolds "must be as easy to visualize as possible".[96] One of the proposed ways to visualize these manifolds, being complex algebraic surfaces, was to take a complex point $(x, y)$ in $\mathbb{C}^2$, and, while considering the two complex coordinates as $x = x_1 + i \cdot x_2$ and $y = y_1 + i \cdot y_2$, to look at the point $(x_1, x_2, y_1)$ in $\mathbb{R}^3$, while "equipping" it with an additional number $y_2$. This way of visualization is compared to how one draws maps: "The procedure is similar to the one used by a surveyor when he presents and investigates points in space on a map: the points are replaced by their projections on the picture plane, and the projections are equipped with an elevation number representing the height over, or the depth under, the plane."[97]

With this way of visualization Heegaard described several objects, such as the line or the plane. However, a second way of visualization is described later. Given a complex algebraic curve in $\mathbb{C}^2$ (when $\mathbb{C}^2$ is considered as a four-dimensional manifold), Heegaard's idea was to cut it with a three-dimensional sphere in $\mathbb{R}^4$, and to consider what happens inside and outside of this sphere, taking into account the points in infinity, that is, the projective part. The aim of Heegaard is stated as follows: "Just as an algebraic function of one independent variable can be spread out in an $n$-valent way on a sphere, it must similarly be possible to spread out the algebraic function $z = f(x, y)$ defined by the [...] equation [$F(x, y, z) = 0$ of degree $n$] in an $n$-valent way on the manifold representing the [complex] plane $(X, Y)$. To each point in the plane there corresponds a value of $z$", i.e. at most $n$ values of $z$. At this point Heegaard describes the numerical properties of a branch curve of an algebraic surface of degree $n$ in $\mathbb{C}^3$: "These"—i.e. the $n$ values of $z$ which correspond to a point in the complex point—"will all be different when the straight line through the

---

[95] Ibid., p. 7–8. The "branching surfaces" are termed in the French translation: "surface de ramification" (Heegaard 1916, p. 169).

[96] Heegaard (1898, p. 8).

[97] Ibid., p. 9.

point and parallel to the Z–axis intersects the surface in $n$ separate points. The collection of points [...] for which this does not happen we shall call the *branching surface* [*forgreningsfladen*]. This will be the surface that represents the boundary [*Konturen*] of the algebraic surface in the plane $(XY)$".[98] This boundary, Heegaard pointed out, is a complex algebraic curve of order $n(n-1)$, having $n(n-1)(n-2)$ cusps and $n(n-1)(n-2)(n-3)$ nodes;[99] however, a proof why this curve has *this* number of cusps and nodes, or why it does not have singularities of higher order, is not given. A possible source of these calculations is Salmon's 1862 book, as Heegaard mentioned it in the bibliography of the thesis.[100]

As described by Epple,[101] the rest of the thesis shifts thematically and focuses on knots and links and the different ways to signify their crossing, and only towards the end of the thesis Heegaard declared, "we shall present some applications of the results stated regarding the theory of algebraic surfaces."[102] In this context, Heegaard focuses only on cusps and nodes of a branch curve. At the end of his thesis Heegaard, similarly to the procedure presented before, considers a sufficiently small four-dimensional ball around a point $P$ of the branch curve $C$. Its boundary is a three-dimensional sphere whose intersection with the branch curve is a knot $K$, which might be eventually unknotted. The restriction of the algebraic surface, branched over $C$, to this three-dimensional sphere, is a three-dimensional Riemannian manifold branched over $K$. One of the examples Heegaard examined, though not calculating explicitly, is the example of $P$ being a cusp;[103] he noted, in contemporary terminology, that the intersection of the (complex) cuspidal curve $x^3 + y^2 = 0$ with the boundary of a small enough four-dimensional ball, whose center is $(0,0)$, is the *trefoil knot*.[104] This is exactly the example that is taken up again by Wirtinger in his 1905 lecture.

* * *

Returning to Wirtinger, in his 1905 lecture, he did not only ignore his former questions, such as those regarding the possible singular points of the branch curve, but he also re-contextualized the problem. Wirtinger now no longer considered the branch curve as a whole, did not mention the total number of cusps and nodes of the branch curve (of a smooth surface), as was noted explicitly by Heegaard, but rather focused only on the *intersection* of a neighborhood of a singular point of the branch curve with the 4-sphere, obviously influenced from Heegaard's treatment in the last part of his thesis, as seen above.

Wirtinger tried to visualize what the singular points of the branch curve look like *locally* —introducing probably in the talk an image of the knot obtained as the intersection of

---

[98] Ibid., p. 31.

[99] Heegaard does term as "cusps" as "Spidser" and "nodes" as "Dobbeltpunkter" ["double points"].

[100] Ibid., p. 96.

[101] Epple (1999, p. 248–250).

[102] Heergard (1898, p. 84).

[103] Another example given is the one of the node.

[104] See also: Epple (1999, p. 251).

$\{x^3 + y^2 = 0\} \cap \{|x|^2 + |y|^2 = c\}$, for $c$ small and positive number, and when $x, y \in \mathbb{C}$.[105] However, he ignored the *global* question that he himself had posed several years before: can every curve with this 'local behavior' be a branch curve? To emphasize—this is not the case of virtual figures, as with Poncelet or Salmon, since in Wirtinger's approach there is no reference to absent figures of the entire branch curve, but rather a reference to an existing figure of a *local* setting of it.

Although being one of the first attempts to actually visualize a branch curve, the problematic issue lays in what Wirtinger chose to illustrate: only a local situation. One may claim that the image of the local behavior in fact hindered the understanding of the global behavior of (the singular points of) the branch curve, and that Wirtinger's visualization may have led also to a dead lock in his research on branch curves.

## 2.4    The End of the Nineteenth Century: A Regression Toward the Local

As the title of this chapter implies, 'separate beginnings' indicate separate local epistemic configurations which were involved in the research on the branch curve; this separatedness, and sometimes even ignorance of other research directions, manifested itself in three ways. First, some authors had no knowledge of the previous work that had been done on the subject. This was the case with Bobillier and Gergonne, who were not aware of Monge's (and Hachette's) earlier research; Gergonne, to recall, reacted cynically to Poncelet's remark on this ignorance, which one might interpret as an underestimation of Bobillier and Gergonne's lack of awareness. Concerning the research that was done later, Salmon, Wirtinger or Heegaard were probably not even aware of Bobillier's results. Second, several authors posited themselves as the first to investigate ramification and branch curves, even when it was not the case. These statements, which can be regarded as performative acts, produce—even if unintentionally—ignorance of the older configuration. Obviously, the phenomenon of discoverers of theorems being forgotten while their results become well known is not rare; but it may well result in a later generation of mathematicians also not knowing the former generation and its results. This was certainly the case of Salmon, who explicitly expressed this, but we will see later other, more subtle forms of ignoration, even when certain authors knew of earlier works. Third, 'branch curve' did not exist as a well-accepted term in the nineteenth century: we saw that various mathematicians gave this curve and the associated surfaces various names, with each reflecting the character of the epistemic configuration in which it was coined—Monge and his "contour apparent" and its corresponding projection, Bobillier's "surface polaire," Salmon's ramification curve as a

---

[105] As Epple notes (ibid., p. 256), Wirtinger supported visualization, functioning as a "productive imagination" done with the help of a drawing, and being not merely a "passive" one, but one, like in a painting of a painter, which produces its own reality (Wirtinger compares the mathematician to a painter in a letter to Klein from 1896; see: Epple 1995, p. 387).

"locus of intersection," and finally Wirtinger's "Verzweigungsmannigfaltigkeit" and the associated notion of "Verzweigung." To that one has to add another term, which I will discuss in the following chapter: the "Uebergangscurve" ("transition curve"), which was employed, in the context of algebraic surfaces, mainly during the 1870s, to designate the ramification curve, but later also the branch curve.

When taking into account how the branch curve was researched during the nineteenth century, one can note a shift toward an investigation of more 'local' objects. As we saw with Wirtinger, there was a regression in the investigation of the entire branch curve (of a complex surface) toward a focus on the local neighborhood of the singular points of the branch curve. This also prompted a shift in configuration, as the investigation was now situated within the theory of knots, and hence explicit visualization techniques were addressing these obtained knots.[106] On the other hand, when the entire branch curve was considered, it was—when concentrating on the visual aspects—by "virtual figures." The investigation itself was mostly analytical (observing that the ramification curve is an intersection of a surface and its first polar) or numerical (counting the number of special tangents). Questions concerning the necessary and sufficient conditions for a plane curve to be a branch curve—though sometimes raised—were impossible to answer with the then available mathematical tools, and were quickly abandoned. However, this global point of view was to be found again in research done by the Italian school of algebraic geometry starting at the end of the nineteenth century, as we will see in the next chapter.

---

[106] Note that during that time, there was still no equivalent research of real 3-dimensional branched coverings, a research which started only during the 1920s (see the Sect. 1.3.1).

# 1900s–1930s: Branch Curves and the Italian School of Algebraic Geometry

3

In contrast to the separate beginnings during the nineteenth century presented in the previous chapter, the twentieth century saw the consolidation of these beginnings—and others—into a more unified research configuration concerning the branch curve. A significant role was played by the Italian school of algebraic geometry which, between the 1890s and 1930s, contributed greatly to the research on algebraic surfaces and, as a result, to the understanding of branch curves.

There are already several thorough accounts of this school and of its main players,[1] and that is precisely why I do not attempt to present an overview of it here. On the contrary, I will concentrate on how several of its main actors—Castelnuovo, Enriques, Beniamino Segre and Zariski—shaped and reshaped the research of algebraic surfaces and of nodal–cuspidal curves, as a result of their research on branch curves. Hence, I will follow the methodology delineated in the introduction: considering the branch curve not as a constant, given, predefined or well-defined and unchanging object, but rather as a research object, which is situated and resituated in various epistemic configurations along the history of the Italian school of algebraic geometry.

Before delineating the structure of the chapter, one question should be addressed: In what sense can one speak of an 'Italian school,' and what does 'school' actually mean? One may define a (mathematical) school as a space for thinking, promoted by a common conceptual understanding and methodology, applied to a wide variety of mathematical problems and disciplines. Schools can be concentrated around one person (famous examples are the Emmy Noether school[2] and the Klein school;[3] a less famous example

---

[1] Brigaglia / Ciliberto (1995), Brigaglia (2001), Conte / Ciliberto (2004), Guerraggio / Nastasi (2006), Casnati et al. (2016), Israel (2004). See also: (Dieudonné 1985, chapter VI and esp. p. 50–54).

[2] Koreuber (2015).

[3] Scharlau (1989, p. 119).

is Chisini's school[4]), around a place (the Göttingen school)[5] or even around a broader, common 'national' theme that unites mathematicians from the same country—such as the Italian school of algebraic geometry. These designations enable us to recognize a school as both a space and a movement within mathematics. While a discussion on the definition of a scientific or mathematical school is outside the scope of this book,[6] the question does arise as to whether the Italian school of algebraic geometry can be considered homogeneous. Norbert Schappacher points out that although "violent polemics have accompanied the history of Italian Algebraic Geometry ever since the golden beginnings [and] [. . .] rival research agendas [existed at the same time]," one may "speak of 'the Italian school of algebraic geometry' in that many Italians have helped create the field [. . .]. However, trying to identify 'typically Italian' notions or methods of research in Algebraic Geometry is problematic, and it can be advisable to ban the epithet 'Italian' from the historical investigation insofar as it may carry unwarranted connotations like 'intuitive,' 'loose,' or worse, charged with national metaphors."[7] Moreover, Aldo Brigaglia claims that "the Italian school was not strictly a national 'school', but rather a working style and a methodology, principally based in Italy, but with representatives to be found elsewhere in the world."[8] We will see below that indeed, the various configurations researching branch curves, while helping to "create the field" of algebraic geometry, were not homogeneous or even necessarily compatible. This heterogeneous character should also be borne in mind when discussing 'classical' algebraic geometry (i.e., the one developed by the Italian school mainly until the 1930s) and its subsequent rewritings by commutative algebra and later, by the theory of sheaves and schemes: as Jeremy Gray points out, "many of the key elementary terms of modern structural algebra can be found in nineteenth-century algebraic number theory and algebraic geometry."[9] This does not necessarily mean that those nineteenth century mathematical configurations were in any way "modern" in the sense that they should be thought of as precursors to "modern structural algebra." The various configurations of the research on branch curves will show exactly this.

This chapter begins in Sect. 3.1 with the first protagonist of the book: Federigo Enriques. During the first decades of the twentieth century, Enriques presented a plurality of approaches to investigate the branch curve, though not all of them had the same degree of success. Several of these various directions culminated in 1923, with the publication of his paper on the 'invariance conditions,' presenting the necessary and sufficient conditions

---

[4] See Sect. 4.2.

[5] Mehrtens (1990, p. 139).

[6] For a thorough account on the notion of 'school' in these contexts as well as a review on the various historical and epistemological approaches to it, see: Koreuber (2015, p. 185–196).

[7] Schappacher (2008, p. 1302); See also (Gario 2014). I will discuss these connotations below. See also: (Brigaglia 2001).

[8] Brigaglia (2001, p. 189).

[9] Gray (2008, p. 21).

for a curve $B$ to be a branch curve, in terms of a map which sends loops circulating the branch curve to the symmetric group, and describes how the surface is ramified. However, already with Enriques' research one can see that not all of the configurations had a common research direction or technique. Apart from the 'invariance conditions,' he also investigated surfaces ramified along a given curve, the moduli of algebraic surfaces with the help of the branch curve,[10] and the numerical invariants deduced from the branch curves. Not all of these configurations culminated in solid results, and some remained underdeveloped or even employed fallible techniques to "prove" erroneous results.

Section 3.2 deals with two of Enriques' followers: Beniamino Segre and Oscar Zariski. One may very well say that Zariski reshaped some of Enriques' results. Zariski took a group-theoretic approach, and concentrated on the calculation of a group termed fundamental group consisting of the set of all loops circulating the branch curve, together with the action of concatenation of those loops. While one may claim—in a sense, retroactively—that Enriques had already implicitly noted the need to research this group (or this set of loops)—hence echoing Gray's above-noted statement, it was only Zariski who explicitly stressed the importance of determining the structure of this group. Zariski's book *Algebraic Surfaces* from 1935 and his work before and after the publication of that book are also analyzed in this section. On the other hand, Segre's configuration pointed in a completely different research direction. Similar to Enriques, he also investigated the necessary and sufficient conditions for a curve $B$ to be a branch curve, but now in terms of the position of the singular points of $B$.

Section 3.3 concludes these various directions, as it might seem that at the end of the 1930s, each of the three configurations met its end: Enriques's 'invariance conditions' marked a natural end to a few of his research directions concerning the branch curve, whereas Beniamino Segre's research was not developed further. Zariski himself shifted from his theoretical topological-group research to another, far more ambitious project: the rewriting of algebraic geometry with the language of commutative algebra. Hence, in Sect. 3.3, I jump to the 1950s, looking at how the views of the different algebraic geometers during this period reshaped the image of algebraic geometry, prompting a retroactive marginalization of the former research configurations of the branch curve. I end the chapter with a short appendix (Sect. 3.4), explaining a few of the mathematical terms (for example, the various genera of an algebraic curve and surface) that appear throughout the chapter.

---

[10]By "the moduli of algebraic surfaces" is meant the variety which parametrizes algebraic surfaces (sometimes of specific type).

## 3.1   Enriques: A Plurality of Methods to Investigate the Branch Curve

Federigo Enriques (1871–1946) was born in Livorno, Italy, to a Jewish family. He was one of the leading mathematicians in Italy in the field of algebraic geometry. His thesis advisor in Pisa was Riccardo De Paolis. After his studies in Pisa, he began collaborating, in the 1890s, with Guido Castelnuovo. This collaboration proved to be incredibly fruitful; it lasted for decades and led to the classification of algebraic surfaces. In 1893, Enriques was also in Turin where he worked with and was also influenced by Corrado Segre, as we will see in the following section. He also worked with Francesco Severi, but had a more strained relationship with him. From 1896 till 1922, Enriques was a professor in Bologna, and after that in Rome. However, the racial laws of 1938 forced Enriques to resign, though he did participate in a clandestine university for Jewish students in Rome that Castelnuovo organized in 1941. As a result of the Nazi occupation in 1943, the school was shut down and both Castelnuovo and Enriques had to go into hiding.[11] His last book, *Le Superficie Algebriche*, was published posthumously, in 1949.[12]

Enriques's research on branch curves, as I claim, should be considered not only in the framework of his mathematical work, but also alongside his philosophical reflections on intuition and visualization—this is why Sect. 3.1.1 delineates Enriques' views on these themes. To summarize briefly, Enriques claims that there are various different forms of intuition that express the 'mathematical spirit,' and hence various ways in which the mathematical object can be researched. In Sects 3.1.2, 3.1.3, 3.1.4, I present the variety of ways that Enriques employed to research the branch curve: within the classification program with Castelnuovo, constructing double (and multiple) planes, investigating maps to the symmetric group[13] and formulating his 'invariance conditions.' I therefore do not attempt to present a complete review of Enriques' work, but rather concentrate on how branch curves appeared in various settings of research on algebraic surfaces, having different roles and functions. What is to be noted is how Enriques slowly considered the branch curve as an increasingly essential component for understanding surfaces, though, as we will see, he did not use just one single method to research this curve, and hence there was not just one 'branch curve' common to all research configurations. To give only one example, as we will see in Sects. 3.1.3 and 3.1.4, along with the classification project of algebraic surfaces, which culminated in 1914, another project emerged, which aimed to

---

[11] See the introduction of Sect. 3.2 below on how the racial laws in Italy reshaped Italian algebraic geometry.

[12] For a detailed biography on Enriques and on his mathematical and philosophical work, see for example (Nastasi 2010; Brigaglia / Ciliberto 1995, chapters 1.5, 4 and 8.1; Simili 1989). For more specific aspects of his work and life, see for example: (Gario 2016; Bussotti / Pisano 2015; Giacardi 2012; Ciliberto / Gario 2012; Bottazzini / Conte / Gario 1996).

[13] Although, as we will see, this did not imply that Enriques employed a group-theoretic approach or promoted it.

classify the branch curves, by finding necessary and sufficient conditions for a curve to be a branch curve.

### 3.1.1 Enriques on Intuition and Visualization

Rather than starting directly with how Enriques dealt with branch curves during his research, I would like to present first the philosophical framework, in which Enriques developed his mathematical ideas, and especially his view on mathematical intuition.

Enriques's views on the role of intuition and visualization of geometric objects should be seen not only in light of his work with Castelnuovo, on which I will elaborate also in the following sections, but also in light of the school in Turin led by Corrado Segre (1863–1924), where Enriques spent a year;[14] Enriques was also a close collaborator with Segre. Segre was one of the main contributors to the early development of algebraic geometry: "Segre's school [. . .] consisted of about a dozen scholars, almost all of whom were university professors [. . .]. The selection of common topics consequently led to a greater agreement, on the part of the Italian algebraic geometers [. . .]."[15] Here one can note that within the larger school of Italian algebraic geometry, emerged other, smaller research configurations, or rather 'schools', concentrated around one person. The rise of the Segre school was possibly mainly due to the participant mathematicians: Castelnuovo and Enriques, among others.[16] How Segre and his school considered the roles of visualization and intuition in mathematics may be seen when examining which opinions Segre and Felix Klein shared. Segre was inspired not only by Felix Klein's approach towards mathematical intuition ("Anschauung"), but also from the physical order of his school in Göttingen. Klein was, starting the 1860s, together with Alexander von Brill and Walther von Dyck,[17] one of the promoters in Germany of the construction of material mathematical models of surfaces and curves from various materials: wood, plaster, metal and strings. The tradition of construction of material mathematical models, which was highly influential at the beginning of the nineteenth century in France (see Sect. 2.1),[18] was later adopted in Germany during the nineteenth century.[19] The construction of models was done not only to merely visualize these objects (as if the goal and usage of these models were *only* pedagogical) but also to enable better understanding of the mathematical object, considering it not only as abstract but also as concrete and sensuous. In that way, the material, visual

---

[14] For a detailed, recent collection of papers of C. Segre work, see: Casnati et al. (2016).

[15] Luciano / Roero (2012, p. 70).

[16] Ibid., p. 71.

[17] On Walther von Dyck, see: (Hashagen 2003).

[18] To recall, this tradition began in France with the rise of descriptive geometry, when Monge and his pupil Thèodore Olivier, who advanced the usage of string models.

[19] See (Sattelmacher 2021).

model could also be seen as an epistemic object. Thus, for example, constructing models prompted new research questions, causing "retrospective investigations on the peculiarities of the presented shape."[20] Indeed, in 1893, Klein indicates that "mathematical models and courses in drawing are calculated to disarm [. . .] the hostility directed against the excessive abstractness of the university instruction [of mathematics]",[21] thus allowing a better approach to mathematics itself. Klein did not deny the pedagogical aspects of these material models, but in addition coupled them with developing spatial intuition.[22] However, this account should be nuanced; as Herbert Mehrtens noted, "models had, for Felix Klein and for a short time, been epistemic things; later he interpreted them as applied mathematics."[23]

Segre, inspired by Klein's views, purchased numerous models and used them during his lectures (see Fig. 3.1); one of his students, Alessandro Terracini, recalls that Segre gave classes "in that lecture hall XVII of the second floor of the university building in Via Po, whose walls were lined with the glass cases with Brill's geometric models".[24] The models were not just to be looked at. Although Segre acknowledges the role of logical reasoning in geometry,[25] he certainly emphasized that "sometimes we even resorted to drawings or

---

[20] Brill (1887, p. 77).

[21] Klein (1911 [1893], p. 109).

[22] This can be seen in his reference to these models in his Erlanger Program from 1872: "If in the text we call the spatial intuition [*Anschauung*] something incidental, then this is meant with reference to the purely mathematical content of the considerations to be formulated. Intuition has for this content only the value of illustration [*Veranschaulichung*], which, however, is very high in pedagogical terms. A geometric model for example is very instructive and interesting from this point of view [Ein geometrisches Modell z. B. ist auf diesem Standpuncte sehr lehrreich und interessant]" (Klein 1872, p. 41–42); He repeats a similar description in 1926: "As is the case today, the purpose of the model in those days was not to compensate for the weakness of the intuition [*Anschauung*], but to develop a lively, clear one, a goal that is best achieved by making models oneself." ["Wie heute, so war auch damals der Zweck des Modells, nicht etwa Schwäche der Anschauung auszugleichen, sondern eine lebendige, deutliche Anschauung zu entwickeln, ein Ziel, das vor allem durch das Selbstanfertigen von Modellen am besten erreicht wird."] (Klein 1926, p. 78). For a general survey on Klein's conception of visualization in mathematics and the role of three-dimensional models, see: (Halverscheid / Labs 2019); especially on the role of models, see: (Tobies 2021, Sects. 2.4.3 and 4. 1.1).

[23] Mehrtens (2004, p. 301).

[24] Terracini (1968, p. 10). translation taken from: (Giacardi 2015, p. 26).

[25] This is to be seen in Segre's introduction to the translation to Italian of Georg Karl Christian von Staudt's book *Geometrie der Lage*, a book known for its treatment of geometry without having a single image. Segre, after emphasizing the "purely geometric reasoning" that students should acquire without relying on "metric consideration", noted that "it was mainly for the same purpose that S. [Staudt] wanted his book to have no figures: by studying the book without using them or by making them himself, during the reading, the reader will perhaps have greater effort to conceive the most difficult figures, but his geometric conception will be strengthened." ["E fu principalmente per lo stesso fine che S. [Staudt] volle che il suo libro non avesse figure: studiandolo senza uso di queste o

**Fig. 3.1**  One of the models
presented in the university of
Turin; the model is of a Kummer
surface with 16 nodes

models of geometric figures in order to *see* certain properties (especially of form or in reality) that could not be obtained with deductive reasoning."[26] Segre also emphasizes the fact that these material models were functioning as epistemic, in a similar fashion how experiments in physics function: "[...] there are other cases in which mathematical results were achieved by using physics. This is also permitted: it is understood in this sense that the result thus achieved, although still scientific (physical), will not be mathematical, but will have only a relative and approximate value; but it will prepare the way to the mathematical result, that is, to what will be established with a complete mathematical deduction."[27]

However, the production of material mathematical models was not as common in Italy as it was in Germany: the construction of material mathematical models in Italy was more of an exception, and Italy depended on other countries, such as Germany, France and England, for the production of these models.[28] A flux in the acquisition of models begins shortly after 1877, the year in which Ludwig Brill's publishing house was opened in Darmstadt.[29] Italian mathematicians participated in those acquisitions, and during the first decades of the twentieth century Castelnuovo and Enriques followed Segre's approach,

---

facendosi solo, man mano che legge, le figure più difficili da concepire, il lettore avrà forse maggior fatica, ma si rafforzerà di più la concezione geometrica."] (von Staudt 1889, p. xii).

[26] Segre (1891, p. 54): "ricorderò che talvolta si è persino avuto ricorso a disegni o modelli di figure geometriche per vedere certe proprietà (specialmente di forma o di realtà) che col. solo ragionamento deduttivo non si sapevano ottenere."

[27] Ibid.: "[...] vi sono altri casi in cui a risultati spettanti alla matematica si giunse valendosi della fisica. Anche ciò è permesso: s'intende in questo senso, che il risultato così conseguito, pur essendo ancora scientifico (fisico), non sarà però matematico, non avrà che un valore relativo ed. approssimato; ma preparerà la via al risultato matematico, cioè a quello che sarà stabilito con una completa deduzione matematica."

[28] In that sense, one may view Beltrami's paper models of the hyperbolic plane, constructed in 1869, as an exception. See: Friedman (2018, p. 141–152); Capelo / Ferrari (1982).

[29] Palladino / Palladino (2009, p. 71–73). For the different usages of material mathematical models in Italy, see also Giacardi (2015).

appreciating this mathematical tradition. There is something remarkable that testifies to this fact in the following citation from Castelnuovo in 1928:

"We had constructed [. . .] a large number of surface models [. . .] [of] regular surfaces [. . .] [and of] the irregular ones [; comparing the surfaces] trouble began and there were exceptions of every kind. In the end, the assiduous study of our models had led us to guess some properties that had to exist, with appropriate modifications, for the surfaces of both showcases; we then put these properties into practice with the construction of new models. If they resisted the test, we were looking for the logical justification for the last phase. With this procedure, which resembles the one carried out in the *experimental sciences*, we have succeeded in establishing some distinctive traits for families of surfaces."[30]

This citation comes on the background of Castelnuovo and Enriques's research on the classification of algebraic surfaces (see Sect. 3.1.2 and 3.1.3). Hence one may suggest that the material models of surfaces were also considered by several mathematicians and in several Italian communities as experimental, as enabling an exploration of the mathematical object. The logical reasoning comes, according to Castelnuovo, only as the last resort.

The above short review of how models were considered sheds light, as we will see now, on how Enriques understood the role of intuition—and in particular visual intuition and illustration in mathematics. However, it is essential to recall, following Schappacher, that there was a general lack of drawings in the Italian mathematical literature of algebraic geometry: for example, "the two famous chapters on algebraic surfaces [. . .] by Guido Castelnuovo and Federigo Enriques [from the *Encyclopädie der mathematischen Wissenschaften*] carry not a single illustration. [. . .] the same is true of Enriques's textbook [from 1949 *Le superficie algebrique*]." Schappacher notes that the Italian algebraic geometers did sketch and use drawings in their correspondences with each other and also during their lectures, but "[. . .] as they were shaping algebraic geometry, the Italian geometers were led to analyzing constellations of objects which are increasingly difficult to visualise adequately".[31] This lack of explicit visualization was already noted with Salmon's work (see Sect. 2.2), and I will return to it when discussing Enriques's works on the branch curve below (see Sect. 3.1.2 resp. Sect. 3.1.4).

Nevertheless, concentrating on Enriques, this is not to claim that the general lack of explicit illustrations is also reflected in his conception of visual intuition. As Enriques

---

[30]Castelnuouvo (1928, p. 194): "Avevamo costruito [. . .] un gran numero di modelli di superficie [. . .]. Una [vetrina] conteneva le superficie regolari [. . .]. Ma quando cercavamo di verificare queste proprietà sulle superficie dell'altra vetrina, le irregolari, cominciavano i guai e si presentavano eccezioni di ogni specie. Alla fine lo studio assiduo dei nostri modelli ci aveva condotto a divinare alcune proprietà che dovevano sussistere, con modificazioni opportune, per le superficie di ambedue le vetrine; mettevamo poi a cimento queste proprietà con la costruzione di nuovi modelli. Se resistevano alla prova, ne cercavamo, ultima fase, la giustificazione logica. Col detto procedimento, che assomiglia a quello tenuto nelle *scienze sperimentali*, siamo riusciti a stabilire alcuni caratteri distintivi tra le famiglie di superficie."

[31]Schappacher (2015, p. 2806).

wrote in 1922 in his review of Klein's *Gesammelte mathematische Abhandlungen*, it was precisely the "tendency to consider the objects to be studied in the light of visual intuition"[32] that brought Klein and the Italian geometers so close together intellectually. The importance that Enriques attached to intuition in scientific research is seen in the following citation from 1938, emphasizing the visual aspects:

> "The faculty, which comes into play in the construction of science and which thus expresses the actual power of the mathematical spirit is intuition. [...] There are in any case *different forms of intuition*. The first is the intuition or imagination of what can be seen. [...] But there is another form of intuition that is more abstract, that—for example—which makes it possible for the geometer to see into higher dimensional space with the eyes of the mind."[33]

The 1938 reference of Enriques to the forms of intuition, among them the imagination (which is certainly visual—e.g. the intuition "of what can be seen"), is not a late development of Enriques's conception of intuition, but rather to be found already in his writings from the 1890s. One form of "experimental intuition" is noted in 1894, when Enriques underlines that "we will seek to establish the postulates derived from *experimental intuition* of the space that appear to be the simplest for defining the object of projective geometry."[34] Moreover, attempting to explain the relations between abstract geometry and intuition, Enriques writes:

> "The importance that we attribute to abstract geometry is not (as may be believed) opposed to the importance attributed to intuition: rather, it lies in the fact that abstract geometry can be interpreted in infinite ways as a concrete (intuitive) geometry by fixing the nature of its elements: so in that way geometry can draw assistance in its development from an infinity of different forms of intuition."[35]

Enriques's remarks on the different types of intuition echo how Klein distinguished between naïve intuition and refined intuition. In 1893, during his Evanston lectures, Klein highlighted the importance of the naïve intuition for the discovery phase of a theory (the example that Klein gave is differential and integral calculus); refined intuition (shown, for example, in Euclid's Elements) on the other hand intervenes in the elaboration of data furnished by naïve intuition, when the foundations of a theory are already established, and in the rigorous logical development of the theory itself: "The naïve intuition is not exact, while the refined intuition is not properly intuition at all, but arises through the logical development from axioms considered as perfectly exact."[36] Similarly Klein presents a division between the "sensuous [sinnliche] *Anschauung*", and the "idealizing spatial

---

[32] Cited from: Giacardi (2012, p. 225).

[33] Enriques (1938, p. 173–174). Translation taken from: Giacardi (2012, p. 231).

[34] Enriques (1894b, p. 551).

[35] Enriques (1894a, p. 9–10).

[36] Klein (1911 [1893], p. 42).

intuition [idealisierende Raumanschauung]", which "goes beyond the inexactness of sense observation".[37] Although a discussion on Klein's conception of the mathematical *Anschauung* (and the possible influence of Kant) is beyond the scope of this book,[38] it is important to note that he and Enriques shared similar views and Klein certainly influenced Enriques.[39] Thus, for example, just as Klein noted that one should always begin with spatial intuition and only then "gradually ascend to more abstract ways", though without giving up mathematical intuition, an opinion he shared with his Italian colleagues,[40] so too did Enriques emphasize the role of intuition in abstract geometry, while objecting to seeing formalism as "an end to achieve, but as a means aimed to use and increase the faculty of intuition. The results themselves, logically established, no matter how far-reaching, must still not be considered as mature achievements until they can be in some way comprehended intuitively. *But in the principles the intuitive evidence must shine brightly*".[41] Moreover, for Enriques, following Klein, the different forms of intuition are not static. Thus, for example, in 1894, in a letter to Castelnuovo, Enriques noted his aims, rather than "invading new fields, [...] to lay the foundations of my [book] *Ricerche* on satisfactory foundations"; and though there are "gaps filled by intuition" in this book,[42] everything is correct and can be "remedied" easily.[43] While intuition here might be thought to be associated with "invading" or discovering "new fields",[44] it can be modified by laying more stable foundations, done with a logical inquiry of the definitions. In 1922 Enriques states similarly, that "abstract geometry is hence based upon a repetition of the logical

---

[37] Klein (1908, p. 88).

[38] Cf. Mattheis (2019); Allmendinger (2014, p. 43–54).

[39] See: Giacardi (2012, p. 223–229).

[40] Luciano / Roero (2012, p. 188): "[...] man [muss] beim Unterrichten von der Anschauung beginnen muss (um erst allmählich zu abstrakteren Auffassungen aufzusteigen) [...]" (a letter from Felix Klein to Mario Pieri, Göttingen, 31 March 1897).

[41] Enriques (1900, p. 12): "Il formalismo logico deve essere concepito, non come un fine da raggiungere, ma come un mezzo atto a svolgere e ad avanzare le facoltà intuitive. Gli stessi resultati più lontani, logicamente stabiliti, non debbono ancora considerarsi come un acquisto maturo, fino a che non possano essere in qualche modo intuitivamente compresi. Ma nei principii l'evidenza intuitiva deve risplendere luminosa." Translation taken from: Giacardi (2012, p. 227). Compare also how the mathematician Fabio Conforto described Enriques's methods, as follows: The algebraic world is "first of all and above all a matter of 'seeing'. Such a conception satisfied deeply the strongly intuitive spirit of Enriques, who often reached the point [...] not to feel the need of a logic proof of some property, because he 'saw'; and that rendered him sure of the truth of the proposition at issue and it fully satisfied him completely [...]. And yet for Enriques, beside the intuitive par excellence, there was also the subtle logic and the highly critical [approach]. [...]. He was for a substantial rigour, more than for a formal one. Finally, he never wanted to adhere to the view that mathematics is a purely logic construction." (Guerraggio / Nastasi 2006, p. 137) I thank Silvia de Toffoli for bringing Conforto's statements to my attention.

[42] See: Bottazzini / Conte / Gario (1996, p. 91). Enriques refers to: (Enriques 1893a).

[43] Bottazzini / Conte / Gario (1996, p. 92).

[44] As we will see, Enriques and Castelnuovo were using often this geographical metaphor.

analysis of the deductive theories, as far as different systems of concepts and forms of intuitions are concerned".[45] According to Paolo Bussotti and Raffaele Pisano, "Enriques stresses that the force of this intuition has been empowered by abstract geometry. [. . .] Certainly it is a faculty of our mind, but it is not given once and forever."[46] Indeed, Bussotti and Pisano stress that for Enriques, the logical models associated to abstract geometries and the physical, material and visual models in physics (and as a possible consequence, in mathematics), were all connected.[47] As Jeremy Gray notes, "Enriques argued that while the high level of abstraction forced geometers to argue more or less formally, he nonetheless believed that there was a core of meaning in the subject without which it was a sterile activity."[48] Moreover, "[i]n Enriques's opinion, geometry is not a matter of writing down some axioms (plausible or not, but in any case mutually consistent) and then reasoning entirely logically. It is concealed talk about physically possible systems, although these need not even remotely be likely to be correct [. . .]. According to Enriques, different systems of postulates, forming various geometries, 'express different physical hypotheses'."[49] For Enriques, this means that abstract geometry as well as material mathematical models were related to each other in the sense that the visual and the more abstract intuition fertilized each other. The lack of drawings did not mean that these visual means were not present or were prohibited, but rather on the contrary—they were considered by Enriques as legitimate objects, which lead to more refined intuition; this underlines not only that the intuitive evidence must "shine brightly" but also the plurality of ways one may research with algebraic geometry.

### 3.1.2 The Turn of the nineteenth Century: First Attempts of Classification of Surfaces

Having seen the philosophical framework of Enriques, concerning the various forms of intuition and visualization, the aim of the following sections is to overview Enriques's work on algebraic surfaces, concentrating on the role which branch curves played. As we will see, this role changed through the years, according to the different techniques, various configurations and modes of argumentation used by Enriques.

---

[45] Enriques (1922, p. 140): "L'istrumento della geometria astratta si basa dunque sopra una costante ripetizione dell'*analisi logica dei principi delle teorie deduttive*, in ordine a diversi sistemi di concetti e a diverse forme d'intuizione."

[46] Bussotti / Pisano (2015, p. 123).

[47] Ibid., p. 126.

[48] Gray (2008, p. 190).

[49] Ibid,. p. 361. For a survey on Enriques's views on the philosophy and epistemology of science, mathematics and geometry, see: (ibid., p. 356–365; Bussotti / Pisano 2015).

One of the main invariants Enriques helped to develop was the genus of an algebraic surface. Here it is important to underline, as François Lê notes, that the consolidation of the term and the invariant 'genus' associated to an algebraic curve was not immediate and took several decades during the second half of the nineteenth century.[50] The situation for algebraic surfaces was even more complicated. The history of the development of various genera associated to algebraic surfaces—e.g. the arithmetic genus $p_a$ or the geometric genus $p_g$– is more convoluted and is outside the scope of this book[51] (see the appendix to this chapter, Sect. 3.4, for a short mathematical explanation regarding the genera associated to algebraic curves and surfaces). Suffice to note that those genera were employed to classify algebraic surfaces. Several mathematicians, among them Enriques and Castelnuovo, had hoped that the classification project would be parallel to the one on algebraic surfaces, but as the examples below will show, this was hardly the case.

In 1894 Enriques discovered an algebraic surface $F$ for which the geometric and the arithmetic genus are zero, that is, $p_g = p_a = 0$, though $F$ was not birationally equivalent to the complex projective plane $\mathbb{CP}^2$; till this discovery it was assumed (wrongfully) exactly the opposite: that any surface with $p_g = p_a = 0$ was indeed birationally equivalent to $\mathbb{CP}^2$.[52] This assumption was based on an equivalent situation for algebraic curves—the rationality of a curve is equivalent to the vanishing of its genus. The surface $F$, which was discovered by Enriques, is of degree 6, has a nodal double curve consisting of the edges of the tetrahedron, whose vertices are triple points for the surface.[53] Such a discovery immediately implied that the task of classifying algebraic complex surfaces would be more complicated than the corresponding task for curves. Indeed, for algebraic curves, it was proved that it is enough to know the geometric genus in order to classify algebraic curves up to birational transformations. It was however clear already in 1891 to Castelnuovo, that one cannot assume by analogy that all the possible properties of curves could be transferred to surfaces: Castelnuovo discovered another example of such an 'impossible' surface: were the analogy to hold, also this surface should not have existed.[54]

This example of the surface $F$ of degree 6 led Enriques, together with Castelnuovo, from the 1890s till 1914 to search for new invariants for algebraic complex surfaces in order to distinguish various birational classes of algebraic surfaces. These invariants were called the *plurigenera* of an algebraic surface, denoted by $P_i$, $i \geq 1$, when $P_1 = p_g$. The 1894 example

---

[50] Lê (2020).

[51] See e.g. Popescu-Pampu (2016, part II).

[52] See: Castelnuovo / Enriques (1896, p. 291–292), Castelnuovo (1896), Enriques (1896b). See the appendix for the definition a rational surface; in short, a rational surface is a surface which is birationally equivalent to the projective plane.

[53] How Enriques describes his discovery to Castelnuovo on 22 July, 1894, see: Bottazzini / Conte / Gario (1996, p. 124–125). See also (Gario, 2016, p. 301).

[54] The example is of a surface whose geometric genus and arithmetic genus are different (such a surface is called *irregular*), a situation which does not have an analog for curves. See also (Gario 2016, p. 296).

motivated Castelnuovo to prove in 1896 that an algebraic surface, having $p_g = p_a = 0$ and that $P_2 = 0$, is indeed a rational surface;[55] the proof marked the importance of taking into account the plurigenera of a surface. This insight led eventually to the culmination of the research: In 1914, Enriques and Castelnuovo published two articles, announcing the classification of algebraic complex surfaces in terms of their plurigenera.[56]

Although the historical development of this classification has already been investigated thoroughly in several works,[57] it is instructive to look at several examples from the beginning of the twentieth century, which show how Enriques and Castelnuovo discovered particular cases of surfaces. To emphasize: the chapter does not aim to be exhaustive when surveying the different approaches to research complex algebraic surfaces during the first decades of the twentieth century, but rather aims to show how the consideration, or rather the couching of the branch curve in different configurations not only prompted novel research directions on complex surfaces, but also led eventually to the emergence of new various mathematical configurations in which the branch curve was considered as the main object of research.

### 3.1.2.1 On Double Covers and Branch Curves

"[B]irational geometry [. . .] is the starting point of the Italian school of algebraic geometry"—this was Corrado Segre's approach to algebraic surfaces, and as a result, also the approach of Castelnuovo and Enriques.[58] One of the first questions asked in this framework, concerning ways to look at surfaces birationally, is to be found in the works of Alfred Clebsch and Max Noether; their approach consisted of representing rational surfaces as double covers of the plane.[59] Here the branch curve, or rather, the ramification curve, is considered as a tool with which a surface can be constructed. The starting point of this approach can be seen with Clebsch's work in 1870, searching for an "injective mapping between a simple and a double plane",[60] when by a "double plane" is meant an algebraic surface which is a cover of degree 2 of $\mathbb{CP}^2$ (such as $z^2 = f(x, y)$, $f$ being an algebraic curve). Clebsch starts an investigation of double planes ("Doppelebene"), which are ramified along a curve denoted by $\Omega = 0$. The curve—according to Clebsch—"contain points, for which [above the 'simple plane'] only one point [on the double plane] corresponds. Along this curve join together the sheets of the double plane; I will call this curve the *transition curve*

---

[55] In the above example of Enriques, the surface $F$ has $P_2 = 1$. See e.g.: Popescu-Pampu (2016, p. 91–92); Gray (1999, p. 62).

[56] Castelnuovo / Enriques (1914); Enriques (1914). See Fig. 3.5.

[57] Brigaglia / Ciliberto (1995, p. 97–123); Gray (1999); Popescu-Pampu (2016, p. 81–96).

[58] Gario (2016, p. 290). As Gray (1999, p. 59) notes, Segre suggested one should study algebraic surfaces birationally, looking also for families of curves which induce an embedding of the surface in a suitable projective space.

[59] See the appendix to this chapter for a definition of rational surfaces and birational mappings.

[60] Clebsch (1870, p. 46): "Eindeutige Beziehung zwischen einer einfachen und einer Doppelebene."

[*Uebergansgcurve*]".[61] As noted at the end of the last chapter, in modern terminology, this curve is the 'ramification curve'.

In Clebsch's paper several cases are presented: the investigation of the case when the "transition curve" is a conic section, when it is of degree 4, or the case when one considers the projection of surfaces of degree 3 from a point on the surface. The question, when the double plane is itself rational, is answered fully only by Max Noether. Noether notes in 1878 that while Clebsch solved this question when the "transition curve" is of degree 4, he intends to develop a general method. Also here Noether employs the term "Uebergangscurve". While it is clear that a double plane is rational if it is birationally equivalent to a double plane branched along a smooth conic (as was already indicated by Clebsch), Noether proves the following: a double plane is rational if it is birationally equivalent to a double plane branched along (1) a smooth quartic; (2) an irreducible sextic curve with two infinitely near triple points;[62] (3) an irreducible curve of degree $2d$ with a point of multiplicity $2d - 2$, with either no other singular points, or at most one additional node.[63]

While Clebsch in 1870 termed the "Uebergangscurve", which was (in modern terminology) the ramification curve—where two sheets came together—, Noether did not define this term in 1878, but one may assume that he employed it in the same way Clebsch does. Noether's result indicates that one way to research surfaces was by considering them as a (double) cover of the complex plane. What became clear from this result is that the characteristics of the surface, once considered as a double cover, are *dependent* on and might even be determined by properties of either the ramification curve or the branch curve.

This conclusion was formulated explicitly in 1896, in a paper by Castelnuovo and Enriques reviewing the recent developments and discoveries in the research of algebraic surfaces, a formulation which also indicated a shift in terminology: Given a "double plane, that is, a surface whose equation can be written under the form of $z^2 = R(x, y)$, $R$ being a rational function [. . .]. It is therefore the question of the rationality of a double plane that we are now facing. A double plane is entirely defined by its *branch curve* [courbe de diramation] (Uebergangscurve) $R(x, y) = 0$, which is the place of all the points where the two corresponding points on the surface coincide. What particularity must the branch curve have in order that the double plane would be rational (as represented above in a simple plane)?"[64] This question was answered by Noether, as is noted immediately afterwards.

---

[61] Ibid.: "Die Curve [. . .] enthält diejenigen Punkte, für welche einem Punkte $y$ nur *ein* Punkt z entspricht. Längs dieser Curve hängen daher die Blätter der Doppelebene zusammen; ich nenne sie die Uebergangscurve."

[62] According to Noether (1878, p. 83), such a point is a point where "three different branches of $\Omega$ [the *Uebergangscurve*] touch."

[63] Those three cases are presented in (ibid, p. 82, 83 and 85).

[64] Castelnuovo / Enriques (1896, p. 305): "[. . .] plan-double, c'est-à-dire sur une surface dont l'équation peut s'écrire sous la forme $z^2 = R(x,y)$, $R$ étant une fonction rationnelle [. . .] C'est donc la question de la rationalité d'un plan double que nous rencontrons maintenant. Un plan double est

Here the question of terminology rises: Castelnuovo and Enriques did not only stress that the "courbe de diramation" is a plane curve (in modern terminology, the branch curve)— this is since it is a curve given by the equation $R(x, y) = 0$, dependent only on two variables—hence, it is indeed a plane curve, which corresponds to another curve "where the two corresponding points on the surface coincide". They also underlined that "courbe de diramation" is the translation of the German term "Uebergangscurve", which was employed by Clebsch and Noether to notate the spatial curve on the surface, where, so Clebsch, "the sheets of the double plane [. . .] join together" (i.e. the ramification curve).[65] As we will see, in their writings Castelnuovo and Enriques often employed the term "courbe de diramation" in French or Italian, although they did use the German term "Verzweigungskurve" later, though less often,[66] and they did not employ the term "Uebergangscurve".[67]

* * *

While an overview of how the result of Noether was considered from the moment it was discovered till the above citation of Castelnuovo and Enriques is outside the scope of this book, I would like to focus on how this citation is to be understood in the context of Castelnuovo's and Enriques's research on algebraic surfaces, and in particular, on their classification project. In order to do that, I will look into the 1896 review more closely, as it uncovers what Castelnuovo's and Enriques's image of algebraic geometry was—and especially of the research of algebraic surfaces—at the end of the nineteenth century.

The review paper, called "Sur quelques récents résultats dans la théorie des surfaces algébriques", begins with a historical overview of the research of algebraic surfaces. The opening statement of Castelnuovo and Enriques is that the origin of the research of "the *Geometry on a general algebraic surface*" is due to the "remark of Clebsch and a

---

entièrement défini par sa courbe de diramation (Uebergangscurve) $R(x, y) = 0$, qui est le lieu d'un point dont les deux points correspondants sur la surface coïncident. Quelle particularité doit présenter la courbe de diramation pour que le plan double soit rationnel (représentable sur le plan simple)?"

[65] Felix Klein also employs in 1882 the term "Uebergangscurve", but in the context of Riemann surfaces, when he defines "*Uebergangscurven*, those curves whose points remain unchanged during the symmetrical transformation in question." ["d. h. derjenigen Curven, deren Puncte bei der in Betracht kommenden symmetrischen Umformung umgeändert bleiben"] (Klein, 1882, p. 72). He gives as an example a sphere: a reflection through a place going through the center of the sphere leaves a great circle fixed –this great circle is an "Uebergangscurve" (ibid., p. 73).

[66] E.g. (Castelnuovo / Federigo 1914, p. 702, 714, 755).

[67] It should be also mentioned that the term "Uebergangscurve" was not only used in the context of algebraic surfaces or of Riemann surfaces (see the previous footnote on Klein's usage), but was also used often as a technical term in other 'applied-mathematics' domains; for example, "Uebergangscurve" (or "Uebergangskurve") was also the term during the nineteenth century for the "track transition curve" used to describe the construction of railroad tracks. See for example: Winckel (1874, p. 128); Boedecker (1882, p. 265); Launhardt (1888, p. 99–112).

fundamental memoire of Nöther".[68] This field of research was developed and supplemented—so Castelnuovo and Enriques—due to "French and Italian mathematicians [...] following two different directions".[69] Here Castelnuovo and Enriques turned to describe the different traditions within which the research on algebraic surfaces was carried out: while in France the geometers preferred the research of transcendental functions, according to Castelnuovo and Enriques, in Italy one concentrated on the study of a linear system of algebraic curves on an algebraic surface. While emphasizing only these two schools, it may seem that Castelnuovo and Enriques considered the Italian community of mathematicians (along with, perhaps, the French one) as replacing German mathematicians who were focusing on algebra, when it came to research on algebraic surfaces. However, as Gray shows, this was not a precise historical description.[70] Nevertheless, it is interesting to note that the adjective "German" was not employed by Castelnuovo and Enriques, in contrast to the explicit mentions of "French" and "Italian" mathematicians. There is here an almost performative act of presenting a unified Italian community of mathematicians—echoing perhaps the unification of Italy in the 1860s, an act which on the one hand points to a geographical differentiation of those mathematicians from France and a mathematical differentiation from the methods of the French mathematicians, and on the other hand from the practices of German mathematicians.[71]

The description of the methods of the Italian school is much more extensive than the one provided by the French school; the review paper notes that while the Italian methods use the "two auxiliary theories"—of the geometry of algebraic curves and the linear system of plane curves—they in fact continue the works of Riemann, Brill and Noether.[72] What is however much more essential, according to Castelnuovo and Enriques, is not to find analogous results for the case of algebraic surfaces, but rather to find "*invariant* characters" of the surfaces, a subject on which "an ensemble of research works, tightly related, [...]

---

[68] Castelnuovo / Enriques (1896, p. 241).

[69] Ibid.

[70] See: Gray (1997). Gray also highlights the differences between the German and the Italian schools of algebraic geometry at the end of the nineteenth century: "in the 1880s and 1890s there were algebraic geometers in two senses of the term—*algebraic* geometers and algebraic *geometers*. For whatever reason, *algebraic* geometers were mostly German and mostly worked before 1890, and algebraic *geometers* were mostly Italian and mostly worked after 1890." (Gray 1994, 154).

[71] While Brigaglia notes that representatives of the Italian school may be also "found elsewhere in world" (2001, p. 189), it is the actors themselves of this school who aimed to emphasize this national point of view. On the relations between French and Italian mathematicians during the 19th and the twentieth century, see (Brechenmacher et al. 2016); in this volume one can note how French mathematicians became aware of the Italian developments in mathematics. For example, "Gaston Darboux [...] wrote in 1870 to Jules Houël [...] that the Germans clearly surpassed the French on the mathematical scene and that if things would continue in that way, the Italians would soon do the same." (Ibid., p. 17).

[72] Castelnuovo / Enriques (1896, p. 242).

[is] published in Italy".[73] Moreover, one should not draw the wrong conclusions with respect to the analogy with curves: "most of the properties of surfaces are far from presenting the characteristics of simplicity one finds with the analogous propositions regarding curves."[74] The introduction of the review ends with a subtle warning, indicating that the "moment for introducing a perfection at the theory of surfaces has not arrived yet"; one must still "overcome the obstacles, which one encounters on the way, for arriving at the summit which shines before us."[75] A solicitation to carry out further research is expressed, a call that abounds with geographical metaphors: the authors anticipate a "disagreeable road", though the "landscape" may be researched with interest, and the different mathematicians may follow "different paths".[76] Indeed, what Castelnuovo and Enriques emphasized is that the "landscape" of the theory of algebraic surfaces is not yet fully understood—and the introduction of the review implies that one has to explore these surfaces with different methods, with what the "landscape offers to our gazes";[77] considering surfaces as covers of the complex projective plane is one way to direct our gaze. What is unfolded here is an image of a mathematical domain which is not yet well researched or fully understood. It is an image of exploration, when certain methods are suggested as more productive (the "invariant characters") and when certain mathematicians are presented as explorers.

As we saw in the citation above, the classification of Noether's rational double covers mentioned by Castelnuovo and Enriques was one of the ways to explore this landscape. Noether proved that there are only three different types of rational double covers up to birational equivalence. However, one of the questions that is asked in the review paper concerns the degeneration of these three cases. By "degeneration"[78] is meant a degeneration of the branch curve into a union of simpler curves; those degenerated branch curves correspond to "double planes, which represent ruled surfaces".[79] This type of question was not raised by Noether and as we will see later (with Chisini or Zappa, for example, see Sect. 4.1.2), such degenerations were a possible way to tackle the question of the classification of surfaces. Another research direction, which also focuses on double covers, is presented at the end of the review, being a summary of results obtained by Enriques in the same year: a classification of every double plane all of whose genera is 1.[80] The classification is again

---

[73] Ibid., p. 243.

[74] Ibid.: "la plupart des propriétés des surfaces sont bien loin de présenter les caractères de simplicité, qu'on trouve en de propositions analogues relatives aux courbes."

[75] Ibid., p. 244: "Ce qui importe à présent, c'est de surmonter les obstacles qu'on rencontre sur le chemin, c'est d'arriver au sommet qui reluit devant nous."

[76] Ibid.

[77] Ibid.: "paysage qui s'offre à nos regards."

[78] Ibid., p. 306.

[79] Ibid.

[80] That is, $p_g = P_1$, $P_2$ and $p^{(1)}$, $p_a$, $P_3$, $P_4$, $\ldots$.

dependent only on the branch curve, though now the singularities of the curve are more complex.[81] It is interesting to note that in another paper from 1896, where Enriques investigated these surfaces, he noted explicitly that his mission is to "classify these double planes [being] [...] analog to the place occupying curves of genus 1 among the hyperelliptic curves".[82] In that sense, in contrast to the image in the review paper, the analogy between curves and surfaces still dominated parts of the research. Moreover, it seems that when Enriques, during the following two years, considered the task of classifying surfaces with specific genera (for example, when $p^{(1)} = 2$ or when $p^{(1)} = 3$), he considered most of the surfaces to be given in the form of a double cover, when their branch curves are explicitly provided.[83]

### 3.1.2.2 End of the 1890s: Enriques's Initial Configurations

At the same time when Enriques was classifying double covers[84] according to their branch curves, he considered also other approaches to construct and characterize algebraic surfaces with the help of branch curves. These approaches—or better yet, explorative reflections on the nature of the branch curve—are to be found in a series of letters sent to Castelnuovo, which were less systematic than what one might have assumed, following his program as presented in his published papers. In that sense, these reflections were indeed less formal, in the sense that they either were formulated as research questions or were denser, less rigorous or just contained unclear arguments—as can be expected from a private communication. Moreover, they show that Enriques employed a variety of techniques in order to explore how different mathematical configurations prompt the discovery of different research results regarding branch curves.

Concerning branch curves, apart from the letters of Enriques on the program to classify double covers,[85] already seen in the last section, he also researched more general questions. In a letter from 22 November 1896, he noted that given a curve, multiple planes (of degree greater than 2) ramified over this curve may not always exist.[86] Two days later, on 24 November 1896 Enriques wrote another letter, which begins with an even less positive tone: "Thinking back to the things I mentioned to you concerning multiple planes, the possible applications, and various examples that I built, I concluded that with respect to

---

[81] See: ibid., p. 316 and: Enriques (1896a). In Enriques (1896a, p. 201), the branch curve is called "curva di diramazione".

[82] Ibid., p. 202.

[83] See: Enriques (1897a, b); Castelnuovo / Enriques (1900).

[84] Note that though I will use either the term "double cover" or "double plane", according to the mathematician who used this term, both terms have the same meaning.

[85] For example, see Bottazzini / Conte / Gario (1996, p. 106), letter from 16 May, 1894: "My studies on double covers are proceeding in a satisfying way."

[86] Ibid., p. 287.

these planes, nothing is understood yet, while it would be very important to understand."[87] The question—*how* to understand, what is not yet understood concerning "multiple planes" and their branch curve—will be at the center of Enriques's research during the following years. It also shows the beginning of a formation of a unique epistemic configuration, when the role which the branch curve would play in it is somewhat unclear.

While Enriques eventually delineated in his letter from 24 November 1896 his plan to concentrate first on double planes, he noted that one "claims to infer invariants" not only from the branch curve, resulting from the intersection of "$f = 0, \frac{df}{dz} = 0$", but also from the intersection of "$f = 0, \frac{df}{dz} = 0, \frac{d^2f}{dz^2} = 0$". This triple intersection was already discussed in Salmon's research (see Sect. 2.2). Salmon noted that this set is the set of the cusps of the branch curve (eliminating $z$), though Salmon, on the one hand, did not term the branch curve with a specific term, and while Enriques on the other hand, did not explicitly term these points as cusps,[88] at least not in this letter. This already indicates that Enriques considered the cusps of the branch curve as carrying additional information.

A series of three letters in June 1897 continues this line of research, but also offers others. Enriques wrote to Castelnuovo, concerning an 1881 article by Leopold Kronecker, that "you yourself asked the question whether 'two multiple planes with the same branch curve [curva di diramazione] can be represented [mapped] one to the other.'"[89] Kronecker's paper that Enriques referred to deals with the discriminant of an algebraic function with *one* variable,[90] which Enriques considered as branch points of an algebraic curve, even though neither the term 'branch point' nor 'ramification point' appear in Kronecker's paper. According to Enriques, a generalization of this theorem leads to a similar question regarding multiple planes: Enriques hence asked about the connection between branch curves and algebraic surfaces. While his question concerned uniqueness, Enriques was also interested in more concrete calculations: he indicates that it is easy to find the numerical invariants of the surface (among them, as he indicates, are the linear genus, the arithmetical genus and the geometrical genus) only by knowing the numerical invariants of a branch curve. Nevertheless, he remarked that he did not have time to deal with this question.[91]

---

[87] Ibid., p. 288: "Ripensando alle cose che ti ho accennato sui p[ia]ni multipli, alle possibili applicazioni, e a varii esempi che mi sono costruito, ho concluso che relativamente ai p[ia]ni multipli ancora non si capisce niente, mentre sarebbe importantissimo di capirne."

[88] Ibid., p. 289.

[89] Letter of June 5, 1897, in: (ibid., p. 340): "Tu stesso anzi ponevi la questione se 'due piani multipli colla stessa curva di diramazione sieno rappresentabili uno sull'altro'."

[90] The paper is called "Ueber die Discriminante algebraischer Functionen einer Variabeln" (Kronecker 1881).

[91] In 1905 a similar approach is indicated. In a letter written to Castelnuovo on 1 February 1905, Enriques noted that once a branch curve ("curva di diramazione"), its degree and its number of nodes and cusps are given, then one can determine the degree of the branched surface from the relations between the various numerical invariants (Bottazzini / Conte / Gario 1996, p. 603).

Almost two years later Enriques presented to Castelnuovo two questions, which develop the above line of investigation regarding whether the branch curve determines uniquely the surface. The first: can a branch curve be arbitrary? I.e., can any curve be a branch curve? The second: given a branch curve is there a *unique* complex surface of degree $n$ branched along it? That is, can one construct a unique multiple plane given the branch curve? This question proved to be important, as Enriques named years later, in 1905, an "old wish".[92] On 26 February 1899 he wrote the following, adding a diagram illustrating what he meant: "Answering an old question proposed by you," and by that Enriques referred to the question of Castelnuovo presented in the letter from June 1897,

> It seems to me that the branch curve of a multiple plane can not be arbitrary, and that if a plane curve $C$ is branch curve of a multiple plane of degree $n$, it will define in general a unique $n$-multiple plane. This is why.

> Take on the plane a pencil of lines passing through $O$ and consider a ($n$ degree) line $a$ through $O$. It defines a certain number $\mu$ of target curves $K_1, \ldots, K_\mu$; choose one curve $K_1$. Rotating the line $a$ around $O$, we continually follow what happens to the curve $K_1$. After a whole turn, the curve $K_1$ generally permutes with $K_2 \ldots$ or $K_\mu$. However, this must be ruled out if $C$ is a branch curve of a degree $n$ multiple plane. But if we impose the condition that $K_1$, for example, separates rationally between the $\mu$ $K$'s represented above the $n$-degree line, generally it will not happen the same for $K_2 \ldots$ or $K_\mu$.[93]

What is the meaning of the method sketched here? It is essential to stress from the outset that the configuration presented here is novel: Enriques presented a new configuration, within which one can investigate branch curves. One is given a plane curve $C$, together with a point $O$, not on $C$, and a pencil of lines passing through $O$, which can be considered as a rotating line: e.g. as the family of lines $\{y = tx : t \in \mathbb{R}\} \cup \{x = 0\}$, when $O = (0,0)$.

---

[92] Ibid., p. 603.

[93] (ibid., p. 400–401): "Ripensando ad una vecchia questione che tu proponevi, mi par di vedere che la curva di diramazione di un piano multiplo non possa darsi ad arbitrio, e che se una curva p[ia]na $C$ è curva di diramazione d'un p[ia]no $n$-plo, essa definirà *in generate* un unico p[ia]no n-plo. Ed ecco perché.

Prendi nel p[ia]no un fascia $O$ e considera una retta ($n$-pla) per $a$ [Lapsus for: una retta [. . .] $a$ per $O$]. Essa definisce un certo numero $\mu$ di curve obiettive $K_1 \ldots K_\mu$; scegliamone una $K_1$. Facciamo ruotare $a$ attorno ad $O$, e seguiamo per continuità ciò che diventa $K_1$. Dopo un giro completo, in generale $K_1$ si permuterà con $K_2 \ldots$ o $K_\mu$. Ciò invece deve escludersi se la $C$ e curva di diramazione di un p[ia]no n-plo. Ma se imponiamo la condizione che $K_1$, ad es[empio], si separi raz[ionalmen]te fra le $\mu$ $K$ rappresentate sulla retta $n$-pla $a$, in generale non accadrà lo stesso per $K_2 \ldots$ o $K_\mu$."

Choosing one line $a$ from this family, it intersects the curve $C$ in several points. Enriques considers the $\mu$ Riemann surfaces of degree $n$ which are ramified over these points, denoting these surfaces by $K_1, \ldots, K_\mu$. Enriques's claim is that once we rotate the line $a$ (inside the pencil of lines) to do a full round, this will induce a permutation of the Riemann surfaces, i.e. a permutation of the set $K_1, \ldots, K_\mu$. However, Enriques claims that if $C$ is a branch curve, then the induced permutation might be in fact a permutation which leaves $K_1$ fixed; explicitly, choosing a Riemann surface $K_1$ it would not be permuted with any of the other Riemann surfaces $K_2, \ldots, K_\mu$.

Enriques's conclusion, being somehow condensed, relies on several other arguments, which appear in a paper of Hurwitz from 1891, a paper which considers a similar setting, of the change of Riemann surfaces caused by a movement of their branch points.[94] Indeed, moving the line $a$ induces also a movement of the intersection points $a \cap C$, above which the Riemann surfaces are ramified. According to Enriques, if $C$ is a branch curve, then (at least) one of the Riemann surfaces $K_1, \ldots, K_\mu$ (branched along $a \cap C$) is a plane section of the complex surface, whose branch curve (under a covering map) is the given branch curve. However, it seems that what Enriques assumed is that there is only one complex surface of degree $n$ branched along $C$ (since the branch curve can "not be arbitrary")—since in this case Enriques hints that the corresponding Riemann surface would not permute with any of the other Riemann surfaces, which are not a plane section of a complex surface.

* * *

One may assume that at the end of the nineteenth century, Enriques still considered branch curves and multiple planes as a not-yet understood landscape, where one can and should explore various mathematical settings, some of them epistemic, in the sense that the ways to research the object—e.g. considering only the singular points, the double covers, or the associated Riemann surfaces to this curve—are unclear or not yet fully developed. However, this approach that adopts a plurality of configurations was not often expressed in the 'official' published papers. This can be seen also in a paper from 1908, which Enriques and Castelnuovo published in the *Encyklopädie der mathematischen Wissenschaften*, a paper titled "Grundeigenschaften der algebraischen Flächen". The paper deals with algebraic surfaces embedded in $\mathbb{CP}^3$, which are given by a homogenous equation $f(x_1, x_2, x_3, x_4) = 0$. Surprisingly, the paper does not present any of the results obtained by the two authors regarding the classification of surfaces according to their different genera. Ramification curves are treated in a somewhat classical way—exactly as Salmon treated them (and indeed, Salmon's work is often cited): as an intersection of the surface and its "first polar".[95] The numerical invariants associated to the ramification curve of a smooth surface of degree $n$—that is, the number of "Inflexions(Haupt)tangenten" (resulting in cusps of the

---

[94] Enriques mentions Hurwitz's research on Riemann surfaces in a letter to Castelnuovo on 7 June 1897 in: (ibid., p. 341); See also: Friedman (2019a, p. 164). Hurwitz's paper is (Hurwitz 1891). Cf. (Epple 1999, p. 185–192) regarding an analysis of Hurwitz's paper and its connection to braid theory.

[95] Castelnuovo / Enriques (1908, p. 652).

branch curve), and the "Doppeltangenten" (resulting in nodes of the branch curve)—are also mentioned,[96] but what is missing is not only any reference to the branch curves as such, but also the variety of explorative questions that Enriques already considered. In the 1908 paper one can note a presentation of an older research direction—that of Salmon's numerical investigation. This changes when considering the published papers of Enriques during the 1910s and the 1920s.

### 3.1.3   Two Papers from 1912 and the Culmination of the Classification Project

In 1912 Enriques published two papers on the subject of branch curves, each paper shows a different approach, which situates the branch curve in a different configuration: the first, presenting a somewhat algebraic approach, concentrating on permutations which can be associated to the branch group; the roots of this approach can be already seen in the 1899 letter. The second paper unfolds a numerical approach, expressing the invariants of any ramified surface given the invariants of its branch curve.

Before examining the two papers, it is important to stress that the first paper, entitled "Sur le théorème d'existence pour les fonctions algébriques de deux variables indépendantes", was later—in 1923—noted by Enriques himself to be not precise enough (see Sect. 3.1.4). Exactly for this reason it is essential to examine it critically, not as a final result, but rather as a first attempt of a construction of a mathematical configuration.

* * *

To introduce the configuration of the first paper, recall from the introduction (Sect. 1.1), that when dealing with Riemann surfaces, already Riemann (among others) noted that every branch point of multiplicity $m$ induces a permutation of the $m + 1$ preimages of points at the neighborhood of the ramification point, when following a loop encircling the branch point. If the branch point is a simple one, then $m = 1$ and only two preimages permute, in this case the induced permutation is called *transposition*, which describes the transposition of the two sheets at the corresponding ramification points. The question that arises is whether one can determine a Riemann surface only from knowing the permutations (or transpositions) themselves. This question was approached algebraically by Adolf Hurwitz in a paper from 1891, a paper Enriques knew well, as was already noted above. As Enriques cited in 1912 this was one of the main results from Hurwitz's paper (though without mentioning his name). It is worthwhile making a detour here to examine Hurwitz approach in more detail, as Enriques attempted to extend it to algebraic surfaces and branch curves.

The goal of Hurwitz's 1891 paper "Ueber Riemann'sche Flächen mit gegebenen Verzweigungspunkten" is "to investigate the totality of the $n$ leafed Riemann surfaces,

---

[96] Ibid., p. 654.

which are branched in a prescribed manner at $w$ points".[97] Hurwitz began by considering Riemann surfaces of degree $n$ branched over $w$ points $a_1, \ldots, a_w$ lying at a two-dimensional real plane $E$. He first chose a point $O$ on $E$ and then drew non intersecting paths $l_1, \ldots, l_w$ from $O$ to $a_1, \ldots, a_w$. Hurwitz then mentioned two other articles that dealt with the question of the construction and representation of Riemann surfaces. The first is Lüroth's "Note über Verzweigungsschnitte und Querschnitte in einer Riemann'schen Fläche",[98] which constructs a set of loops in a *certain order* on the complex line (in Hurwitz's setting, the plane $E$), encircling the branch points. The permutations induced from the loops indicate how the Riemann surface can be constructed. The second is Clebsch's "Zur Theorie der Riemann'schen Flächen",[99] which continues Lüroth's work by proving that in fact any set of loops can be taken when considering effects on the induced permutations while changing the set of loops. Following Clebsch's and Lüroth's works, Hurwitz constructed the Riemann surface itself: "The Riemann surface is now formed by connecting the $n$ sheets along the cuts $l_1, \ldots, l_w$ in the following manner."[100]

While Hurwitz drew several drawings of this system of paths,[101] he follows a more algebraic-combinatorial practice, as he describes the problem using algebraic terms and conditions. For each of the paths $l_1, \ldots, l_w$ Hurwitz assigned a permutation: $S_i$, when $1 \leq i \leq w$. If $S_i = \begin{pmatrix} 1, & 2, & \ldots, & n \\ \alpha_1 & \alpha_2, & \ldots, & \alpha_n \end{pmatrix}$ is the permutation sending 1 to $\alpha_1$, 2 to $\alpha_2$ etc. (when the $\alpha_i$ are integers between 1 and $n$), then along the path $l_i$ the sheets numbered $1, \ldots, n$ of the Riemann surface are transformed to the sheets $\alpha_1, \ldots, \alpha_n$. Here Hurwitz expressed the necessary and sufficient conditions for the existence and construction of a Riemann surface in algebraic terms:

> The substitutions $S_1, S_2 \ldots, S_w$, which are chosen for the construction of the surface, should only satisfy the following two conditions: (I) The transition from any element to any other is to be possible by substitutions. (II) The composition of all substitutions is to give the identity, i.e. $S_1 S_2 \ldots S_w = 1$.[102]

---

[97] Hurwitz (1891, p. 2): "Die Gesammtheit der $n$ blättrigen Riemann'schen Flächen zu untersuchen, welche an $w$ gegebenen Stellen in vorgeschriebener Weise verzweigt sind". I follow here also the analysis given in (Epple 1999, p. 186–192).

[98] Lüroth (1871).

[99] Clebsch (1872).

[100] Hurwitz (1891, p. 4) "Die Riemann'sche Fläche entsteht jetzt, indem man die $n$ Blätter längs der Schnitte $l_1, \ldots, l_w$ in folgender Weise mit einander verbindet."

[101] Ibid., p. 34, 36.

[102] Ibid., p. 4: "Die Substitutionen $S_1, S_2 \ldots, S_w$, welche man zur Herstellung der Flache wählt, sollen nur folgenden beiden Bedingungen genügen: (I) Vermöge der Substitutionen soll ein Uebergang von jedem Element zu jedem andern möglich sein. (II) Die Zusammensetzung aller Substitutionen soll die Identität ergeben, es soll also $S_1 S_2 \ldots S_w = 1$ sein."

In modern language, the first condition forces the group generated by the $S_i$'s to be transitive, which is equivalent to saying that the Riemann surface constructed would be connected. The second condition ensures that while circling the point $O$ (which is not a branch point), the order of the sheets will not be permuted. What is essential to note is that via this algebraic formulation, expressed in terms of conditions on permutations, it becomes irrelevant *where* the branch points $a_1, \ldots, a_w$ are located on the complex line $\mathbb{C}$, and the initial question is "reduced to a purely group theoretic question".[103] Therefore Hurwitz has proved that while one can construct the Riemann surface topologically when giving a number of sheets, the position of its branch points, and the permutations describing the number of sheets, this construction gives the algebraic data *equivalently*—which is clearly to be seen in the way Hurwitz considers a Riemann surface, as an *w-tuple* of permutations: $(S_1, S_2, \ldots, S_w)$, where "obviously, the substitution $S_i$ immediately gives the 'type' of the branching at the point $a_i$."[104]

Returning to Enriques's 1912 paper, how did Enriques reconsider Hurwitz's construction? At the beginning of his paper, Enriques recounted the Hurwitz's construction: given a set $M$ of $2m$ simple branch points, $n$ the degree of the to be constructed Riemann surface, and $2m$ transpositions corresponding to the branch points, forming a transitive group inside the symmetric group of $n$ elements, a Riemann surface exists branched over these $2m$ points. Enriques formulated it succinctly as follows: One "fixes the corresponding transpositions between the branches [...], so that these transpositions form a transitive group and their product is the identity; [therefore] there will exist a corresponding class of algebraic functions $y(x)$, or if one likes better an irreducible function [...]."[105] The question that Enriques posed is whether the same situation can be generalized to branch curves and the construction of complex surfaces.

Enriques emphasized his claim (he already made in 1899) that the branch curve cannot be arbitrary and is usually singular: "given the invariants $(p_a, p^{(1)})$ of the [surface given by the] equation $F(xyz) = 0$ [of degree $n$] one finds that in general the [branch] curve $f(xy) = 0$ has a certain number of nodes which are $> 0$ for $n > 3$ and a certain number of cusps which are $> 0$ for $n > 2$, when these numbers can be calculated with the help of the known formulas."[106] Enriques then stated—without proof—that branch curves do not have more complicated singularities.[107] A more surprising claim—also stated without giving an

---

[103] Epple (1999, p. 187).

[104] Hurwitz (1891, p. 6): "Offenbar ergiebt die Substitution $S_i$ sofort die 'Art' der Verzweigung in dem Punkte $a_i$."

[105] Enriques (1912a, p. 419): "[...] fixer aussi les transpositions correspondant entre les branches [...], de façon que ces transpositions forment un groupe transitif et que leur produit soit l'identité; [donc] il existera une classe correspondante de fonctions algébriques $y(x)$, ou si l'on aime mieux une fonction irréductible [...]."

[106] Ibid., p. 419–420.

[107] Ibid.: "on exclura qu'il y ait des singularités plus élevées." This was only proved in 2011, in: (Ciliberto / Flamini 2011).

example—is that given two plane curves of the same degree, with the same number of nodes and cusps, it may be that one would be a branch curve, while the other not.[108]

It seems that what led Enriques to make the second claim are the necessary and sufficient conditions that he proposed for a curve as a branch curve. These conditions are stated in a completely algebraic way, and, as Enriques claimed, due to their complexity, do not hold for every curve. Specifically, denoting the degree of the branch curve $B$ by $2m$, the question is how to assign transpositions for each of the $2m$ branches (of the branch curve), transpositions that would describe the way the complex algebraic surface is branched. Formulated similarly to how Hurwitz formulated his theorem for Riemann surfaces, these conditions deal with the properties of the map sending a loop around one of the $2m$ branches to a transposition. In order to find the sufficient and necessary conditions, Enriques considered two types of critical points of the branch curve: simple ones[109] and cusps. He then posed three purely algebraic conditions concerning the associated transpositions, indicating that if they are satisfied then the curve is a branch curve and one can construct a complex surface branched along it.[110] The claim, however, is made without proving that those conditions are necessary and sufficient. Moreover, the nodes of the branch curve and the conditions they might have implied regarding their associated transpositions go unmentioned. Here one can note another indication to Enriques's explorative configuration. This is also to be seen in the fact that this pure algebraic consideration is re-formulated in a 1923 paper, where it is re-situated in a more topological context. But before turning to the 1923 paper, it is essential to examine another work of Enriques: the second paper published in 1912.

The paper deals with one of the famous questions concerning algebraic varieties: on how many continuous parameters does an algebraic variety depend? For curves, it was already established in 1857 by Riemann, that if the genus of a smooth curve is $p > 1$, then the curve depends on $3p - 3$ continuous parameters, named also moduli.[111] This computation was done with the help of counting parameters, related to branch points, and the problem was known as the 'moduli problem'. For surfaces the situation was more complicated, and in 1912 a coherent conclusion was not yet to be found. While Noether made some progress regarding regular surfaces in 1888,[112] Enriques aimed to show the

---

[108] Enriques (1912a, p. 420). This is an important statement, as it indicates—in modern terminology—that the variety which parameterizes all plane curves with degree $n$ with $\delta$ nodes and $\kappa$ cusps, with no other singularities (as a subvariety in the projective complex space $\mathbb{CP}^{\frac{n(n+3)}{2}}$), may be, for given $n$, $\delta$, and $\kappa$, reducible. The first example of this reducibility involving branch curves is an example of Zariski from 1929 (see Sect. 3.2.1).

[109] These points "correspond to parallel lines to the $y$ axis, tangent to $f = 0$ at a simple point" (ibid.).

[110] Ibid., p. 420–421. One of the conditions is for example that the product of all of the propositions would be the identity—a condition similar to the one formulated for Riemann surfaces.

[111] See: Riemann (1857, p. 120). Note however that Riemann did not use the term "genus"; see Sect. 1.1. for a short discussion on this.

[112] See: Noether (1888). A surface is regular if its arithmetic genus $p_a$ is equal to its geometric genus $p_g$.

connection of this problem with the calculation of numerical invariants of the branch curve. $M$ being the moduli, Noether suggested in 1888 the formula $M > 10p_a - 2p^{(1)} + 12$, though only for regular surfaces with $p_a = p_g > 3$ and $p^{(1)} > 5$; his argument consisted of computations done for family of surfaces in $\mathbb{CP}^3$ with ordinary singularities and a double curve. As I will elaborate below, Enriques improved Noether's formula.[113] Already in 1908 Enriques proves that the surfaces with given genera $p_a$, $p_g$ and $p^{(1)}$ depend on $M = 10p_a - p_g - 2p^{(1)} + 12 + \theta$ continuous parameters, where $\theta$ is a non-negative integer.[114] To prove that $\theta$ is indeed non-negative, he introduced in 1912 an assumption— given such a surface $F$, varying in an irreducible continuous system $\{F\}$ of surfaces of the same degree, Enriques assumed that the system $\{F\}$ is *complete* (in the sense that it is not contained in a larger continuous system of surfaces of degree $n$, having the same numerical invariants).[115] To show that this assumption is plausible, Enriques computed the same number of continuous parameters on which a (family of) surface(s) depends, but now considering the surface as a ramified cover of degree $n$, and inspecting the numerical invariants of the branch curve. Arriving at the same numerical result concerning $M$, he concluded that if a "plane curve $K_{2m}$ of degree $2m$, having $d$ nodes and $e$ cusps, is a branch curve [curva di diramazione] of an $n$-cover [...] [then] all the plane curves of the same degree $2m$ with the same number of nodes and cusps, belonging with $K_{2m}$ to the same continuous family, are branch curves of analogous $n$-plane covers".[116] This means that the continuous system of branch curves is complete, that is, as Zariski formulated it in 1935, "it is not contained in a larger continuous system of curves of the same order and with the same number of nodes and cusps as a generic [branch curve]."[117] However, even if this conclusion would have been true (and in 1974 it was proven that it is not),[118] another problem was—as Enriques stated in his 1912 paper explicitly in a section called "around a theorem of existence"—proving this theorem, that is: proving that a surface depends only on the branch curve, its number of nodes and cusps, and can be constructed uniquely from it. The title of this section is remarkable—one finds oneself "around" the theorem, but a proof is not yet offered. Enriques himself noted that the presented mathematical configuration is itself not yet clear, and this theorem may be clarified if one looks at other mathematical configurations. This is seen at the conclusion of the paper where Enriques hinted at the difficulties of constructing such a surface or proving its existence: it is not enough that a curve would have the right number of cusps and nodes, it should also "satisfy *certain conditions of arithmetical nature*, which do not diminish the number of moduli of the class of corresponding surfaces. [...] a first look at the research of this theorem would

---

[113] Cf. also Brigaglia / Ciliberto (1995, p. 101–102).

[114] Enriques (1908).

[115] Enriques (1912b, p. 303).

[116] Ibid., p. 306.

[117] Zariski (1935, p. 100).

[118] By Wahl (1974). See also Sect. 5.1.

seem very difficult [. . .] [but one may] overcome the difficulty with the help of an analysis of the monodromy group [. . .] as I will show in another work"[119]—in this way he referred to his first paper from 1912. The "conditions of arithmetical nature" are those being associated to the symmetric group, which will be formulated more clearly only in 1923.

These two results from 1912 are mentioned in the second contribution of Enriques and Castelnuovo to the *Encyklopädie der mathematischen Wissenschaften* in 1914, called "Die algebraischen Flächen vom Standpunkt der birationalen Transformationen". In contrast to the contribution of 1908, this contribution no longer considers (only) surfaces embedded in $\mathbb{CP}^3$ given by an equation $f = 0$, but inspects them from the point of view of birational transformations. The above results are summarized shortly: "the conditions of existence of a $n$-cover plane with [a given] branch curve [*Verzweigungskurve*] [. . .] result from an investigation which concerns certain groups of substitutions [permutations] that play a fundamental role."[120] The crux of the paper however is not Enriques's results from 1912, but rather the presentation of the outcome of the classification project, according to their genera.[121] The shift in this point of view, from the 1908 paper to the 1914 paper, is also to be seen in how the 1914 paper ignores the variety of approaches Enriques considered regarding the role of the branch curve in the research of surfaces, and reduces (or presents) them as having had only one aim, being the moduli problem, resp. contributing to the classification project.

\* \* \*

Putting aside for a moment what constitutes one of the foci of Castelnuovo and Enriques's project—i.e. the classification project of algebraic surfaces, it seems that not all of the mathematicians who were working on algebraic geometry were satisfied with how the research was presented. The plurality of approaches to research surfaces was not only to be found in Enriques's work, but also across different mathematical communities. The different mathematical approaches for dealing with algebraic functions at the beginning of the twentieth century were described by Felix Klein in 1926: the Italian school of algebraic geometry (or, as Klein noted, the "Italians") during the first decades of the twentieth century with its "geometric thinking", and the Germans with their "arithmetic procedures".[122] These two approaches, according to Klein, led to an almost unsolvable

---

[119] Enriques (1912b, p. 307): "soddisfi a certe condizioni di natura aritmetica che non diminuiscono il numero dei moduli della classe di superficie corrispondente. [. . .] A prima vista la ricerca d'un teorema siffatto apparirebbe molto difficile. [. . .] [È possibile] superare la difficoltà mediante l'analisi del gruppo di monodromia [. . .] come mostrerò in un altro lavoro."

[120] Castelnuovo / Enriques (1914, p. 714): "[. . .] die Existenzbedingungen einer $n$-fachen Ebene mit $C$ als Verzweigungskurve [. . .] ergeben sich übrigens aus einer Untersuchung, welche gewisse Gruppen von Substitutionen betrifft, die dort eine fundamentale Rolle spielen."

[121] Ibid., p. 760–761.

[122] Klein (1926, p. 327). In contrast to the explicit general designation of the "Italians", Klein does not assign any nationality to the ones who preferred "arithmetic procedures"; However, all of the

dissonance within the mathematical community. Mentioning also the contribution of Castelnuovo and Enriques in 1914, Klein noted the following:

> "It is like the Tower of Babel that the different languages soon no longer understand each other. And one wants, because it is so uncomfortable to get used to, to not understand each other anymore. In any case, in the *Enzyklopädie der mathematischen Wissenschaften* we had to present two contributions set alongside each other. The 'geometrical' contribution of Castelnuovo-Enriques is in Vol. III."[123]

Exactly because of this "Tower of Babel", a table 'translating' the terms from one language to another was added already at the beginning of Castelnuovo and Enriques paper of 1914, a table situated at the end of a list of books and articles, which were considered of as fundamental to the theory. After a paper by Heinrich W. E. Jung from 1910 is mentioned, the table at Fig. 3.2 is provided. Klein mentioned this table in 1926 immediately after the above citation: "The remarks about [the essay by] Jung in the table of contents have only been enforced by me with great effort."[124] He then continues, lamenting on the splintering of mathematical languages: "This tendency not only to divide science into ever more numerous individual chapters, but to create school distinctions according to the way in which they are treated, would, if unilaterally enforced, bring about the death of the science."[125]

### 3.1.4   1923: After the Classification Project

While the 1914 paper of Castelnuovo and Enriques presented the existence result as based on the "fundamental role"[126] of the symmetry group, a role which was explicated in 1912, in 1923 Enriques expressed his own dissatisfaction from the rigor of his own proof in 1912, where these group-theoretic considerations were presented. This is done in a paper entitled

---

mathematicians he named who were employing those "arithmetical procedures" (Hensel, Landsberg, Jung) were Germans. See also: Slembek (2002, p. 113–115).

[123] Klein (1926, p. 327): "Es geht dann wie bei dem Turmbau von Babel, daß sich die verschiedenen Sprachen bald nicht mehr verstehen. Und man will, weil es so unbequem ist sich umzugewöhnen, sich vielfach nicht mehr verstehen. Jedenfalls haben wir bei der Enzyklopädie der mathematischen Wissenschaften zwei Referate neben einander ansetzen müssen. Das 'geometrische' Referat von Castelnuovo-Enriques liegt in Bd. III (C 6b) vor."

[124] Ibid.: "Die Bemerkungen über [den Aufsatz von] Jung in dem Inhaltsverzeichnis sind nur mit großer Muhe von mir durchgesetzt worden."

[125] Ibid.:"Diese Tendenz, die Wissenschaft nicht nur in immer zahlreichere Einzelkapitel zu zerlegen, sondern Schulunterschiede nach der Art der Behandlung zu schaffen, würde, wenn sie einseitig zur Geltung käme, den Tod der Wissenschaft herbeiführen."

[126] Castelnuovo / Enriques (1914, p. 714).

| *Algebraisch-geometrische Theorie* | *Arithmetische Theorie* |
|---|---|
| Irreduzibele Kurve auf *F* | Primteiler |
| Beliebige Kurve auf *F* | Divisor |
| Lineares Kurvensystem auf *F* | Divisorenklasse |
| Gesamtheit der Kurven auf *F*, die vollständige Schnitte sind | Hauptklasse |
| Dimension, Grad, adjungierte Systeme, Defekt, charakteristische Schar usw. eines linearen Kurvensystems | ebenso, einer Klasse |
| Summe und Residualsystem | Produkt und Quotient zweier Klassen). |

**Fig. 3.2** A table of the different concepts used in the different theories (the "algebro-geometric" and the "arithmetic") for investigating algebraic functions was added in the *Enzyklopädie der mathematischen Wissenschaften*, at the beginning of the article "Die algebraischen Flächen vom Gesichtspunkte der birationalen Transformationen aus" of Castelnuovo and Enriques. The editors, Meyer and Mohrmann, noted: "Here is a short index of the basic terms which correspond to each other" ["Hier ein kurzes Verzeichnis der einander entsprechenden Grundbegriffe"] (in: Castelnuovo/ Enriques 1914, p. 676)

"Sulla costruzione delle funzioni algebriche di due variabili possedenti una data curva di diramazione", where Enriques declared at the outset that while submitting the 1912 paper, he already encountered several difficulties. As he stated, it was precisely due to such difficulties that he now wished to revise and resubmit the paper.[127]

As in the previous papers, Enriques noted also here that not just any plane curve can be a branch curve[128]—but once more he failed to give even a single example. Similar to the 1912 paper, Enriques's aim is to provide a proof for the necessary and sufficient conditions for a curve being a branch curve, formulating more precisely and more extensively the algebraic conditions given eleven years ago. The 1923 paper begins with the same setting of the 1899 letter: a plane curve $C$ is given in the plane $P = \mathbb{C}^2$ (or $\mathbb{CP}^2$) assuming it is a branch curve of a surface $F$ of degree $n$; a point $O \in P$ not on $C$ is also given, and a pencil of lines $\{y = tx\} \subset P$ on this plane, passing through $O$. The parameter $t$ is a complex one, whose values vary in the (complex) plane denoted by $\tau$. Taking $t = 0$, the line $y = 0$ cuts the curve $C$ in $m$ points: $A_1, \ldots, A_m$. When considering the section of the surface $F$ above this line,[129] one obtains a (smooth) Riemann surface, denoted by $K_0$, which is considered as a branched cover of the line $y = 0$, branched over $A_1, \ldots, A_m$, which are simple branch points.

---

[127] Enriques (1923, p. 185).

[128] Ibid.

[129] Denoting by $O'$ the point from which one projects $F$ to the complex plane, one considers a plane $P$ passing through $O'$ and $\{y = 0\}$. The above-mentioned section is the intersection of $P$ and $F$.

Enriques's key move is to note that the line $y = 0$ is a complex line, homeomorphic to a two-dimensional real plane. Hence one can look at a "system of loops"[130] which are denoted by $l_i$, in the two-dimensional real plane $y = 0$, going out from $O$ and encircling the points $A_1, \ldots, A_m$. Every loop $l_i$ corresponds to a permutation $S_i = (a\ b)$,[131] a permutation which describes the permutation of the sheets $a$ and $b$ of the Riemann surface $K_0$, while moving along $l_i$; this permutation eventually describes the permutation of the sheets of the algebraic complex surface $F$ itself. Following the arguments from 1899, Enriques then moves the line $y = 0$ (as a member in the family of lines $\{y = tx\}$), claiming that after moving the parameter $t$ along a loop at the plane $\tau$, the resulting permutation should be the identity permutation. This results in conditions of invariance concerning these permutations, which Enriques claimed are true for every branch curve. To show what these conditions of invariance are, Enriques started by noting that while rotating the line $y = 0$ in the pencil of lines, the rotated line might intersect three types of critical points of the curve $C$: simple tangent (branch) point, nodes and cusps. Before analyzing what happens to the above-mentioned permutations $S_i$ while approaching one of these critical points, Enriques asked the following question: when moving $t$ along a loop at the plane $\tau$,[132] what happens to the loops $l_i$ when approaching one of these critical points? Assuming that $y = t_c x$ is a line which intersects $C$ at a critical point, then there are two intersection points $A_{r_i}$ and $A_{s_i}$ which are merged (see Fig. 3.3).[133] Renumbering the points of intersection, one can assume that the points $A_1$ and $A_2$ are merged. Taking a value $t_0$ very close to $t_c$,[134] Enriques noted that "for simplicity [...] we can demonstrate that one can transform the loops, permitting us to reduce it to the case when $l_1$ and $l_2$ would be fairly close."[135]

---

[130] Ibid., p. 187.

[131] The notation $(a\ b)$ stands for a permutation, which permutes between $a$ and $b$ and leaves all the other numbers as they are.

[132] That is, when for example, rotating the line $y = 0$ in the family of lines $y = tx$.

[133] The fact that there are *two* points that coincide is due to the nature of the critical points (tangent point, node or cusp). Were there other types of singular points, this might not have been the situation. However, Enriques did not prove—also in this paper—that the only singular points of a branch curve are nodes and cusps.

[134] Denoting by $L_0$ the complex line $y = t_0 x$, the question that Enriques poses concerns, in *modern* terminology, the finding of a well-ordered basis for the fundamental group of the complement of the intersection points $A_1, A_2, \ldots$ of $L_0$ with $C$: $\pi_1(L_0 - \{L_0 \cap C\}) \simeq \pi_1(\mathbb{C} - \{A_1, A_2, \ldots\})$. Enriques (ibid., p. 189) asked if one can always rearrange the loops in this group such that $l_1$ and $l_2$ would be "fairly close", meaning that they "tend to merge without including any other point $A$ or crossing other loops." ["[...] può accadere che $l_1$ e $l_2$, per $t = t_c$, diventino o possan farsi diventare *onestamente vicini*, cioè tendenti a confondersi senza includere alcun altro punto $A$ o attraversare altri cappi."]

[135] "Ma, riferendoci per semplicità a questo case elementare, possiamo dimostrare che, in ogni case, una conveniente trasformazione dei cappi, permette di ridursi al caso in cui $l_1$ e $l_2$, diventino onestamente vicini." (ibid., p. 190) In modern terminology, since the loops are elements in $\pi_1(L_0 - \{L_0 \cap C\})$, one can reformulate Enriques's question as one about automorphism of this

**Fig. 3.3** The complex line $y = t_0 x$ intersects the curve $C$ near a singular point $N$, and the points of intersection $A_{r_i}$ and $A_{s_i}$ will coincide at this point, once the line gets closer and closer. This complex line $y = t_0 x$ is homeomorphic to a two-dimensional real plane— which is also drawn in order to depict where the loops $l_{r_i}$ and $l_{s_i}$ are to be found; these loops surround the intersection points and exit from the point $O$ (which is also on the line). © Graphics: M.F.

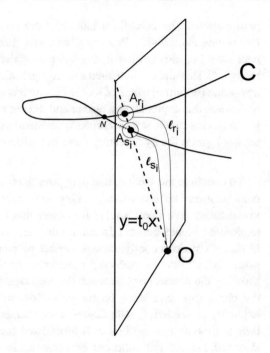

With these assumptions Enriques dealt with the three types of critical points, investigating what the relations between the induced permutation $S_1$ and $S_2$ are. He started by investigating what happens in the neighborhood of a critical point, noting that "the permutations $S_1$ and $S_2$ change respectively to $S_1' = S_1 S_2 S_1^{-1}, S_2' = S_1$."[136] Enriques's argument relies mainly on visual reasoning and diagrams of loops, giving no explicit algebraic proof with the tools of group theory. He then employed algebraic inference methods, proving that in this case of the tangency point, $S_1' = S_1, S_2' = S_2$. Using the same argumentation, he subsequently investigated what happens in the neighborhood of a node and a cusp. In the case of a node, Enriques proved that locally the corresponding permutations should be disjoint (e.g. $S_1 = (1\ 2)$, $S_2 = (3\ 4)$, such that $S_1 S_2 = S_2 S_1$) while in the case of a cusp, the corresponding permutations should have only one index in common (e.g. $S_1 = (1\ 2)$, $S_2 = (2\ 3)$, such that $S_1 S_2 S_1 = S_2 S_1 S_2$).

The recapitulation of the theorem, appearing at section IV of Enriques's paper, summarizes the necessary conditions for a curve to be a branch curve, i.e. what results when it is known that $C$ is a branch curve. The theorem presents the map sending the loops (in the complex line $y = tx$, encircling the points $A_1, \ldots, A_m$) to their corresponding

---

fundamental group. To stress: Enriques does not use the terms 'automorphism' or 'fundamental group', but rather talks about how the induced permutations 'change'.

[136]"vediamo cosi che le sostituzioni $S_1$ e $S_2$, si cambiano rispettivamente in $S_1' = S_1 S_2 S_1^{-1}, S_2' = S_1$." (Ibid., p. 192)

permutations. The conditions induced from this map would later be called 'Enriques's *invariance conditions*'. Proving afterwards that the conditions presented are not only necessary but also sufficient, Enriques formulated his theorem in terms of "elementary loops".[137] This already indicated a more algebraic formulation, which might have indicated a possible (or future) usage of tools of group theory. Nevertheless, Enriques did not attempt to develop this approach any further and did not term his theorem algebraic. As will we see in the next sections, it was Zariski who aimed at a re-formulation of Enriques's results in terms of group theory, situating them in a different epistemic configuration.

* * *

To conclude the section, one may pose the following question: How can these research configurations be considered, taking into account Enriques's views on intuition and visualization (see Sect. 3.1.1)? It is clear that Enriques considered a variety of ways to explore the branch curve. I do not use the verb 'explore' incidentally, as it is to be noted in Enriques's various configurations—either by constructing double covers branched over a given (branch) curve, deducing numerical conclusions regarding moduli of surfaces, or showing the connections between the topological and the algebraic objects associated to the curve (the invariance conditions). How this exploration took place is to be seen explicitly in his letters with Castelnuovo, sometimes in form of reflections, which from time to time do not end in a well formulated conclusion, when directions of research are sketched but are not followed or taken again only several years later. The image of exploration as an image of knowledge was moreover articulated by Castelnuovo and Enriques themselves in 1899, as they described an exploration of an uncharted territory (being algebraic geometry in general and algebraic surfaces in particular), or in their 1896 review.[138] The "obstacles" in this landscape were also noted by Klein, when he mentioned the "Tower of Babel" of algebraic geometry.

This exploration can also be seen as a way to refine the geometric intuition, which was also partially visual. As seen above, Enriques called to develop "different forms of intuition", when "abstract geometry can be interpreted in infinite ways."[139] This plurality of epistemic configurations, in which the branch curve is situated, may be considered under this 'infinite interpretation'. It sometimes led to wrong assumptions or to dead ends, but nevertheless it was essential for Enriques's research.

---

[137] "giri elementari" or "sistema primitive di cappi" (ibid., p. 198).

[138] A similar geographical metaphor appears in Castelnuovo's preface to the 1949 book of Enriques *Le superficie algebriche* (see Sect. 3.3).

[139] Enriques (1894a, p. 9–10).

## 3.2    Zariski and Segre: Novel Approaches

Having seen how Enriques treated branch curves, a project which culminated in 1923 with the precise formulation of the invariance rules, I would like to turn to two mathematicians who not only continued Enriques's research, but also opened new horizons concerning how one should characterize branch curves. The first mathematician is Oscar Zariski. Zariski (1899–1986), born as Ascher Zaritsky to a Jewish family in Kobrin (now in Bellarus), was one of the influential mathematicians in the field of algebraic geometry during the twentieth century. Starting in Pisa, he then moved to Rome and was working as a student with three of the senior members of the Italian school of algebraic geometry: Castelnuovo, Enriques, and Severi. Castelnuovo was his thesis advisor, and Zariski completed his doctorate in 1924. The ascent of Mussolini to power in 1922 and the increasingly alarming situation in fascist Italy—especially for Jews and non-Italians—forced Zariski to immigrate to the USA, where he worked at John Hopkins University, between 1927 and 1947. Zariski's research between the years between 1927 and 1937 is characterized by using various topological techniques in the domain of algebraic geometry, a method warmly supported by Solomon Lefschetz, who enthusiastically supported Zariski's work, and with whom Zariski was in close contact over those years; Zariski was also influenced by his topological methods.[140] During his stay at John Hopkins University he also published his influential book *Algebraic Surfaces*, whose writing prompted Zariski to establish algebraic geometry on the foundation of modern commutative algebra. During the years 1946–1947 he was at the University of Illinois; he finally became professor at Harvard University in 1947, where he remained until his retirement. It was during these years that another influential book was written by Zariski, together with Pierre Samuel: *Commutative Algebra*, published in two volumes, in 1958 and 1960.[141]

The second mathematician, with whom this section will deal, is Beniamino Segre. Segre (1903–1977), the nephew of the mathematician Corrado Segre, was a Jewish Italian mathematician, specializing in algebraic geometry. After graduating in Turin, doing his thesis under the guidance of Corrado Segre, he was Francesco Severi's assistant in Rome and from 1931 a professor in Bologna. Due to the racial laws of the fascist Italian government, he was expelled from the university, and stayed in England till 1946, getting a teaching position only in 1942 in Manchester. He continued researching in the field of algebraic geometry, but researched also diophantine equations and arithmetic of algebraic varieties. Returning to Bologna in 1946, he succeeded Severi in 1950 in Rome, a position from which he retired in 1973. Segre was an extremely productive mathematician,[142] he had more than 300 publications during his life, on a wide variety of subjects.[143]

---

[140] See: Parikh (2009, p. 38–46).

[141] For a detailed biography of Zariski, see: (Parikh 2009).

[142] See: Segre (1987, p. xi–xxxi).

[143] For a detailed biography on B. Segre, see: (Sernesi 2012; Vesentini 2005; du Val 1979).

Already these two short biographical accounts show the influence Mussolini's fascist regime had on Jewish mathematicians starting the 1920s; this was already noted above with Enriques, who was forced to resign in 1938. As Angelo Guerraggio and Pietro Natasi note, "the shameful support of [the non-Jewish] Italian mathematicians to this further wickedness of the regime only made this detachment deeper."[144] The extent of the influence of the racial laws on the Jewish mathematicians has been analyzed thoroughly in several publications;[145] to recall shortly, the racial laws of 1938 included, among others, not only the a process of 'purge' and exclusion of Jews from universities, but also the prohibition of the usage and publication of books and papers written by Jewish authors. Giorgio Israel underlines that these laws "made irreversible a crisis in Italian mathematics [...]. The isolation and decline that had been showing for a long time, both in terms of internal and institutional dynamics, were accelerated and made worse by the (material and intellectual) autarchy and the policy for the purity of the race."[146] We will see during the following sections the results of this isolation.

Returning to the mathematical work of Zariski, as Michael Artin and Barry Mazur note concerning his work between 1927 and 1937, Zariski led "an energetic attack on topological problems in algebraic geometry—the fundamental group, mainly."[147] To state the obvious—it was due to the political circumstances in Italy that Zariski was forced to emigrate. Had he not done that, the influence of Lefschetz on him might have been not so decisive. Moreover, one can speculate, that his "attack on topological problems" and on the "fundamental group" may not have taken place, had he not emigrated. Concerning the term 'fundamental group', I will explain below precisely what this group is; it suffices to note that in this context it is the set (more precisely, the group) of all loops in the (projective) plane $\mathbb{C}^2$ (or $\mathbb{P}^2$) surrounding the branch curve. In this respect Zariski was influenced by Enriques's work,[148] on the one hand, and by Lefschetz's work, on the other. The "attack" on the fundamental group consisted mainly in understanding its structure and properties, and went beyond the research delineated by Enriques, that is, beyond the mere description of the map sending those loops to the symmetric group, as was described in Sect. 3.1.3. During the same years, Segre published only one paper on the subject of branch curves in 1930, to be found within the setting of his investigation on moduli of nodal-

---

[144] Guerraggio / Natasi (2006, p. 281). Moreover, they also note that publishing in English or making contact with mathematicians outside of Italy may have been practically impossible (ibid., p. 277–278).

[145] On the situation of Jews in Mussolini's Italy, see for example: (Sarfatti 2006; Zimmerman 2005). On Jewish mathematicians and scientists during the fascist regime, see: (Guerraggio / Nastasi 2018; Capristo 2005; Finzi 2005; Israel 2004). Cf. also: (Luciano 2018; Guerraggio / Nastasi 2006, p. 243–281; Zappa 1997).

[146] Israel (2004, p. 47)

[147] Artin / Mazur (2009, p. 137).

[148] See: Brigaglia / Ciliberto (1995, p. 117–123).

cuspidal curves.[149] In this paper he showed explicitly another way to think about branch curves—in terms of the location of its singular points, and in this he opened another epistemic configuration to research those curves, which was—due to reasons I will describe below—less accepted then Zariski's.

I will begin therefore with Zariski's early work during the late 1920s (Sect. 3.2.1), then reviewing Segre's paper from 1930 (Sect. 3.2.2). Section 3.2.3 will unfold Zariski's work during the 1930s, focusing mainly on his 1935 book *Algebraic Surfaces*.

### 3.2.1   The Late 1920s: Zariski on Existence Theorems and the Beginning of a Group-Theoretic Approach

The period in Zariski's research between 1927 and 1937 is characterized by an intensive investigation of nodal-cuspidal curves, also of branch curves, being also nodal-cuspidal curves.[150] Zariski began his research on those curves in 1927 and their associated fundamental group of the complement. The group theoretic approach, which was accompanied with a topological one, is to be seen in two papers, written in 1928 and 1929 respectively. In these papers Zariski asserted that he was continuing Enriques's research.

In the 1928 paper Zariski already noted that while Enriques was the first to pose the problem of the necessary and sufficient conditions of a curve to be a branch curve, the answer he gave did not explicitly introduce "the concept of the fundamental group".[151] Before examining Zariksi's approach in more detail, I would like to explain what the *fundamental group* is.

Given a topological space $X$, the fundamental group $\pi_1(X)$ of this space is the group[152] formed by the set of all loops under the equivalence relation of homotopy, equipped with the action of concatenation of loops; that is, the elements of this group are paths in $X$ with the same initial and final point $p \in X$, under the following equivalence: if one loop can be continuously deformed to another within $X$, then these two loops are considered the same. The identity element of this group is the set of all paths homotopic to the degenerate path consisting (only) of the point $p$. This group, developed by Henri Poincaré at the beginning of the 1890s, was consolidated in the work of the generation after Poincaré.[153] Indeed, Zariski used it extensively during the 1920s and the 1930s, as we will see, to calculate the fundamental group $\pi_1(\mathbb{C}^2 - C)$, for $C$ being a nodal-cuspidal curve, and especially for

---

[149] See: Sernesi (2012).

[150] As was (correctly) assumed at that time, but not yet proved.

[151] Zariski (1928, p. 134).

[152] A group is a set $A$ equipped with an operation (denoted here by *), such that the following four axioms are fulfilled: closure, associativity, unit element and inverse.

[153] For the history of the development and research of the fundamental group, see for example: (Vanden Eynde 1999; Krömer 2013).

*C* being a branch curve. But here it is essential to stress that the consideration of this group of loops of the complement of a curve was not Zariski's novel idea; already Wirtinger developed in 1905 a procedure to find the generators and relations of the fundamental group of the complement of a knot.[154] Zariski was well aware of these techniques of knot theory, as he mentions them in both papers from 1928 and 1929.[155]

Returning to the 1928 paper, by the "concept of the fundamental group" Zariski meant, as noted above, the group of loops in the "residual space $S_4 - D$",[156] when $S_4$ is a real 4-dimensional space $\mathbb{R}^4$ (being homeomorphic to $\mathbb{C}^2$), and $D$ is the branch curve.[157] The same formulation appears in the 1929 paper "On the Problem of Existence of Algebraic Functions of Two Variables Possessing a Given Branch Curve", a translation of the 1928 paper with new results also added. The title of the paper already indicates Zariski's main question, to formulate a parallel existence theorem to the one formulated by Riemann for algebraic functions with one variable: According to Zariski, in 1857 "Riemann announced the following existence-theorem", which associates permutations to the branch points of "an algebraic function $y(x)$", permutations which fulfill several conditions.[158] Zariski noted that Enriques's approach found a similar existence theorem for algebraic surfaces,[159] referring to his 1923 paper on the invariance rules, but at the same time Zariski called for a new approach, or rather, for extracting what was—at least in Zariski's eyes—implicit in Enriques's approach. The critique of Zariski is however explicit:

> Concerning Enriques's problem of "conditions of invariance [. . .] it was pointed out to me by S. Lefschetz that [this] problem is intimately bound up with the Poincaré group of *f* relative to its carrying complex projective plane, the conditions given by Enriques merely expressing the fact that the assigned substitutions must satisfy the generating relations of the fundamental group of the curve *f*. [. . .] Enriques' result, important as a principle, is nevertheless incomplete, since we do not know anything about the fundamental group [of the complement] of an algebraic curve. The complete solution of the existence problem depends upon the solution of the following purely topological problem: *Given an algebraic curve, to find its fundamental group.* In this paper we attempt to throw some light upon the structure of the fundamental group."[160]

---

[154] On the history of the research of this knot group (i.e. the fundamental group of the complement of a knot in $\mathbb{R}^3$) during the first decades of the twentieth century, see Epple (1996, p. 252–255, 267–281, 304–329).

[155] Zariski (1928, p. 136–7; 1929a, p. 311–2).

[156] Zariski (1928, p. 133).

[157] Note that Zariski did not use the notation $\pi_1(\ldots)$ in his papers from 1928 and 1929.

[158] Zariski (1929a, p. 306)

[159] Ibid., p. 308. The two existence theorems will be consequently termed in the literature as the "Riemann-Enriques existence theorem" (see Sect. 5.1).

[160] Zariski (1929a, p. 305).

This passage cited above is important, since it is the first time the investigation of the fundamental group of the complement of a plane (algebraic) curve appears in relation to the branch curve. While Zariski was aware that a similar group was researched for knots (i.e. the knot group), he presented his results as a restructuring of Enriques's configuration. The investigation of branch curves should be focused on finding "the structure" of this associated group and not on the curve itself. Zariski's investigation is aimed therefore to be much more algebraic, and more specifically, group theoretic. The question that arises is whether one may term Zariski's approach a 'structural' one. The answer must be rather negative. The term "structure" here functions as an actor's term, which Zariski uses in 1928–1929 much more in order to differentiate himself from Enriques's work.[161] The usage of this term is certainly not to be seen as a precursor of the algebraic rewriting of algebraic geometry, led by Zariski, which only started in 1937.

That being said, Zariski reformulated Enriques's results explicitly in a group-theoretic way. According to Zariski, "[a] finite set of generators of $G$ [the fundamental group] can easily be constructed", using Lefschetz's method of finding a more convenient basis to work with,[162] when "the generators $g_i$ [of the group $G$] satisfy several relations, called *generating relations*."[163] These generators are what Enriques called a "primitive system of loops",[164] but Zariski took this somewhat vague formulation and made it precise using group theory. By looking at the relations between the generators, relations which are induced from the critical points of the branch curve, Zariski noted that the relations of the corresponding permutations turn into algebraic relations between the different generators $g_i$.[165] Zariski noted that for a tangent critical point, "$g_1$ is transformed into $g_2$ and $g_2$ is transformed into $g_2^{-1} g_1 g_2$ [. . .] This leads to the generating relation: $g_1 = g_2$."[166] Stating also the relations induced from a node and a cusp of the branch curve, Zariski did stress that "the above transformation of the loops can also be deduced *rigorously* [. . .] by a careful inspection of the expansion of the function $y$ in the three considered cases";[167] Zariski then reformulated Enriques's results in an algebraic way:

---

[161] See also Corry (2004, p. 305–306, esp. footnote 46).

[162] Zariski (1929a, p. 307): "It can be shown,* that any circuit $g$ is equivalent to a circuit $g'$ belonging to a generic 'line' $l$ [. . .] through $O$ (a two-dimensional manifold, homeomorphic to a sphere)." In the footnote * Zariski refers to: Lefschetz (1924, p. 33).

[163] Zariski (1929a, p. 307).

[164] "sistema primitivo di cappi" (Enriques, 1923, p. 198).

[165] Zariski (1929a, p. 310–311).

[166] ibid., p. 310.

[167] ibid., p. 311. I will return in Sect. 3.2.3 to how Zariski in his book *Algebraic Surfaces* (and afterwards) considered rigor in the Italian school of algebraic geometry.

"The following theorem is an implicit consequence of the existence theorem, as it is stated by Enriques: THEOREM 4. The elementary generating relations together with the relation $g_1 g_2 \ldots g_n = 1$, form a complete set of generating relations, i.e. every relation between the generators is a consequence of them."[168]

Before continuing the comparison between Enriques and Zariski and respectively their configurations, it is essential to note another theorem of Zariski from the 1929 paper: if $C$ is a reducible curve, having only nodes as singular points, then the fundamental group of the complement of it is abelian.[169] Zariski himself relied, in the course of his proof, on a result of Severi concerning the irreducibility of the moduli space of nodal curves.[170] However, it was discovered that Severi's proof contained a gap, hence Zariski's proof was not valid anymore; the theorem was proved correctly only by Fulton and Deligne in 1980, without using Severi's result.[171] Zariski's "proof" shows, just as was noted above for Enriques, the fallibility of his assumptions in his own configuration, even though he stressed, as we will see later, how rigor is essential for the development of algebraic geometry.

Returning to how Zariski not only restructured Enriques's research configuration but also went beyond his methods, yet another development can be seen in Zariski's papers from the late 1920s. While Enriques only hinted at the fact that there are curves which are not branch curves but have the same numeric invariants as branch curves, Zariski clearly formulated and proved this claim. In 1928, Zariski asked "is the number of the cusps enough for determining the fundamental group of a curve of a given curve, or does this group depend also on the position of the cusps?"[172] Zariski gave an example, which, also years later, would serve as an exemplary one: two non-isotopic curves of order 6 with 6 cusps.

The first curve is a branch curve of a complex surface of degree 3:[173] the branch curve is of degree 6, has 6 cusps all of them lying on a conic (see Fig. 1.3b). This result was in no way obvious since only a unique conic passes through 5 points in a generic position, i.e., there is no conic that passes through 6 points in a *generic* position. Zariski also gave its explicit equation:

---

[168] ibid., p. 312

[169] Ibid., p. 319.

[170] See: ibid., p. 318, footnote *. Cf. Severi (1921, Anhang F).

[171] Fulton (1980); Deligne (1981). See (Fulton 1980, p. 407): "Zariski's proof relied on the assertion of Enriques and Severi that any node curve can be degenerated to lines in general position. This assertion remains unproved, however."; cf. also (Deligne 1981, p. 1): "[Zariski's] démonstration utilisait le théorème de Severi, selon lequel l'espace des courbes planes irréductibles de degré donné, ayant un nombre donné de points doubles, est irréductible [...]. On ignore toujours si l'assertion de Severi est vraie ou non."

[172] Zariski (1928, p. 137).

[173] The branch curve is obtained under a generic projection from a point not on a surface.

"Let $F(x, y, z) = 0$ be the equation of the cubic surface. The branch curve $f$ of the function $z$ is a sextic with 6 cusps *on a conic*, the equation of which has the following form:

$$f(x, y) = \phi_2^3(x, y) + \phi_3^2(x, y) = 0$$

where $\phi_2 = 0$ and $\phi_3 = 0$ represent a conic and a cubic curve respectively."[174]

He then computes the fundamental group of the complement of this curve, proving that it is isomorphic to the non-abelian group generated by two elements $g_1$ and $g_2$ such that $(g_1)^3 = 1$ and $(g_2)^2 = 1$.[175]

The second curve is also of degree 6, when the 6 cusps are in a general position.[176] Zariski asked whether the two curves have the same fundamental group.[177] Initially there is only a negative answer to this question,[178] but only in few years later Zariski proved that "the fundamental group of a sextic with six cusps not on a conic is cyclic", that is, this group is generated by one element $g$, with only one relation $g^6 = 1$.[179] However, this is done without finding the explicit equation of the curve. The computations that Zariski performed in both cases were completely algebraic, and moreover, in order to compute the fundamental group of a sextic with six cusps not on a conic, Zariski used a deformation argument by "remov[ing] a certain number of cusps" from a generic sextic with 9 cusps.[180] I will describe in more detail in Sect. 3.2.3 the techniques Zariski used in this later work. Moreover, as we will see in the next section, in 1930, by using completely different methods, Segre proved that the second curve is *not* a branch curve. This example showed that nodal curves and nodal-cuspidal curves are different objects not only with respect to their singularities, but also and more essentially with respect to their associated fundamental group, and with respect to the decomposition of the variety parametrizing those curves.[181]

To conclude: Zariski shifted in the late 1920s the focus of the research to be more group-theoretic, also pointing towards a positional approach concerning the branch curve, by noting the importance of taking into account the position of the singular points of the branch curve.[182] Before I continue surveying the work of Zariski during the 1930s, I will

---

[174] Zariski (1929a, p. 320).

[175] (Ibid., p. 325). A different yet equivalent description for the relations is given by Zariski in 1928: the relations presented there are $g_2 g_1 g_2 = g_1 g_2 g_1$ and $(g_1 g_2)^3 = 1$.

[176] Zariski (1928, p. 137).

[177] Ibid., p. 138. More precisely, he asked whether the two groups are isomorphic.

[178] Zariski (1929a, p. 320).

[179] Zariski (1937, p. 357).

[180] ibid., p. 356.

[181] Since for nodal curves this variety is irreducible (as was (wrongly) proved by Severi, but later correctly proved) and the fundamental group of the complement of a nodal curve is abelian.

[182] To emphasize: the expression 'positional approach' is an expression I use and it is not an actors' term.

examine the work of Beniamino Segre, who attempted to characterize branch curves by generalizing Zariski's positional approach, but also to situate it in a different configuration, with different research methods.

## 3.2.2   1930: Segre and Special Position of the Singular Points

A year after the publication of Zariski's 1929 paper, which had researched the branch curve of a smooth surface of degree 3, Beniamino Segre generalized Zariski's result. The 1930 paper—the only paper Segre published on branch curves—pointed towards a different direction for research and generalized Zariski's results in two ways. Firstly, while Zariski showed that the cusps of the branch curve of a (smooth) complex surface of degree 3 are in a special position (i.e. all of them lie on a conic), Segre showed the singular points of any branch curve of a (smooth) complex surface of degree $n$ – *for any n*, when $n \geq 3$—are in a special position, when one projects the surface from a generic point not lying on the surface. This means that the position of these singular points is not generic. Secondly, Segre also pursued the questions posed by Enriques and Zariski: what are the necessary and sufficient conditions for a nodal cuspidal plane curve to be a branch curve of a smooth complex surface embedded in the (projective complex) three-dimensional space $\mathbb{CP}^3$. Segre, however, does not follow their methods. Here it should be stressed that while the first result is a result in the body of the mathematical configuration, the second goal pursued by Segre can be considered to be found at the intersection of the body and the image of this configuration, since it not only asks about the necessary and sufficient conditions, but also claims which methods should be employed. When taking Zariski's statements regarding the usage of the fundamental group, one can claim that these 'double' statements were not rare in mathematical discourse.

Indeed, Segre realized a shift was needed not only in the methods of inference used, but also in the mathematical configuration, in which the problem was located. Segre noted at the beginning of his paper that the approaches of Zariski and Enriques "do not exhaust the argument. The difficulties encountered in the above approaches depend on the fact that not every algebraic plane curve is a branch curve of a (non-cyclic) multiple plane: it is a matter of *characterizing* branch curves of such [multiple] planes",[183] implying that the other methods were not "characterizing" these curves in a good enough way. On the one hand, this statement might indicate that the research of Enriques done during 1890s (at least as it appears in his published work) was only a particular investigation of special families of surfaces, and as a result, an investigation of specific curves, sometimes more singular than nodal-cuspidal ones, which might be branch curves of double planes. On the other hand, the more general investigation, as in Enriques's 1923 paper or in Zariski's 1929 paper,

---

[183] Segre (1930, p. 97): "[...] non esauriscono l'argomento. Le difficoltà che s'incontrano nella suddetta estensione, dipendon dal fatto che non ogni curva piana algebrica è curva di diramazione d'un piano multiplo (non ciclico): si tratta dunque di *caratterizzare* le curve di diramazione di tali piani."

concentrated on a derived object associated to the (branch) curve: the map to the permutation group or the fundamental group. This might suggest that Segre saw his 'characterization' as a more precise one, as he used this term also as a rhetorical means to differentiate his methods from those of Zariski and Enriques.

How did Segre characterize branch curves? Firstly, Segre concentrated only on complex smooth surfaces embedded in $\mathbb{CP}^3$. As Salmon already knew the numerical invariants of a branch curve of a surface of degree $n$ (see Sect. 2.2), Segre referenced his work and his results:[184] The degree of the branch curve is $n(n - 1)$, the number of nodes is $\frac{1}{2}n(n - 1)(n - 2)(n - 3)$ and the number of cusps is $n(n - 1)(n - 2)$. Segre then used the machinery of *adjoint curves*: Given a plane curve $C$, a second curve $A$ is said to be *adjoint* to $C$ if it contains each singular point of $C$ of multiplicity $r$ with multiplicity at least $r - 1$. In particular, $A$ is adjoint to a nodal-cuspidal curve $C$ if it passes through all nodes and all cusps of $C$. For example, for the branch curve of a surface of degree 3, Zariski showed that the six cusps lie on a conic; hence the conic is an adjoint curve to this branch curve. One of Segre's results involved proving that,[185] for example, the following adjoint curves to the branch curve exist:

(1) Two adjoint curves of degrees $(n - 1)(n - 2)$ and $(n - 1)(n - 2) + 1$ passing smoothly through the nodes and the cusps of the branch curve.[186]
(2) An adjoint curve of degree $n(n - 1) - 2$, having nodes at the cusps of the branch curve and passing smoothly through the nodes of the branch curve.[187]

Nevertheless, Segre's main result runs in the opposite direction: that is, he proved the following theorem:

A nodal-cuspidal plane curve $B$ of degree $n(n - 1)$ with $\frac{1}{2}n(n - 1)(n - 2)(n - 3)$ nodes and $n(n - 1)(n - 2)$ cusps is the branch curve of a generic projection of a smooth surface of degree $n$ in $\mathbb{CP}^3$ *if and only if* there are two adjoint curves of degrees $(n - 1)(n - 2)$ and $(n - 1)(n - 2) + 1$ passing through the nodes and the cusps of the curve.[188]

Taking into account these necessary and sufficient conditions, it is clear why Segre considered his method as a complete 'characterization'. However, one should ask whether this 'characterization' was indeed complete. The answer is negative, and for two reasons. Firstly, Segre's results concerned surfaces without singularities in $\mathbb{CP}^3$. Other surfaces, which might have a double curve or isolated singularities, were not even mentioned or

---

[184] Ibid., p. 99.

[185] Segre used during the proof algebraic-geometric methods, such as the existence and properties of linear series, equivalence of divisors and Noether's $AF + BG$ theorem.

[186] Ibid., p. 100, 102.

[187] Ibid., p. 101.

[188] Ibid., p. 111.

taken into account.[189] One can claim that a general treatment of these surfaces—especially for surfaces in $\mathbb{CP}^3$ with a generic (i.e. singular) double curve (being a curve with pinch and triple points)—was not even possible during the 1930s, since the mathematical machinery to 'imitate' Segre's proof did not yet exist for generic or singular surfaces in $\mathbb{CP}^3$. Nevertheless, a second reason might be given, why after his 1930 paper Segre did not investigate branch curves any further. As Edoardo Vesentini remarks,

> the "memoir by B. Segre on the characterization of the branch curve of a multiple plane, that appeared in 1930 and was inspired by a paper of Enriques, followed shortly by a paper by O. Zariski on the same topic. But a critical remark made by O. Zariski on an infinitesimal method used by Enriques in earlier papers on the moduli of an algebraic surface, set some doubts on the validity of Enriques' argument and, consequently on the papers that Segre devoted to this topics."[190]

Zariski's "critical remark" concerned the (wrong) assumption Enriques made regarding the completeness of the "characteristic system of a complete continuous system of surfaces" (see Sect. 3.1.3), a proof of which, according to Zariski, "is not likely to be an easy undertaking".[191] I will discuss this assumption later.

In addition, as Edoardo Sernesi notes, other mistakes in some of Segre's related research were discovered. Another paper by Segre[192] dealt with the construction of 'new components' of the variety parameterizing plane curves of degree $n$ with $d$ nodes and $c$ cusps; Segre started from given components, aiming to establish for which values of $n$, $d$ and $c$ the variety of these curves is not empty.[193] But as Sernesi notes, though "Segre's procedure seems to be correct", "his conclusions, as they stand, are incorrect."[194] Again we meet here another occurrence of fallible practices, which were considered as valid and correct by a certain community. Sernesi adds that due to the mathematical tools used, the problem of investigating the variety itself—for example, whether it has singular points— has remained out of reach for Segre and his colleagues: "Only with the help of scheme theory [. . .] [t]he first example of a singularity of [this variety of nodal-cuspidal curves] [. . .] has been given by Wahl [in 1974]: it is a curve of degree 104 with 3,636 nodes and 900 cusps."[195] This is Wahl's example that disproved Enriques's assumption, which was mentioned above. These mistakes as well the incomplete proofs may explain why Segre failed to develop his research on branch curves further and led to the fact that the research

---

[189] Other projections, such as a projection from a point on the surface, were also not taken into account.

[190] Vesentini (2005, p. 188).

[191] Zariski (1935, p. 99).

[192] Segre (1929).

[193] See especially: ibid., p. 37–38. See also: Sernesi (2012, p. 446).

[194] Ibid., p. 447.

[195] Ibid., p. 447. Sernesi refers to (Wahl 1974).

on this moduli variety "suffered the [...] fate [...] [of being] incomplete and inconclusive."[196]

### 3.2.3   1930–1937: Before and After Zariski's *Algebraic Surfaces*

While Segre published only one paper on the subject of branch curves, Zariski had certainly published more on this topic: this is seen not only with his article from the end of the 1920s, but also with his famous book *Algebraic Surfaces*, which was published in 1935, and which contained an entire chapter on branch curves. The book was regarded as presenting a modern *state of the art* treatment on the subject of algebraic surfaces, making rigorous proofs and claims which were done by his colleagues, while also presenting new material, ranging from the beginning of the twentieth century till the 1930s. I will discuss the book later in this section, but for now one should already take into consideration that chapter VIII of the book is called "Branch Curves of Multiple Planes and Continuous Systems of Plane Algebraic Curves". The chapter, although shorter than the other chapters in the book, shows that Zariski considered branch curves unique: not only should they be researched separately, but also one should understand their relations with other nodal-cuspidal curves, which are not branch curves.

As Michael Artin and Barry Mazur already described the mathematical background of Zariski's results during these years,[197] I will concentrate mainly on how Zariski himself considered his own research on branch curves, as well as how he re-conceptualized older research results. As we will see, Zariski attempted not only to bring new methods to the research on branch curves, but also to situate the research configurations in a new context, at the same time taking into account the research done in Italy.

These new directions are already to be seen not only in the 1929 paper discussed above (see Sect. 3.2.1), but also in another paper published in 1929: in this paper Zariski proved that given an irreducible curve $C$ defined by $f(x, y) = 0$, and a multiple plane, defined by $z^n = f(x, y)$ and ramified along $C$, when $n = p^a$ with $p$ a prime number, then the first Betti number and the geometrical genus of the surface are both zero.[198] The proof, using topological tools, has as a key step the investigation of the generators of the fundamental group of the complement of the branch curve $C$.[199] Seeing the fundamental group as an essential component to understand algebraic surfaces appears in another paper two years later. The proofs in this paper underline not only properties of the branch curve, or of the fundamental group, but also how these properties reflect and impose conditions on

---

[196] Brigaglia / Ciliberto (1995, p. 102).

[197] Artin / Mazur (2009, p. 137–148).

[198] The first Betti number of the space $X$ is defined as the rank of the abelian group $H_1(X)$, the first homology group of $X$.

[199] Zariski (1929b, p. 495).

invariants of the ramified surface; Zariski noted that this research was instigated by the theorem

> "that if $f(x, y) = 0$ is a plane algebraic curve the fundamental group of which [. . .] is cyclic, then the algebraic surface $z^n = f(x, y)$ is regular for any value of the positive integer $n$. This suggests the problem of characterizing the plane algebraic curves $f = 0$, which give rise to irregular cyclic multiple planes $z^n = f(x, y)$ and which *eo ipso* can be considered as branch curves of non-cyclic multiple planes, since by the above theorem the fundamental group of such curves is not cyclic."[200]

Here Zariski drew attention to what Segre, Enriques and he himself emphasized also in earlier papers: curves "*eo ipso* can be considered as branch curves" if these curves satisfy special conditions. These conditions can be expressed in terms of the fundamental group of the complement, but also, as was seen with Segre, in terms of the position of the singular points of the curve. In that sense, in this paper Zariski brought together two mathematical configurations and put them into a closer relation: the positional approach (concerning the singular points of branch curves), and the group-theoretic approach. This is to be seen in another theorem proved in the paper. There Zariski proved that given a nodal-cuspidal curve of order $m = 6j$, its fundamental group of the complement was non-cyclic if the set of all curves of order $m - 3 - j$ passing through the cusps (of the curve of order $m$) contained more curves than expected.[201] To give an example, when $j = 1$, then $m = 6$ and $m - 3 - j = 2$. Zariski's claim was that the fundamental group of the complement of a nodal-cuspidal curve of degree 6 was not cyclic, if the set of conics (i.e. curves of degree $m - 3 - j$) passing through the cusps contained 'non-generic' curves. If one takes Zariski's famous example—the branch curve of a cubic surface, being a sextic curve with six cusps, then indeed there is a conic passing through the cusps. This means that the associated fundamental group is not cyclic.

The proof of the theorem relies on properties of multiple planes ramified along this curve. As Zariski notes, the theorem "puts in evidence for the first time [. . .] the connection which exists between the structure of the fundamental group of a plane algebraic curve and the special position of its cusps [. . .]."[202] Zariski did not claim to be the first to note the possibility of such a connection, but he certainly intertwined the group-theoretic research configuration with the positional one regarding plane curves, presenting the branch curve as to be found at an intersection of these two mathematical configurations.

---

[200]Zariski (1931a, p. 485). As Zariski notes: irregularity "is the difference $p_g - p_a$ between the geometric and the arithmetic genus of the surface [. . .]." (ibid.). The name 'irregularity' comes from the fact that for smooth complex surfaces in $\mathbb{P}^3$ the irregularity vanishes. For precise definition, see the appendix in Sect. 3.4.

[201]Zariski's formulation is as follows: "the linear system $|C_{m-3-j}|$ of curves of order $m - 3 - j$ passing through the cusps of $f$ should be superabundant." (ibid., p. 508).

[202]Ibid.

Cyclic multiple planes appear also in two other configurations underlined by Zariski: First, in 1932, he noted the analogy between algebraic surfaces as finite covering of the complex plane (ramified over a singular algebraic curve) and "$k$-sheeted [3-dimensional] manifolds with the knot as branch curve".[203] As was seen in Sect. 3.2.1, Zariski noted already in 1928 that the technique of computing the fundamental group of the complement was employed both for knots and for nodal-cuspidal curves. But mentioning the analogy between ramified 3-dimensional manifold and ramified algebraic surfaces shows that Zariski only considered this analogy in 1932 with 3-dimensional manifolds ramified over knots as "branch curve"—as James Waddell Alexander also termed them (although this theme was already researched at the beginning of the twentieth century, as I noted briefly in Sect. 1.3.1).[204] Second, in 1931, Zariski used his former results to prove the non-existence of curves of order 8 with 16 cusps. Though the paper proving this result did not directly deal with branch curves, it used the results on cyclic multiple planes to prove this non-existence. Zariski noted that this non-existence stands in "contradiction with a statement made by B. Segre [. . .]."[205] This was not the only critique made during these years regarding the methods of the various actors mentioned so far, as B. Segre. Egbert R. Van Kampen, who collaborated with Zariski during the early 1930s, expressed another critique.

Van Kampen (1908–1942), a Dutch mathematician, arrived at Johns Hopkins University in 1931, where Zariski was working.[206] Van Kampen helped Zariski to prove that the relations that he found for the fundamental group of the complement of an algebraic curve were sufficient, and that the methods Enriques used for his proofs were not exact.[207] The formulations of van Kampen in 1933 and of Zariski in 1935 are telling, as they rewrite the

---

[203] Zariski (1932, p. 453, footnote §).

[204] Alexander (1928, p. 303). The work on coverings ramified over knots led Alexander to the discovery of the Alexander polynomial associated to a knot, already during the early 1920s. With respect to this Libgober notes (2011, p. 4): "Though Zariski was under the strong influence of Lefschetz and met Alexander in November 1927 [. . .], he either learned about the Alexander polynomial only in March 1932 from a rather roundabout source or preferred to avoid its use", and adds that this "was four years after the appearance of Alexander's paper [on the Alexander polynomial] and after Zariski most fundamental papers on the topology of singular curves."

[205] Zariski (1931b, p. 309). Zariski refers to (Segre 1929).

[206] See: Parikh (2009, p. 49): "Most valuable to Zariski was the hiring of E. R. van Kampen, a gifted topologist from Holland. Warm and charming, part Indonesian, he shared with Zariski a lively interest in fundamental groups."

[207] For a survey on van Kampen's work, see: Fokkink (2004, p. 12). As Fokkink notes, "Zariski presented the problem [of computing the fundamental group of a complement of an algebraic curve] to Van Kampen who then solved it and wrote three consecutive papers [. . .] on the problem [. . .]. All three articles have become standard references, although they are rarely mentioned together. The algebraic geometers refer to the first article, the topologists to the second ["On the connection between the fundamental groups of some related spaces,"], and the group theorists to the third ["On some lemmas in the theory of groups"]. This is remarkable as in fact only the first article stands alone, and even then only barely." Also remarkable is that none of the papers contain any figures or diagrams.

history of research on branch curves. In his paper "On the Fundamental Group of an Algebraic Curve" Van Kampen notes:

> "The relations [of the fundamental group of the complement of an algebraic curve] have been determined implicitly by Enriques. Zariski pointed out that Enriques' results imply that a set of relations for these generators can be found on determining their transforms when the line containing them is moved round all singularities of the curve and round all the tangents from the origin to the curve. As the resulting proof seemed too algebraic for this simple and nearly purely topological question, Dr. Zariski asked me to publish a topological proof which is contained in this paper".[208]

As we saw, Enriques did not determine these relations implicitly. He did not even suggest that he was working with groups. Moreover, the method of rotating a line originated from Enriques, and not from Zariski. It is however not clear whether van Kampen had access to all of Enriques's papers. Additionally, the fact that Zariski asked van Kampen to publish a "topological proof" shows that Zariski did not think yet of a complete rewriting of algebraic geometry in terms of algebra. Zariski gave in 1935 a more correct account of Enriques's description of the relations of this fundamental group: "from this [the 1923 paper of Enriques] does not follow immediately the completeness of the set of generating relations [...] for the fundamental group $G$, proved by van Kampen."[209] To stress: there is indeed one aspect in van Kampen's method which did not appear in Enriques's papers, and which concerns how to compute the desired fundamental group with the help of an explicit procedure. Van Kampen formulated it as follows:

> "To determine the fundamental group of a projective plane $P$ minus an algebraic curve $C$, take a point $A$ not on $C$ and $\alpha$ line a not containing $A$. Determine in $\alpha$ [being the line] $m$ points $A_i$ by means of the lines through $A$ and tangent to $C$ or through singular points of $C$ [the $A_i$ are hence the branch points of $C$ under a projection from $A$]. Take in a line through $A$ [denoted by $l$], but not through any of the $A_i$, $n$ loops $g_i$ from $A$ round the $n$ points of $C$ in that line, capable of generating the group of $P - C$."

Van Kampen then concluded that the "relations between, those elements are" induced from the branch points as well as the relation $g_1 g_2 \cdot \cdots \cdot g_n = 1$.[210]

In modern language, van Kampen claims that if $l$ is a line, passing from the projection point $A$, cutting the curve $C$, which is of degree $n$, in $n$ points: $c_1, \ldots, c_n$, then the generators of $\pi_1(\mathbb{P}^2 - C)$ are the $n$ loops $g_i$ in $\pi_1(l - \{c_1, \ldots, c_n\})$, each $g_i$ encircling $c_i$ only once.[211] The relations between the $g_i$ are obtained when one inspects the change of the $g_i$ when

---

[208] Van Kampen (1933, p. 255).

[209] Zariski (1935, p. 162).

[210] Van Kampen (1933, p. 260).

[211] Recall that the line $l$ is a complex line and hence homeomorphic to the real plane.

**Fig. 3.4** Depiction of the curve
$C$, the line $l$ cutting $C$ in $c_1$, $c_2$,
$c_3$, . ., the projection from $A$ to
the line $\alpha$, and the branch points
$A_1, \ldots, A_m$ © Graphics: M.F.

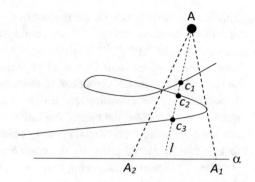

following the $m$ loops on $\alpha$ which surround the $m$ points $A_i$ (see Fig. 3.4)[212]—i.e. the
elements in $\pi_1(\alpha - \{A_1, \ldots, A_m\})$; van Kampen described in his paper a general method
how to obtain these relations.[213] This method, of first setting a basis to $\pi_1(l - \{c_1, \ldots, c_n\})$
and then to see how the loops in $\pi_1(\alpha - \{A_1, \ldots, A_m\})$ act on it (i.e. induce the relations)
will become the standard method of computing the fundamental group of the complement
of an algebraic curve, as we will see in the following chapters.

At this point I would like to turn to Zariski's book *Algebraic surfaces*. Indeed, as Parikh
notes: "From 1932 to 1935 Zariski published nothing. In 1932 he was promoted to
associate professor with tenure, and that winter he began work on *Algebraic Surfaces*, a
definitive account of the classical theory of algebraic surfaces that would convince him of
the need to rewrite the entire foundations of algebraic geometry."[214] Whether this book
contains the seeds which convinced Zariski "to rewrite the entire foundations of algebraic
geometry"—this will be considered later within the framework of his treatment of branch
curves.

### 3.2.3.1  1935: Zariski's *Algebraic Surfaces*

"The aim of the present monograph is to give a *systematic exposition* of the theory of algebraic
surfaces emphasizing the interrelations between the various aspects of the theory: algebro-
geometric, topological and transcendental. [...] In many instances, due to exigencies of
simplicity and *rigor*, the proofs given in the text differ, to a greater or less extent, from the
proofs given in the original papers."[215]

---

[212] As the line $l$, the line $\alpha$ is a complex line and hence homeomorphic to the real plane.

[213] For example, for two lines intersection at a node, the induced relation is $g_1 = g_2 \, g_1 \, g_2^{-1}$. For a cusp
one induces the relation $g_2 \, g_1 \, g_2 = g_1 \, g_2 \, g_1$. Note that these examples do not appear in van Kampen's
paper, but can be induced from Enriques's computations.

[214] Parikh (2009, p. 51).

[215] Zariski (1935, preface) (cursive by M.F.).

This is what Zariski wrote at the preface of his 1935 book *Algebraic Surfaces*. Two aspects, or rather "epistemic values",[216] are emphasized already from the beginning: "systematic exposition" and "rigor".[217] The first aspect is expressed by the fact that Zariski gave the state of the art of the main themes of algebraic geometry, though he did acknowledge that he discussed neither the classification of surfaces, nor the theory of real algebraic surfaces. However, that the classification project was not discussed does not mean that Zariski ignored or minimized the role of the Italian school of algebraic geometry. Silke Slembek analyzed the sources that Zariski used to write *Algebraic Surfaces*, reaching the conclusion that Zariski discussed primarily the contribution of the Italians to the theory of algebraic surfaces.[218] However, analyzing these sources more accurately, she notes that following the years and the languages of the research papers Zariski cited, one may conclude that "between 1890 and 1910, the theory of algebraic surfaces seems to have been mainly written by Italian authors, in Italian and published in Italian journals. [...] Between 1920 and 1935 this has shifted: the publications of non-Italian authors in American or English journals and in English increased to the same extent as the Italian literature considered decreased."[219]

Zariski's systematic exposition of the Italian literature, creating indeed an impression of a unified Italian school of algebraic geometry operating at the turn of the century, was not without critique of that very same school. This critique was expressed using two modi. The first emphasized which theorems were not yet proven, or what the proven theorem implied, in contrast to what the mathematicians who proved these theorems considered. Thus, for example, Zariski noted that one of Enriques's theorems relies on another theorem whose "proof, however, was not available";[220] Zariski underlined again the differences between Enriques's and van Kampen's methods to calculate relations in the fundamental group of a complement of a branch curve,[221] and he noted uncertainties regarding certain theorems on nodal-cuspidal curves.[222] The second modus concerns the epistemic value of rigor.[223]

---

[216] Goldstein (2018, p. 491). As Goldstein notes, "What are called 'epistemic values,' that is the internalised criteria of what constitutes good mathematics at one time, have [...] been studied: rigour is the most obvious perhaps and has a complex history [...]" (ibid., p. 491).

[217] On the history of rigor in development of analysis between the 17th and the 19th century, see: (Schubring 2005). Schubring concludes his study by emphasizing that rigor "is the strongest subjective moment in mathematics" (ibid., p. 617).

[218] Slembek (2002, p. 128). Slembek also notes that this is to be seen also in Zariski's course on algebraic geometry given in 1932 at John Hopkins University (ibid., p. 129).

[219] Ibid., p. 129–130.

[220] Zariski (1935, p. 124).

[221] Ibid., p. 162–163.

[222] Ibid., p. 173.

[223] Another example for the rigorous approach which Zariski insists on in *Algebraic Surfaces* is brought by Brigaglia and Ciliberto, when they discuss the question of reduction of singularities. They note that while "Enriques and Campedelli dispose of the matter in a few lines [...] Zariski's attitude

Already during the first pages of the first chapter, Zariski presented the "rigorous development of the notion of infinitely near multiple points, [which] [...] provide a rigorous foundation for the general theory of linear systems on an algebraic surface [...]."[224] Indeed, for several proofs Zariski explicitly noted that they are not "entirely rigorous",[225] or that "[i]t seems [...] that [a certain] proof [of Enriques] makes use of an *assumption*, which raises an interesting question and which it would be highly important to establish with complete rigor."[226] However, several other proofs were praised for being rigorous: Zariski described a proof of Chisini to be "outstanding for the standard of rigor achieved [...] leaving little, if anything, to intuition."[227] This citation already highlights Zariski's views on intuition, as what should not be overly trusted, or as what may serve as a hindrance to rigor. Discussing another theorem, he noted "[m]ost authors seem to have missed the point in the belief that the compositions of [two types of mappings] [...] are identical, as a matter of intuition."[228] He then listed Enriques and Chisini as two of these authors.

How these views, on clarification of theorems on the one hand, and on rigor on the other hand, come into expression with Zariski's treatment on branch curves in the book? A short discussion on branch curves is to be found already before Chapter VIII, which is the chapter dealing mainly with branch curves. Section 4 of chapter VI is called "The representation of $F$ upon a multiple plane". Taking the "projective models of $F$ a surface $f(x, y, z) = 0$ of order $n$ with ordinary singularities", Zariski considered its projection to the "plane $(x, y)$". He then defined that the "branch curve $D$ [...] is obtained by eliminating $z$ between $[f = 0]$ and $\partial/\partial z = 0$ and by neglecting in the resultant $R(x, y)$ the factor corresponding to the apparent branch curve, i.e. the projection of the double curve of $F$."[229] Zariski explained that nodes and cusps of the branch curve correspond to special tangents to $F$ or to the ramification curve, and also added that the branch curve "possesses only ordinary double points and cusps [...]."[230] No proof was given to this claim, not even a reference. Assuming that this claim is clear—as we saw, it was indeed assumed during the beginning of the twentieth century that branch curves obtained from generic projections of surfaces with only ordinary singularities have only nodes and cusps—Zariski immediately continued to the next subject—the fundamental group: "The number of cusps can be evaluated and shown to be always $> 0$, if $n > 2$. This also follows from a theorem on

---

was quite different [...] dedicat[ing] six whole pages [to this theorem] [...]. Zariski placed exactly those 'details of the proofs' at the center of his attention, those to which [...] Enriques remain supremely indifferent." (Brigaglia / Ciliberto 1995, p. 121)

[224] Zariski (1935, p. 6).

[225] Ibid., p. 83.

[226] Ibid., p. 96.

[227] Ibid., p. 23.

[228] Ibid., p. 12.

[229] Ibid., p. 103–104.

[230] Ibid., p. 104.

the fundamental group $G$ of [the complement of the branch curve] [. . .] to the effect that if $D$ is irreducible and possesses only ordinary double points then $G$ is cyclic."[231] Zariski referred to Chapter VIII in the book, but at this point the question arises: can it not be that a fundamental group of a complement to a plane curve with nodes and cusps would be cyclic? Obviously the reasoning presented at Chapter VI is vague, and not so rigorous itself; I will return to this question below.

Chapter VIII, on the other hand, presents a more systematic investigation. The chapter is mainly an overview of the results obtained so far, mostly by Zariski and by other Italian mathematicians, on the subject of the branch curves. The chapter begins again with the definition of the branch curve of a surface of degree $n$, and then discusses the results obtained by Enriques and van Kampen regarding the structure of the fundamental group $G$, emphasizing the difference between them,[232] and noting the calculation of Enriques raises the question of finding normal subgroups of $G$ with finite index.[233] A discussion on properties of a fundamental group of a complement of a nodal curve ensues: "The theorems [presented above] [. . .] show that in order that the group $G$ be noncyclic it is necessary that $f$ possess singularities other than nodes. The curves with nodes and cusps are of special theoretical interest, since the branch curve of a surface with ordinary singularities is in general a curve of this type."[234] There are two aspects that should be emphasized here; both concern the rigor that Zariski attempted to convey with this statement.

The first aspect is the logical connection between branch curves and the non-cyclicity of the associated fundamental group, already mentioned above. Indeed, branch curves have nodes and cusps, but this does not directly imply that $G$ would be non-cyclic or non-abelian. Zariski's own example, of a sextic with six cusps not a conic, shows exactly that this logical connection should be more explicit (as $G$ in this case is cyclic; the curve in this case is not a branch curve).[235]

The second aspect concerns a footnote Zariski added after the statement cited above: "The branch curves of general surfaces of a given order have been characterized by

---

[231] Ibid.

[232] Ibid., p. 162: "from this [Enriques's theorem] does not follow immediately the completeness of the set of generating relations [he found] for the fundamental group $G$ [. . .] proved by van Kampen." See also Libgober (2014, p. 480).

[233] Zariski (1935, p. 163).

[234] Ibid., p. 164.

[235] The argument is (or should be) finer; for the sake of completeness, I present it here (in a more modern formulation), though it does not appear in Zariski's book. If $D$ is a branch curve, then there exists an epimorphism from $G = \pi_1(\mathbb{P}^2 - D)$ to the symmetric group $Sym_n$. This epimorphism was already described by Enriques, as the map sending every loop to its corresponding transposition. As the symmetric group is non-abelian for $n > 2$, and since a quotient of $G$ is isomorphic to $Sym_n$ (as a direct result from the first isomorphism theorem in group theory), one obtains that $G$ itself is non-abelian (and hence non-cyclic). Although the epimorphism is described at the beginning of chapter VIII of *Algebraic Surfaces*, it is not clear whether Zariski knew group theory well enough to explicitly note this, notwithstanding his usage of group theory in earlier papers.

B. Segre in terms of the position of their nodes and cusps on algebraic curves of lower orders." Zariski referred to Segre's 1930 paper; this not only shows the "special theoretical interest"[236] which branch curves prompted but also other ways (and other epistemic configurations) to characterize them; however, the footnote may create the impression that Segre's paper dealt (also) with fundamental groups, which is definitely not correct. Moreover, taking into account that Zariski mentioned Segre's paper on branch curves this one time, we may certainly assume that it was one of the reasons for Segre's result being forgotten. Indeed, the chapter ends with mentioning the famous example of Zariski, of two disjoint systems (i.e. families) of curves of the same degree and having the same number of singularities (i.e. sextics with 6 cusps in a conic, and sextics with 6 cusps not on a conic). Mentioning Segre's result again might have prompted research on other pairs of such non-isotopic curves, having the same degree and the number of singularities.[237]

The section that follows in Chapter VIII deals with properties of a cyclic multiple plane $F$ of the form $z^k = f(x, y)$ and summarizes earlier results. It is necessary, according to Zariski "to bear in mind that we use the surface $F$ [ramified above $f$] only as a *tool* for deriving properties of the curve $f$. Thus the *true significance* of [the obtained] theorem lies in [deriving a] [. . .] non-trivial invariant [. . .]."[238] This remark is telling. Branch curves and the associated ramified covers are not only, as one might have gotten the impression, an object for research. They can be also considered as a tool to investigate other objects, and to derive other invariants which are the "true significance" of the theorem in question. Zariski shifted here the center of research configuration: branch curves not only can be considered with various techniques (of topology, group theory or algebraic geometry), but can have different roles in the mathematical investigation.

The rest of the chapter deals with nodal curves and with nodal-cuspidal curves. Regarding the latter, Zariski noted "our information on curves with cusps and nodes consists only of a few partial results, which [. . .] give only an indication of the difficulties involved."[239] While not dealing directly with branch curves, Zariski's following statement is worth taking into account, referring to Enriques's assumption: "It is not known if there exist complete continuous systems of curves with nodes and cusps whose characteristic series is incomplete",[240] which means that the dimension of the system is larger than the expected one. Though not pointing explicitly that such an example would come from a family of branch curves, it might be possible that Zariski was hinting in this direction; and indeed, as was noted above, it was proven in 1974 by Wahl that for a certain family of branch curves, the characteristic series is incomplete.[241]

---

[236] Ibid.

[237] Shustin found in 2011 other pairs of curves having this property, one of the curves in this pair being a branch curve (Friedman/Leyenson/Shustin 2011).

[238] Zariski (1935, p. 166–167) (cursive by M.F.).

[239] Ibid., p. 172.

[240] Ibid., p. 173.

[241] Wahl (1974).

### 3.2.3.2 After *Algebraic Surfaces*

Contrary to what Carol Parikh assumes, I claim that the book *Algebraic surfaces* was *not* a representative or even a precursor of Zariski's algebraic re-writing project of algebraic geometry. Aldo Brigaglia and Ciro Ciliberto emphasize that "despite the fact that Zariski makes reference to [...] the new algebraic methods [...], all these references remain somewhat superficial and episodic. [...] [More essential is that] Zariski makes use as much as possible of the transcendental methods filtered through the topological approach of Lefschetz."[242] However, the transcendental methods did not ever come into expression in Chapter VIII, whereas the algebraic methods were also serving topologically,[243] and this in a very limited way, when researching fundamental groups of complements of branch curves.

After completing *Algebraic Surfaces*, Zariski received an invitation from the newly formed Institute for Advanced Study at Princeton for the 1934–35 academic year, where he worked with Lefschetz. Slembek notes that in the yearly reports of John Hopkins University it is mentioned that Zariski took part in the courses given by Emmy Noether (on class fields), Alexander and Lefschetz (on topology), as well as by Carl Ludwig Siegel (on number theory).[244] It is also at the same time that he began teaching himself modern algebra and ideal theory from the books of B. L. van der Waerden *Modern Algebra* and Wolfgang Krull *The Theory of Ideals*. As Zariski noted, while he was writing *Algebraic Surfaces,* it became clear to him that "I found I didn't even know the elementary facts of modern algebra. If somebody had asked me, 'What's a ring?' I couldn't have defined it. I spent the next couple of years just studying modern algebra."[245] The process of studying modern algebra was followed by joining the lectures at Princeton, as was noted above. Emmy Noether writes to Helmut Hasse on 28.11.1934 the following: "He [Henry Morgan Ward][246] belongs with Zariski to the professors among the listeners; and the latter begins to plunge into the arithmetic theory of algebraic functions!"[247]

---

[242] Brigaglia / Ciliberto (1995, p. 120–121).

[243] This is also Zariski's formulation during the late 1980s: "I wouldn't underestimate the influence of algebra [...] but I wouldn't exaggerate the influence of Emmy Noether.[...] I was always interested in the algebra which throws light on geometry, and I never did develop the sense for pure algebra. Never." (Parikh 2009, p. 57).

[244] Slembek (2002, p. 138).

[245] Parikh (2009, p. 51–52).

[246] Henry Morgan Ward (1901–1963) was an American mathematician; among his research interest were Diophantine equations, abstract algebra and lattice theory.

[247] "Er [Henry Morgan Ward] gehört mit Zariski zu den Professoren unter den Zuhörern; und letzterer fängt an sich in die arithmetische Theorie der algebraischen Funktionen zu stürzen!" In: (Lemmermeyer / Roquette 2006, p. 218–219). Zariski recalls his own encounter with Noether as having a motherly character, noting that "[s]he spoke about ideal theory in algebraic number theory. [...] I was trying to learn ideal theory, so I went faithfully [to her lectures] even if I didn't understand everything [...]. She was very motherly to me, although I didn't learn ideal theory from her, but from her papers." (Parikh 2009, p. 57)

What these accounts underlined is that Zariski's rewriting of algebraic geometry with the language of algebra only happened from 1935 onwards. Artin and Mazur note that around 1937, "Zariski was struck by the realization that the classical language was inadequate and that a thorough reconsideration of the field in terms of the new commutative algebra was needed."[248] However, in the years 1936 and 1937, one still finds Zariski publishing topological papers dealing with nodal-cuspidal curves, as an echo of his interest on these questions.[249] The paper from 1937 is of fundamental importance to the research of branch curves. It discusses the example that Zariski gave in 1929 of two curves having the same degree and number of cusps, yet not being isotopic to one another—the example of the two curves of degree 6 with 6 cusps, one being a branch curve, the other not.

One of the fundamental results in this paper is the computation of the fundamental group of the complement of the curve "of order $2n$, [which] possesses $k = 3n$ cusps and $d = 2n(n - 3)$ nodes"; this curve is "the dual of a general elliptic plane curve [. . .] of order $n$".[250] Inspecting the case of $n = 3$, one obtains a sextic with nine cusps. The arguments here lead Zariski to the computation of the fundamental groups of the complement of (1) a sextic curve with 6 cusps on a conic; and (2) a sextic with 6 cusps not on a conic. While Zariski already computed the fundamental group of the complement of the first curve in 1929, the second group was not yet computed. The computation is done as follows: considering the fundamental group of the complement of the sextic curve with 9 cusps, there are, according to Zariski, "9 relations [which] are typical cuspidal relations, and one may conjecture that they correspond to the 9 cusps of the curve. [. . .] [by previous calculations] only 7 of these relations are group-theoretically independent [. . .] [h]ence if we remove a certain number of cusps of $C$, it is immaterial which cusps are removed, as long as the number of removed cusps does not exceed 2."[251] Removing two cusps, one obtains that the resulting fundamental group is "a group generated by the two elements $[u, v]$ [. . .] satisfying the relations $u^2 = v^3 = 1$."[252] At this point comes the crucial part of the argument: "[T]his group is also the fundamental group of a sextic with six cusps on a conic [. . .]. Hence there must be among the seven cusps left, a third cusp whose removal has no effect on the fundamental group. [However,] [. . .] if the removal of an additional cusp yields [a certain relation] [. . .], the group becomes cyclic."[253]

---

[248] Artin / Mazur (2009, p. 137).

[249] The paper (Zariski 1936) deals with curves of degree $2(n - 1)$ with $3(n - 2)$ cusps and $2(n - 2)(n - 3)$ nodes, being dual to rational nodal curves. Zariski proves that the fundamental group of the complement of these nodal-cuspidal curves is a quotient of the braid group ("'Zopfgruppe' of Artin" as Zariski called it (ibid., p. 607)) of order $n$ with the additional relation $g_1 g_2 \cdot \cdots \cdot g_{n-2} g_{n-1}^2 g_{n-2} \cdot \cdots \cdot g_2 g_1 = 1$. Note that the fundamental group of the complement of this curve is the $d$−th braid group of the Riemann sphere.

[250] Zariski (1937, p. 355).

[251] Ibid., p. 356.

[252] Ibid., p. 357.

[253] Ibid.

This shows that there are two options for the fundamental group—but the calculation is done without finding explicitly (the equation of) the sextic with 6 cusps not on a conic. The computation in fact bypasses any need to find the explicit object, for which one finds the fundamental group. Moreover, there are no more options, as Zariski summarized: "It is known that the sextics with six cusps distribute themselves into two distinct continuous systems, according as the six cusps lie or do not lie on a conic. [Hence] [...] the fundamental group of a sextic with six cusps not on a conic is cyclic (of period 6)."[254]

This paper would be the last paper Zariski publishes for a long time—till 1958—on the subject of branch curves. It presents the subject of branch curves as a side-result, as what stems from the main results of the paper and not as the center of it. In the 1937 paper neither Segre nor Enriques are mentioned regarding the research of the moduli space of nodal-cuspidal curves, and it seems that at least for Zariski, the subject did not offer any more insights worthy of investing in. Why this occurred is described by Mumford as follows:

"In the period 1937–1947, [...] Zariski completely reoriented his research and began to introduce ideas from abstract algebra into algebraic geometry. Along with B. L. van der Waerden and André Weil, he undertook to completely rewrite the foundations of algebraic geometry without making any use of topological or analytical methods. There were two motivations for this: first, it became clear to Zariski, particularly after writing his Ergebnissebericht *Algebraic Surfaces* that many of the classical Italian 'proofs' were not merely controversial but were really incomplete and imprecise at certain points. Second, it had become clear that it was both logical and useful to develop an 'abstract' theory of algebraic geometry valid over an arbitrary ground field."[255]

Above I discussed whether the first "motivation" which Mumford points is indeed histori-cally correct; as we saw, when inspecting the theme of branch curves, Zariski himself was sometimes not rigorous enough, at least in *Algebraic Surfaces*. What is true, however, is that the arguments used by Zariski in the research of branch curves were highly topological, though they were pointing towards also a much more group-theoretic approach, as one could have guessed from the above-mentioned account.

## 3.3   Reflections on Rigor: Reassessment and New Definitions in the 1950s

The period from the 1930s till the 1950s marks the rise of a secluded self-conception within the Italian mathematical community. One of the results of the seclusion of this community is the relative blossom of another Italian group of researchers dealing with branch curves— the group around Oscar Chisini—on which I will elaborate in the next chapter. This group, however, remained 'outside' the algebraic rewriting project, which was led among others

---

[254] Ibid.

[255] Mumford (2009, p. 151).

by Zariski, and was not taken into consideration within this project. However, at the end of the 1940s a reassessment of the research done during the first decades of the twentieth century was taking place from Castelnuovo's and Enriques's side as well as from Zariski's. It is with this reassessment that I will also conclude this chapter.

In 1949 Enriques's book *Le superficie algebriche* was published posthumously by Castelnuovo. Castelnuovo added a preface to the book, praising Enriques's "genius intuition."[256] While surveying the content of the book and Enriques's discoveries, Castelnuovo expressed in this preface his critique on the new methods of mathematics, employed during the "present century". In one of the last paragraphs of the preface, he doubted that these methods would lead to a perfection of the theory of algebraic surfaces, as was done for algebraic curves:

> "My doubts are fueled by the observation that mathematics has taken a very different direction in the present century from that which dominated in the last century. The imagination, the intuition that guided the research of that time are today looked upon with suspicion, for the terror of the errors to which they can lead. [...] It was [during the former century] the exploration of a vast territory glimpsed from a distant peak. [...] Today, more than the terrain to be explored, it is the path that leads to it that interests us, and this path is now sown with artificial obstacles, now it hovers in the clouds."[257]

Castelnuovo's lamentation on the state of algebraic geometry, formulated with geographical metaphors: the "vast territory", "the terrain to be explored" or the "artificial obstacles"—is also a critique, and can be read on the background of the rise of the new algebraic methods of algebraic geometry, which are neither discussed nor used in Enriques's 1949 book. Another way to react to these new methods was expressed by Enriques himself. Throughout the book, Enriques insisted that his methods are rigorous, as indicated already at the first paragraph of Enriques's preface.[258] The adjective "rigorous" or the adverb "rigorously" are used often, to describe, for example, the works by Enriques or by Enriques and Castelnuovo.[259] However, when it comes, for example, to his assumption, whether the continuous systems of irreducible surfaces in $\mathbb{CP}^3$ with ordinary singularities

---

[256] Enriques (1949, p. vi).

[257] Ibid., p. vii: "A nutrire I miei dubbi m'induce l'osservazione che la matematica ha preso nel secolo attuale un indirizzo ben diverso da quello che dominava nel secolo scorso. La fantasia, la intuizione che guidavano la ricerca di allora sono oggi guardate con sospetto per il terrore degli errori a cui possono condurre. [...] Era l'esplorazione di un ampio territorio intravisto da una cima lontana. [...] Oggi più che il terreno da esplorare interessa la via che vi conduce, e questa via ora vien seminata di ostacoli artificiali, ora si libra tra le nuvole."

[258] Ibid., p. ix: "For many years we have elaborated such an exposition [...] in such a way as to give the theory itself the most rigorous and simple framework." ["Per lunghi anni abbiamo elaborato tale esposizione [...] in guisa da conferire alla teoria stessa l'assetto più rigoroso e più semplice."]

[259] For example, see: ibid., p. 58, 124, 251, 269, 347.

are complete, Enriques insisted on calling it a "working hypothesis".[260] This might show that he acknowledged the hypothetical status of this assumption. Nevertheless, on few accounts Enriques did justify a usage of "bold intuition",[261] regarding which Castelnuovo comments that the "arguments presented [by Enriques] contain several gaps."[262]

While the results presented on branch and ramification curves are mainly a survey of the research done till the publication of the book, it is clear that the highlight of the book presented the results of the classification project; the classifying table (see Fig. 3.5) is presented on the last two pages. The book's ending paragraph couples this table with the call to reveal the hidden, divine harmonies of the theory of surfaces: "[. . .] it used to be said that while algebraic curves (already composed in a harmonic theory) are created by God, surfaces, instead, are the work of the devil. Now, on the contrary, it is clear that God chose to create for surfaces an order of more hidden harmonies, where a wonderful beauty shines forth [. . .]".[263]

* * *

While the book of Enriques represents the classical algebraic geometry before its rewriting with commutative algebra, Zariski and André Weil represent in the 1950s the modern algebraic geometry. The book of Enriques was published three years after the publication of André Weil's book *Foundations of Algebraic Geometry*. Working over fields of any characteristic, Weil concentrates on intersection theory. While the book does not mention branch or ramification curves (or branch varieties), it certainly shaped retroactively a certain image of the Italian school of classical algebraic geometry. Weil wrote:

"Algebraic geometry, in spite of its beauty and importance, has long been held in disrepute by many mathematicians as lacking proper foundations. [. . .] at times when vast territories are being opened up, nothing could be more harmful to the progress of mathematics than a literal observance of strict standards of rigor. Nor should one forget, when discussing such subjects as algebraic geometry, and in particular the work of the Italian school, that the so-called 'intuition' of earlier mathematicians, reckless as their use of it may sometimes appear to us, often rested on a most painstaking study of numerous special examples, from which they gained an insight not always found among modern exponents of the axiomatic creed. At the same time, it should always be remembered that it is the duty [. . .] of the mathematician to prove theorems, and that this duty can never be disregarded for long without fatal effects. [. . .] [Moreover] in this field [of algebraic geometry], the work of consolidation has so long been overdue that the delay is now seriously hampering progress in this and other branches of mathematics."[264]

---

[260] Ibid., p. 209, 210.

[261] Ibid., p. 333.

[262] Ibid., p. 333, footnote 1.

[263] Ibid., p. 464: "[. . .] si soleva dire che, mentre le curve algebriche (già composte in una teoria armonica) sono create da Dio, le superfici sono opera del Demonio. Ora si palesa invece che piacque a Dio di creare per le superficie un ordine di armonie più riposte ove rifulge una meravigliosa bellezza [. . .]"

[264] Weil (1946, p. vii).

$$P_{12} > 1 \atop (n < 2\pi - 2)$$

$p^{(1)} = 1$

$p_a = -1$, superficie ellittiche con un fascio di genere $p_\sigma$ di curve ellittiche $(p_a = -1, \ P_4 \neq 1)$.

$p_a \geq 0$, superficie le cui curve canoniche e pluricanoniche sono composte colle curve ellittiche d'un fascio di genere $p_\sigma - p_a$: dipendono da più caratteri interi arbitrarii.

$p^{(1)} > 1$

$p_a \geq 0$, superficie che ammettono un modello proiettivo canonico $(p_\sigma > 3)$ o bicanonico $(p^{(1)} > 3)$ ecc.: un numero finito di tipi per un dato valore del $p^{(1)}$.

$P_{12} = 0$: rigate $n > 2\pi - 2$, curve eccezionali non eliminabili, schiera continua di trasformazioni non formanti gruppo.

$p_a = 0$ superficie razionali $(p_a = P_2 = 0)$
$p_a = -1$ rigate ellittiche $(p_a = -1, \ P_4 = P_6 = 0)$
$p_a < -1$ rigate di genere $p = -p_a > 1$.

$P_{12} = 1$: curva canonica virtuale d'ord. 0, $n = 2\pi - 2$ $(p^{(1)} = 1)$

$p_\sigma = 1$

$p_a = 1$, infinite famiglie di superficie $(p_a = P_2 = 1)$
$p_a = -1$, infinite famiglie di superficie iperellittiche di divisore $\delta = 1, 2, \ldots$ $(p_a = -1, \ p_\sigma = P_4 = 1)$.

$p_\sigma = 0$

$p_a = 0$, superficie del $6^\circ$ ordine passanti doppiamente per gli spigoli d'un tetraedro $(p_a = P_3 = 0, \ P_2 = 1)$.
$p_a = -1$, superficie ellittiche $I_a I_b \ II_a II_b \ III_a III_b III_c$ $(P_2 = 0,1, \ P_4 = 0,1, \ P_3 = 0,1, \ P_4 = 0,1)$.

**Fig. 3.5** The classifying table of algebraic surface according to their genera, in: Enriques (1949, p. 463–464)

Just as Castelnuovo, Weil employed here the geographical metaphor regarding the "vast territories" which are opened up, but he immediately emphasized that what lacks is not a further exploration of this "territory" but rather "proper foundations", employing an architectural metaphor. This critique of the "Italian school" is obvious, though Weil did attempt to relativize it. However, the point is clear: one should leave aside the "intuition" coming from special examples, and fulfill the "duty" of the mathematician: proving theorems, which, as implicitly hinted, was only partially fulfilled by the Italian school (or by several of its members). Weil compared his own efforts to reconstruct an edifice, where all mathematicians (or at least those who deal with algebraic geometry) would speak one language: "[...] attention must be and has been given to the language and the definitions. Of course every mathematician has a right to his own language — at the risk of not being understood [...] [leading to the] same fate [...] for mathematics as once befell, at Babel [...] in such a subject as algebraic geometry, where earlier authors left many terms incompletely defined [...], all terms have to be defined anew, and to attach precise meanings to them is a task not unworthy of our most solicitous attention."[265] The analogy with the 'Tower of Babel' is clear, and it refers implicitly to what Klein already noted in 1926 about the different languages used in algebraic geometry (see Sect. 3.1.4).[266] What Weil proposed, restricting the algebraic language to the "simplest facts about abstract fields [...] [and] the theory of ideals", is "to return at the earliest possible moment to the palaces which are ours by birthright, to consolidate shaky foundations, to provide roofs where they are missing, to finish, in harmony with the portions already existing, what has been left undone."[267] This argument of a "return" to more ancient "palaces" is highly unlikely, as there was no historical continuity to be found in Weil's book. This is underlined also by Zariski's review of the book, published in 1948. Zariski sarcastically noted that if an algebraic geometer would have to choose between Weil's language, consisting of "constructions full of fields, linearly disjoint fields, regular extensions, independent extensions, generic specializations, finite specializations and specializations of specializations"[268] and between "constructions full of rings, ideals and valuations",[269] then this geometer would probably "decline the choice and say: 'A plague on both your houses!'"[270] Nevertheless, Zariski emphasized that Weil's book "is presented with that absolute rigor to which we are becoming accustomed in algebraic geometry."[271]

Similar images of classical and modern approaches, concerning how one should research algebraic geometry, were presented two years later, again by Zariski. In 1947 Zariski moved to Harvard University, and in the same year he delivered there a general lecture on algebraic geometry at the international congress for mathematics. The image Zariski presented concerns the new methods in algebraic geometry, having being

---

[265] Ibid., p. viii.

[266] Interesting to note is that this metaphor appears again in 1948 in the manifesto published by the Bourbaki group 'The Architecture of Mathematics': Bourbaki warns that the "exuberant proliferation" of mathematics might lead to a situation that mathematics might become "a tower of Babel, in which autonomous disciplines are being more and more widely separated from one another, not only in their aims, but also in their methods and even in their language." (Bourbaki 1950, p. 221). Recalling that Weil was an early member of the Bourbaki group, one may assume that the members of the group knew his book.

[267] Weil (1946, p. viii).

[268] Zariski (1948, p. 671).

[269] Weil (1946, p. viii)

[270] Zariski (1948, p. 672).

[271] Ibid., p. 671.

'arithmetized' with the tools of modern algebra. Zariski did not only present the developments in this new domain, he also compared it with the classical Italian algebraic geometry; this image which Zariski presented became well accepted. Zariski opens his lecture as follows:

> "The past 25 years have witnessed a remarkable change in the field of algebraic geometry, a change due to the impact of the ideas and methods of modern algebra. What has happened is that this old and venerable sector of pure geometry underwent (and is still undergoing) a process of arithmetization [which] [. . .] was criticized either as a desertion of geometry or as a subordination of discovery to rigor. I submit that this criticism is unjustified and arises from some misunderstanding of the object of modern algebraic geometry. This object is not to banish geometry or geometric intuition, but to equip the geometer with the sharpest possible tools and effective controls. It is true that the lack of rigor in algebraic geometry has created a state of affairs that could not be tolerated indefinitely. Effective controls over the free flight of geometric imagination were badly needed, and a complete overhauling and arithmetization of the foundations of algebraic geometry was the only possible solution. This preliminary foundational task of modern algebraic geometry can now be regarded as accomplished in all its essentials. [. . .]
>
> Modern algebra, with its precise formalism and abstract concepts, provided [more precise] tools. [. . .] the modern developments in algebraic geometry are characterized by great generality. They mark the transition from classical algebraic geometry, rooted in the complex domain, to what we may now properly designate as *abstract algebraic geometry*, where the emphasis is on abstract ground fields."[272]

I bring this long quotation in order to show the almost too well accepted narrative on the Italian school of algebraic geometry that Zariski definitely helped to shape.[273] The "free flight of geometric imagination" must be tamed, with the "sharpest"—rigorous—"possible tools". Obviously this call to rigor was already present in *Algebraic Surfaces*, but nowhere Zariski wrote in 1935 what he expressed in 1950, that "[t]he Italian geometers have erected, on somewhat shaky foundations, a stupendous edifice."[274] Obviously what could have been read as a complement regarding the "stupendous edifice", is undermined by the

---

[272]Zariski (1950, p. 77)

[273]I bring here two later accounts which echo Zariski's narrative from 1950. First, a similar image regarding the role of Weil and Zariski is depicted by Mumford (2009, p. xxii), noting that "in the twenties and thirties [of the 20th century], they [the Italian geometers] began to go astray. It was Zariski and, at about the same time, Weil who set about to tame their intuition, to find the principles and techniques that could truly express the geometry while embodying the rigor without which mathematics eventually must degenerate to fantasy." Second, note also Dieudonné's account regarding the image of knowledge of algebraic geometry since the 1950s: "[. . .] many of the Italian proofs [of theorems concerned with projective curves and surfaces] were inconclusive, or based on 'intuitive' arguments, and it was necessary to put them on more secure foundations, chiefly based on sheaf cohomology and other modern techniques [. . .]". (Dieudonné 1985, p. 114).

[274]Zariski (1950, p. 88)

reference to "shaky foundations", "lack of rigor" and free "imagination". As we saw above, Castelnuovo reacted to this kind of allegation with a certain annoyance, naming irritated the new level of rigor and the new language as "obstacles". Zariski's reference to the "free flight of imagination" might have also misled the reader (resp. the listener) to think that "geometric intuition" was associated with erroneous visual thinking, which was definitely not the case, as Schappacher showed.[275] To stick to a more faithful description of the events, one may follow Slembek here, who analyzes Zariski's project of the reduction of singularities of algebraic surfaces at the end of the 1930s. According to Slembek, although Zariski generalized old concepts and invents new ones, which had no parallel in the classical algebraic geometry, Zariski's "motivation was not exclusively to increase rigor" at that time.[276] This is also to be seen in his treatment of branch curves, also after the publication of *Algebraic Surfaces* and until 1937. There was, one may claim, no increase in rigor, also when considering the treatment of Zariski in his last paper from the 1930s on the subject. The account of Slembek highlights how the 'rigorous' proofs of the main actors presented here: Enriques, Zariski, B. Segre—might have also been not as rigorous as they seemed, employing fallible practices sometimes. Following Gert Schubring,[277] "rigorous" is a very much subjective value and is dependents on both techniques and tools employed as well as on the image of the configuration which the mathematical community has.

Compared to the 1930s, the research on branch curves had a different character (with respect to the methods employed) and status during the end of the 1950s. Zariski, in his lecture series "An Introduction to the Theory of Algebraic Surfaces" given in 1957–1958 at Harvard University, did not mention branch curves in a word. Indeed, he mentioned in the preface that "[t]he purpose of these notes is to acquaint the reader with some basic facts of the theory of algebraic varieties, and to do that by self-contained, direct and I would almost say—ad hoc methods of Commutative Algebra [. . .] [however], the title of these lecture notes is somewhat misleading, for only three of the sixteen sections [. . .] deal specifically with algebraic surfaces [. . .]."[278] Taking this into account, this implies that branch curves were too specific to be dealt with within the framework of what is called "the basic facts"

---

[275] Schappacher 2015.

[276] Slembek (2002, p. 194).

[277] Schubring (2005, esp. p. 616–617).

[278] Zariski (1969, preface) (underlined in the original). How the ramification of varieties as covers became somehow marginal in the work of Zariski during the 1950s is also to be seen the book *Commutative Algebra*, vol. 1, published in 1958 (by Zariski and P. Samuel). Though the book introduces thoroughly the subjects of ramification in Dedekind domains (Zariski/Samuel 1958, p. 284ff.), there is no mention of algebraic surfaces; indeed, Zariski notes in the preface that while the original plan was to write a manuscript on algebraic geometry and "from time to time [to insert] algebraic digressions [. . .] it soon became apparent that such a parenthetical treatment of the purely algebraic topics, covering a wide range of commutative algebra, would impose artificial bounds [. . .]." (ibid., p. v) This also underlines the shift which algebraic geometry went through—heavily based now on commutative algebra.

concerning algebraic varieties, which obviously stands in contrast to Chapter VIII in Zariski's book from 1935 *Algebraic Surfaces*. This point of view changed though in a paper published in 1958, called "On the Purity of the Branch Locus of Algebraic Functions". In this short paper Zariski proved that given a variety $V$ of dimension $r$, and its normalization $V^*$, then the "the set of points of $V^*$ which are ramified with respect to $V$ is an algebraic variety [...]. This variety is called the *branch locus* [...] denoted by $\Delta$ [in modern language, the ramification locus]. The main object of this note is to prove the following [...] [that] if $V$ is a non-singular variety, then $\Delta$ is a pure $(r - 1)$-dimensional subvariety of $V^*$".[279]

One could have expected that along with the proof of the above theorem, the case of complex algebraic surfaces embedded in $\mathbb{CP}^3$ would be presented or at least given as an example, since the ramification locus is calculated as the intersection of the surfaces $f(x, y, z) = 0$ and $df/dz = 0$, a computation Zariski presented in *Algebraic Surfaces*. However, Zariski's proof is completely different: Zariski situated the "branch locus" in the research setting of commutative algebra. He defined first in a completely algebraic way what is an unramified point over a field of any characteristics, without considering whether the variety $V$ is affine or embedded in a projective space. The criterion for a point being unramified is given in the language of commutative algebra: via primary and prime ideals and separable extensions of fields,[280] and when taking an affine part of $V$, in the language of $k$-derivations, when $k$ is an "arbitrary ground field". None of these terms which appear in the paper: primary ideal, normal varieties, fields of characteristics different than zero, $k$-derivations and more—appear in the classical theory of algebraic geometry. Indeed, this definition was introduced in Sect. 1.2, and at first glance it can be considered as a sharp break with the classical definition of the ramification curve. However, Zariski then proved that the algebraic definition of a non-ramified point is equivalent to proving that a certain matrix of derivatives at the point has a certain rank; one can claim, though Zariski did not state it, that for the case of an (affine) algebraic surface in $\mathbb{C}^3$ one obtained the classical definition of a non-ramified point. Hence, this new definition points towards an algebraic shift, but not a complete rupture between the two mathematical configurations, as the case of algebraic surfaces in $\mathbb{C}^3$ or in $\mathbb{CP}^3$ is deduced from the general theorem Zariski presented. Moreover, Zariski reduced the proof to "the case in which the ground field is algebraically closed",[281] and proved the claim for the cases for which "$k$ is either of characteristic zero or is a perfect field of characteristic $p \neq 0$."[282] However, notwithstanding the link to the older definition of ramification curve, there was not a single hint in Zariski's paper from 1958 that during

---

[279] Zariski (1958, p. 793).

[280] Ibid., p. 791.

[281] Ibid., p. 793.

[282] Ibid. Zariski also notes that "The generalization to nonperfect ground fields will be found in the note of N. Nagata which immediately follows the present note." (ibid.)

the 1930s he worked extensively on a similar subject within the 'classical' configuration of the research on algebraic surfaces. That is: neither the methods used by Zariski himself during the late 1920s and the 1930s (topological or group-theoretical) nor the various mathematicians who worked on the subject or the approaches used to characterize the branch curve (group-wise of positional-wise) are even mentioned.

* * *

To conclude, concentrating on the period between the end of the nineteenth century and the end of the 1930s, it is clear that the narrative that was shaped during the 1950s about the Italian school of algebraic geometry was somewhat misleading. While some of Enriques's assumptions were proven to be incorrect, other research directions which he delineated did eventually enable the rise of various epistemic configurations. These directions were not necessarily lacking rigor, but rather were not written in an algebraic language (which might have led to discovering other results, such as in the case of the fundamental group of the complement of the curve). Rather than dismissing the results of this school as merely 'intuitive', which would be sometimes a projection of images of later configurations on the methods of former ones, a more proper way to view how the branch curve functioned during this period is to consider it as having changing roles and functions, dependent on the mathematical configuration: It was situated at an intersection of various approaches (topological, group-theoretical, researched via adjoint curves or defined with algebraic methods); but it also functioned as a tool to construct other objects (as Enriques's double covers or Zariski's cyclic covers $z^n = f(x, y)$). Indeed, it is essential to note that the branch curve was researched within the project of classification of surfaces, but afterwards in the context of classifying nodal-cuspidal curves. This ongoing research shows how several configurations in algebraic geometry were integrated; the various research configurations on branch curves indicate how the Italian school shaped algebraic geometry, employing a variety of disciplines and methods. Moreover, those research configurations and their transformations point out how the research interests of algebraic geometers shifted: While Zariski's 1935 book dedicated an entire chapter to coverings of the plane, such a treatment is missing in his lecture series in 1957–1958 on algebraic surfaces, which indicates also how the Italian school of algebraic geometry and the importance of its results were later conceived.

## 3.4    Appendix to Chap. 3: Birational Maps and Genera of Curves and Surfaces

The following Appendix aims to give a short summery of the main concepts used in Chap. 3: rational maps and the various genera of curves and surfaces. Its function is neither to give fully developed definitions of the concepts nor to present here a historical analysis, how these concepts were developed,[283] but rather to give the reader a review of the variety of the tools, techniques and terms with which the various algebraic geometers at the beginning of the twentieth century worked.

Let $X$ and $Y$ be varieties;[284] a *rational map* $f$ from $X$ to $Y$ is defined as a morphism from a nonempty open subset $U$ of $X$ to $Y$. A *birational* map from $X$ to $Y$ is a rational map with an inverse, i.e. it is a rational map $f : X \rightarrow Y$ such that there is a rational map $g : Y \rightarrow X$ with $g \circ f = id_X$ and $f \circ g = id_Y$, as rational maps. Two varieties $X$ and $Y$ are called *birational* if there is a birational map between them. For example, generic projections of projective varieties in $\mathbb{P}^n$ onto $\mathbb{P}^{n-1}$ (either from a point on or not on the variety) are an example of a rational mapping. To give a specific example, the circle is birationally equivalent to an affine line, when the map $f$ is the projection from a point on the circle to a line. Another example of a birational morphism is a blow-up of a variety. A *rational variety* is birationally equivalent to a projective space of some dimension.

Before discussing the notion of genera of surfaces, let us recall quickly the definition of a *genus of a curve*: taking into consideration curves up to birational equivalence, one can define the genus of a smooth algebraic curve to be to the dimension of the space of regular differential 1-forms on this curve. The genus of a smooth, algebraic plane curve of degree $n$ is $\frac{1}{2}(n-1)(n-2)$, which can be proven by, for example, the Riemann–Hurwitz formula. Rational curves have genus 0. If the algebraic plane curve is singular, then the genus decreases according to the singular points. For example, if the curve has $d$ nodes and $c$ cusps, then its genus is $\frac{1}{2}(n-1)(n-2) - d - 2c$. Two curves having the same genus are birational. This also means that the genus is an invariant classifying curves up to birational equivalence.

Contrary to the situation with algebraic curves, for algebraic surfaces one has a plurality of genera. To begin with, one has the *geometric genus*: Brill and M. Noether showed that the genus of a singular plane curves of degree $n$ can be also defined as the number of linearly independent adjoint curves of degree $n - 3$, passing $(m - 1)$ times through the singular points of order $m$. The geometric genus $p_g$ generalizes this definition to the case of surfaces: given a (possibly singular) surface in $\mathbb{P}^3$ of degree $n$, $p_g$ is the number of linearly independent adjoint surfaces of degree $n - 4$ which are linearly distinct, passing once through the double curve of the surface and with multiplicity $r - 2$ through any singular point of multiplicity $r$ of the surface. The geometric genus $p_g$ can also be defined as the number of independent holomorphic 2-forms (following an analogous definition for the case of curves). However, it may be that the surface does not have a non-zero holomorphic 2-form, but that there exists such a form of weight 2. To account for such phenomena, one denotes by $K$ the canonical line bundle, whose sections are the holomorphic 2-forms, and by $P_n$ the dimension $\dim H^0(K^n)$

---

[283] For an excellent historical overview regarding the development of the various genera, both for curves and surfaces, see: (Popescu-Pampu 2016).

[284] Generally speaking and working over the complex numbers, an affine variety is defined as the zero locus of a set of polynomials $S \subset \mathbb{C}[x_1, \ldots, x_n]$, i.e. as the set $\{x \in \mathbb{C}^n : f(x) = 0 \; \forall f \in S\}$. Similarly one may define quasi-projective or projective varieties.

when the $P_n$ are called *plurigenera*. Then according to the definition, $p_g = P_1$. The above described situation means that while for certain surfaces $p_g = P_1 = 0$, it can be that $P_2 > 0$. The plurigenera (and hence the geometric genus) are also birational invariants.

One can also define the *arithmetical genus $p_a$*. Recall that the genus of a curve can be obtained by counting coefficients of the curve, that is, by a numerical formula dependent only on certain invariants of the curve (e.g. the degree and the number of singular points). In a similar way one may define the arithmetical genus of a surface,[285] which is also a birational invariant. But this number is not always equal to the geometric genus, i.e. to the number of linearly independent adjoint surfaces of degree $n - 4$ (note that other equivalent definitions were also given, for example, via cohomology sequences). While for a smooth surface in $\mathbb{P}^3$, the arithmetic and the geometric genus are equal, it can be that $p_g > p_a$. The difference $q = p_g - p_a$ is called the *irregularity*. In contrast to algebraic curves, there are algebraic surfaces which are not rational (i.e. birationally equivalent to a plane) but have vanishing geometric and arithmetic genera. Last but not least in this list of genera, one can also define the *linear genus $p^{(1)}$* being the virtual arithmetic genus of the divisors in the canonical system of the surface, that is, the genus of the intersection of the surface with one of its adjoint.

---

[285] For example, with the postulation formula of Noether one obtains the arithmetical genus in terms of the degree of the surfaces, and the degree, the number of double points, the genus and the number of triple points of the double curve of this surface.

# 1930s–1950s: Chisini's Branch Curves: The Decline of the Classical Approach

**4**

We saw in the last chapter the influence that Enriques and Castelnuovo (and as a consequence, Zariski and B. Segre) had on the research on branch curves. Moreover, the importance of their works to the rise of Italian algebraic geometry during the first decades of the twentieth century is studied in several works.[1] Though both died later, in 1946 and 1952, respectively, the period of greatest creativity of the Italian school of algebraic geometry ended in the late 1920s. As we already saw with Klein's reference to the Tower of Babel of algebraic geometry, the algebraic treatment, mostly led by German mathematicians, had no equivalent in Italy during those years.[2] In the subsequent period, from the 1930s to 1950s, the most original ideas of this school can be attributed mostly to contributions by what were considered "minor" authors.[3] Partially isolated from other mathematical communities through their own choice—i.e., the choice not to follow the discussions and developements in modern algebra, partially due to the fascist regime, which isolated and in fact forbade the scientific and hence mathematical community from having contact with other communities outside of Italy—the mathematical research of these 'minor' authors and groups sometimes developed into unique configurations which did not, or could not, take into account the rise of modern algebraic geometry. By 'minor,'

---

[1] Guerraggio / Nastasi (2006, p. 125–158); Brigaglia / Ciliberto (1995, p. 24–33, 97–124); Dieudonne (1985, p. 52–54). See also: Gario (2014).

[2] On the "difficult presence of Algebra" (Guerraggio / Natasi 2006, p. 119) during the first decades of the twentieth century in Italian algebraic geometry, see: Ibid., p. 119–125.

[3] The expression is taken from: Conte / Ciliberto (2004, Sect. 6): "Around the end of the 1940s some of the most notable figures disappear, such as Castelnuovo, Enriques, Fano. This highlights, due to the lack of equally authoritative continuators [. . .] the decadence [in Italian algebraic geometry]. [. . .] the period of greatest creativity of the Italian school ends around the end of the 1920s, and the most interesting and original ideas of the period from the 1930s to the 1950s are to be attributed more to contributions by authors considered 'minor' [. . .]." (translation by M.F.)

M. Friedman, *Ramified Surfaces*, Frontiers in the History of Science,
https://doi.org/10.1007/978-3-031-05720-5_4

however, I wish to avoid any connotation that these groups were unimportant or that their mathematical configuration was not leading to any new research questions. Rather, 'minor' refers to an epistemic configuration which deals with objects and techniques that were considered—either at that time or later—as representing marginalized, older, perhaps outdated or no longer significant research programs. Such is the work of Oscar Chisini and his students, who chose not to take into consideration other, more algebraic research directions when developing their own research on branch curves. This might be considered an example of the general phenomenon of how, between the 1930s and 1950s, "the awareness [of the Italian school] of their own means had, over time, transformed into an over-reliance on its methods and problems. This had led to the neglect of a critical analysis,"[4] a neglect we will observe when inspecting Chisini's theorems (see Sects. 4.1.3 or 4.2.2).

As already mentioned, the isolation of the Italian school of algebraic geometry was induced by the autarkic conception of the fascist regime, which rose to power in 1922. Even before the legislation of racial laws, the special relations that existed at the end of the nineteenth century between the French mathematicians and their Italian colleagues were damaged with the rise of Mussolini.[5] How Fascism deprived the field of algebraic geometry in Italy of the works of some of its major contributors, especially after the 1938 racial laws, was already briefly discussed in Chap. 3; to recall, Enriques and Beniamino Segre had to resign their positions, whereas Zariski had already left Italy in 1927. After the fall of the fascist regime and the end of the war, it gradually became painfully clear that the Italian school of algebraic geometry was not aware (or did not wish to be aware) of the new methods and fields of research that had been developed during the previous decades, such as algebraic topology or commutative algebra. As we will see, into the 1950s, Oscar Chisini and his students continued their research as if still under the conception of an autarkic, self-sufficient Italian mathematical community.

This chapter begins in Sect. 4.1 with a thorough investigation of Chisini's work, which has remained, till now, largely unknown in the historical literature on algebraic geometry, while the existing accounts are somewhat terse.[6] When working on the branch curve, Chisini followed Enriques's methodological footsteps, presenting a variety of configurations to research this curve: associating it to a bundle of braids, degenerating the ramified surface or the branch curve, or conjecturing the uniqueness of a ramified

---

[4] Ibid.

[5] See: Brechenmacher et al. (2016).

[6] Thus for example, we find the following short description at (Brigaglia / Ciliberto 1995, p. 113–114): "The model [of Chisini] [. . .] is [. . .] called [by him] the *characteristic braid* of [. . .] algebraic curve, and allowed him to place in evidence the topological-combinatorial aspects of the theory of curve singularities and multiple planes." A similar dense account is to be found in (Guerraggio / Nastasi 2006, p. 131): "Chisini [. . .] introduce[d] in 1933 the concept of characteristic braid of an algebraic curve, composed of a finite number of tracts of spatial curves that (twirling with each other as in a braid) point out—in the particular way of twirling—the essential numerical traits of a curve's singular points."

surface when a branch curve is given. The work of Chisini is characterized by a turn to more visual and material reasoning and practices, and this during a period when algebraic geometry was starting to be written with the tools of commutative algebra. Thus, for example, Chisini worked with material models of strings to present theorems about singular curves and braids. Moreover, while continuing to develop and research Enriques's 'invariance rules' with the help of degenerations of (branch) curves, those methods led Chisini to 'prove' erroneous theorems about the possible equivalence between various modes of presentation of algebraic curves. At the same time, while ignoring Zariski's novel research directions, Chisini and his research group were becoming minor with respect to their relevance to other practices which were used more commonly during this period.

Section 4.2 continues to survey the group of students around Chisini and concentrates on the work of two of them: Modesto Dedò and Cesarina Tibiletti. Continuing Chisini's work, theirs showed how the 'minor' traditions in the Italian school of algebraic geometry were consolidated. This consolidation consisted, from Dedò's side, of an attempt to introduce more algebraic techniques into Chisini's work, while also keeping its visual and diagrammatical aspects, as seen in the work of Tibiletti. Whereas this might have been considered an attempt to take into account the turn to algebraic structures in the 1940s in algebraic geometry, it is clear from the works of Dedò and Tibiletti that their conception of 'algebra' was quite different from this turn, and reached its limits very quickly. How those research directions were eventually abandoned will be concluded in Sect. 4.3.

The Appendix to this chapter (Sect. 4.4) deals with a short introduction to the *braid group*, and especially with its presentation by Emil Artin. I bring this appendix as this group appears in several contexts, not only in this chapter, but also in the research done during the 1970s onward, as we will see in Chap. 5.

## 4.1   The 1930s and Chisini's First Conjecture

Oscar Chisini (1889–1967) was an assistant of Federigo Enriques, being the coauthor (together with Enriques) of the four volumes of *Lezioni sulla teoria geometrica delle equazioni e delle funzioni algebriche*, which were published between 1915 and 1934. He was therefore well acquainted with the works of the members of the Italian school of algebraic geometry, such as Luigi Cremona, C. Segre, Castelnuovo and, of course, Enriques himself. He was a professor in 1923 in Cagliari and from 1925 in Milan, where he remained until his retirement in 1959. Staying in these positions during the fascist regime, he had chosen to agree with the actions taken by this regime against Jewish mathematicians (see also Sect. 4.1.2). Chisini was also the editor of the educational journal for mathematics, *Il periodico di matematiche* from 1946 to 1967. Carlo Felice Manara, one of Chisini's students, underlines in the following citation a few characteristics of Chisini's approach to mathematics: "the vividness of the imagination, understood as spatial imagination, the ability to guess and grasp the analogies between different situations and to generalize, and finally the inexorable and radical critical sense. Regarding the latter, that is to say of the mistrust in front of every formal reasoning and every algorithmic acrobatics, one could bring many testimonies."[7] While one may assume that Chisini was influenced by Enriques concerning his approach to "imagination", and hence to intuition, one cannot say that Enriques mistrusted "every formal reasoning" but rather considered it as helping to shape a more refined form of intuition, as we saw in Sect. 3.1.1. Chisini, as we will see, had a slightly different approach.

During the 1940s, Chisini had several students, namely Modesto Dedò, Carlo Felice Manara, Cesarina Tibiletti Marchionna and Ermanno Marchionna. They "developed their scientific activity starting from the Maestro [Chisini] but with originality and personal choices according to their peculiarities and all, sooner or later, arrived at the Universities of Milan as full professors [...]".[8] Chisini researched a variety of subjects, among them complex algebraic surfaces as coverings, branch curves and singular curves in general. As I described in previous works Chisini's research on new modes of presentation of plane singular curves with the help of braids,[9] I will review this presentation only shortly, concentrating in the following section on his research on branch curves.

### 4.1.1    The "Characteristic Bundle"

In 1933 Chisini suggested a new way of "representing" a plane algebraic (complex) curve, as can be seen clearly in the title of his 1933 article "A Suggestive Real Representation for Plane Algebraic Curves" ("Una suggestive rappresentazione reale per le curve algebriche piane"). What was this new way? What was the nature of this representation? Chisini began with a complex plane curve of degree $n$: $f(x, y) = 0$ which might be singular. Examining its projection to the $x$-axis $p : (x, y) \longmapsto x$, he looked at the $N$ branch points on the $x$-axis, considering also singular points. The $x$-axis is a complex line, hence homeomorphic to the real plane, and Chisini drew $N$ loops on this plane denoted as $\gamma_1, \ldots, \gamma_N$, all exiting from a given point $O$, when each loop encircles another branch point. The point $O$, chosen not to be a branch point, has $n$ preimages (see Fig. 4.1a for an example of a curve of degree 3, with three preimages $y_1, y_2, y_3$). Taking one of the loops $\gamma_i$ starting (and ending) at $O$, Chisini looked at what happens to the $n$ preimages of $O$ as one goes along this loop. Considering the complex points $y_j$ as points in $\mathbb{R}^2$ having coordinates $(Re(y_j), Im(y_j))$, what one obtains is a movement of $n$ points in $\mathbb{R}^3$, having coordinates $(t, Re(y), Im(y))$, when $0 \leq t \leq 1$.[10] When one looks at the movement of these points as delineating curves in three-dimensional space, one obtains an image of a braid, composed of $n$ strings; see Fig. 4.1b for an example

---

[7]Manara (1987, p. 15).

[8]Tibiletti (1999, p. 198).

[9]See: Friedman (2019b).

[10]The parameter $t$ corresponds to the function describing the loop: $t : [0, 1] \to x -$ axis; hence Chisini considers $t$ as ranging from 0 to 1.

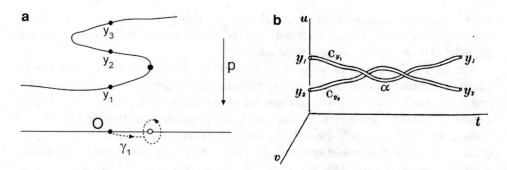

**Fig. 4.1**  (**a**) A depiction of a loop $\gamma_1$ encircling a branch point and the three preimages $y_1, y_2, y_3$ of the point $O$ for a curve of degree 3 (figure drawn by M.F.). (**b**) Chisini's depiction of a braid induced from a node (Chisini 1933, p. 1150)

of a depiction by Chisini of a braid composed of two 'strings', induced from a loop encircling the branch point of a "double point" of a curve $f$—i.e. a node.

Chisini called this construction and similar ones a "model" in several of his papers; he declared that his proposed "model" leads to "remarkable results",[11] although during the 1930s what was modelled was usually not called a "braid" ["treccia"] but rather a "bundle" ["fascio"].[12] This kind of depiction—a drawing of a braid to describe the motion of the points in the neighborhood of a ramification point—was certainly not new (it was used, for example, by Riemann, Carl Gottfried Neumann, Felice Casorati, Gustav Holzmüller, Enriques and Severi).[13] Moreover, following the preimages of a point along a loop was also a procedure which was employed often during these years, as we saw in the last chapter with Enriques and Zariski. The novelty in Chisini's "model" lies not only in the terming of this description (also) as a "braid". It is also due to the fact that while the above treatment presents what occurs when following one loop on the $x$-axis, the "bundle" is in fact associated to a concatenation of all of the braids obtained from a *concatenation* of all the loops $\gamma_1, \ldots, \gamma_N$, encircling each one a different branch point. In this way, what Chisini called the "characteristic bundle [fascio carrateristico]"[14] (i.e. the concatenation of all the obtained braids) presents not only how ramification and singular points of the curve 'look like', but also their respective position in relation to each other. In addition, Chisini emphasized that one not only can choose a different base point $O$, but also may vary the

---

[11] Chisini (1933, p. 1142).

[12] However, for specific examples Chisini did use the term "braid", for example, for a "braid of three lines" or for a "braid of $r$ lines" (ibid., p. 1151, 1152), for describing the braid deduced from encircling the image of an $r$-fold point.

[13] See: (Friedman 2019a, p. 113 (footnote 5), 119, 124, 141, 143); In his book *Teoria Geometria delle Equazioni e delle funzioni algebriche*, Enriques (1915, p. 361), for example, draws a braid depicting a ramification point of multiplicity 3 of a degree 5 Riemann surface; the book was edited by Chisini, hence he must have seen this visualization (also) already during the 1910s.

[14] Chisini (1933, p. 1146).

loops themselves, taking note "of the variations for continuity that give equivalent systems of loops."[15] Moreover, Chisini preferred "models" which are symmetric and simple, and the point $O$ is chosen according to these criteria.[16]

The question arises: what is meant by "model" in Chisini's research configuration? Chisini emphasized throughout the 1933 article that this model becomes more accessible and convenient to work with once one uses "material models", with "material threads with notable thickness".[17] Both Piera Manara and Maria Dedò, the daughters of Carlo Felice Manara and Modesto Dedò respectively, recall the string models of Chisini, being of different colors used to model curves.[18] The fact that Chisini considered his models as material is also to be seen in the thickness of the curves depicted in the Fig. 4.1b, as if the curves were real strings. In another paper, from 1952, Chisini discussed coloring the braids, as it will be easier that way to determine the torsion of each braid.[19] This material construction of models of braids is not surprising, since, as we have seen in Sect. 3.1.1, Enriques, together with Guido Castelnuovo, constructed material models of algebraic surfaces. Chisini, just as Enriques, certainly supported a visual and material thinking of mathematics. Manara recalls that for Chisini, one had to verify the results obtained regarding complex curves "which led him to persist in the construction of material models, not to be satisfied with drawings and formulas; those who met him in those years remember that one of the most frequently repeated phrases was '. . . I do not trust'. And this distrust led him to want to build tangible and material models."[20] This skepticism led Chisini also to ignore Emil Artin's research (and the subsequent research) on the braid group, as we will see later, and may have prompted a disconnect from other research directions which were on the rise at that period, for example, commutative algebra.[21]

In 1937 Chisini presented two explicit examples of the 'characteristic bundle', presenting them as two examples of a "canonical form".[22] The first "canonical form" is of the smooth curve $y^n - ny + x^n = 0$, a curve which has $n(n-1)$ simple ramification points. The second "canonical form" is of $n$ generic lines, intersecting at $\frac{n(n-1)}{2}$ nodes. For the last

---

[15] Chisini (1937, p. 60):"[. . .] dalle variazioni per continuità che danno sistemi di cappi equivalent." Though this is clear that Chisini considered implicitly homotopy of paths, an explicit discussion is only to be found in 1952 Chisini (1952b, p. 18).

[16] Chisini (1933, p. 1151). However, Chisini did note how the variations of the system of loops will affect the "characteristic bundle", and pointed that these variations can be decomposed into a series of transformations between two neighboring loops (see also: Friedman 2019b, p. 12–13).

[17] Chisini (1933, p. 1146–1147).

[18] Private communication with Piera Manara (Email from 21.2.2018), and private communication with Maria Dedò (Email from 27.2.2018).

[19] Chisini (1952a, p. 360).

[20] Manara (1987, p. 23).

[21] Moreover, the tradition of material mathematical models was declining during this time: those models were hardly produced anymore in Germany during the 1920s and the 1930s, and were turning slowly into a museum object (Sattelmacher 2014, p. 140–142).

[22] Chisini (1937, p. 50).

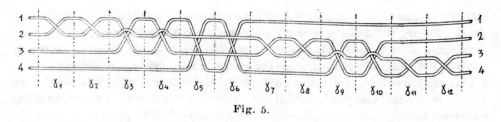

Fig. 5.

**Fig. 4.2** An example of the second "canonical form": The "characteristic bundle" ["fascio caratteristico"] associated to four generic lines, intersecting at six nodes. Each consecutive pair of crossing (denoted as $\gamma_1$ and $\gamma_2$; $\gamma_3$ and $\gamma_4$ etc.) corresponds to a node. (Chisini 1937, p. 58)

example, Chisini drew a diagram, what the associated "characteristic bundle" ought to look like (see Fig. 4.2), when one takes into account all of the loops encircling the images of the nodes on the $x$-axis.

### 4.1.2 On Braids, Branch Curves and Degenerations

After publishing his first paper in 1933 on the "characteristic bundle", Chisini sought to show how his method can "characterize" branch curves, as he described in 1934.[23] Chisini noted that all of the previous attempts—of Enriques, Zariski and B. Segre—analyzed the problem, "without, however, arriving at a response that could be applied in practice."[24] What Chisini proposes is that with the help of his "real model"[25] of the "characteristic bundle [...] of the branch curve of [a surface] $\varphi$", the "invariance conditions of Enriques can be translated into a condition with a topological character".[26] The usage of Chisini in the verb "translate" implies a transfer between two configurations: from Enriques's invariance conditions to Chisini's research on more specific class of surfaces and branch curves; those results in Chisini's configuration would have a "topological character".

How this can be done is by employing another method, which is less material or sensuous than the material models: Chisini's method consisted of taking a curve $\overline{\varphi}$ with $n-1$ components $C_1, \ldots, C_{n-1}$, when the curve itself is given by $\overline{\varphi} = C_1 \cdot C_2 \cdot \cdots \cdot C_{n-1}$. Chisini then looked at the square of this curve $\overline{\varphi}^2$, and claimed that with "infinitesimal variations" of $\overline{\varphi}^2$,[27] which satisfy certain conditions, one can obtain another (irreducible) curve $\varphi$, which is a branch curve of an $n$-degree cover. The "infinitesimal variations" are

---

[23] Chisini (1934, p. 688).

[24] Ibid.: "[...] senza tuttavia pervenire ad una risposta suscettibile d'applicazione pratica."

[25] Here Chisini presumably referred also to his material model of strings.

[26] Ibid., p. 690: "La condizione di invarianze di Enriques si traduce cosi in una condizione di carattere topologico per il fascio caratteristico rappresentativo della curva di diramazione $\varphi$."

[27] Ibid., p. 691: "variazione infinitesima".

done not in order to find explicitly the curve φ (or its equation), but rather to examine how the singularities of the degenerated curve deform. Chisini concludes that a node of the degenerated curve is varied into four nodes and two simple branch points of the branch curve φ; and a tangency point between two components of the degenerated curve (or an intersection point of two consecutive curves) is varied into three cusps and a simple branch point.[28] With these "variations", starting with a reduced curve with $n - 1$ components, one can obtain branch curves of only certain "multiple covers" of the plane of degree $n$ (for example, of smooth surfaces in $\mathbb{CP}^3$ ); this limitation was later criticized by Bernard d'Orgeval (1909–2005), as we will see later.

It is important to stress that this was not the first time that Chisini worked on degenerations of surfaces and of curves. In the 1934 paper Chisini referred to another paper which he wrote in 1917, called "Sulla riducibilità della equazione tangenziale di una superficie dotata di curva doppia".[29] The motivation of the 1917 paper is as follows: taking a algebraic surface $f_c(x_1, \ldots, x_4) = 0$ in the projective space $\mathbb{CP}^3$, it may have a double curve $C$ (or order $n$); if one takes the intersection of $f_c = 0$ with $df_c/dx_i$ ($1 \leq i \leq 4$), and projects the intersection to the projective plane, one obtains two curves: the branch curve $h = 0$ (of order $r$) and the image of the double curve $c = 0$ (of order $n$), counted twice, i.e. $c^2 h = 0$. Chisini suggested to consider the surface $f_c$ as a limit of functions in the family of surfaces $f_c + \lambda f_0 = 0$, when the other members in this family (i.e. when $\lambda \neq 0$), are smooth surfaces in the projective space $\mathbb{P}^3$, and hence the branch curves of those surfaces are irreducible (i.e. they do not have two components).[30] This leads Chisini to state that the "problem of degenerations of an envelope of an algebraic surface $f$, having a nodal [double] curve $C$, is intimately connected with the problem of degenerations of an envelope of a plane curve, of which [one component] is separated and counted twice."[31] Chisini then stated that the degenerations of branch curves is to be considered by taking "$c^2 h = 0$ as a limit of a curve $k$ varying in a bundle $c^2 h - \lambda l = 0$, when $l = 0$ is a generic curve of order $2n + r$."[32] First considering such limits in a general family of curves (when $h$ and $c$ are not necessarily curves related to any surface as above), Chisini focused on how the singular points of $k$ are transformed when passing to the limit. Only afterwards did he consider the more special case: what happens when $c^2 h = 0$ is the branch curve of a surface with a double curve. The last part of the paper is a list of cases, describing the transformations of singular points in the degenerated curve $c^2 h = 0$ in the family;[33] exactly those cases are listed again in the 1934 paper. While Chisini declared in the 1917 paper that his aim with

---

[28] In: (ibid., p. 766–771) other cases of degenerated branch curves are considered, also dependent on the intersection between the double curve of the surface and branch curve.

[29] In: Chisini (1934, p. 691) Chisini refers to (Chisini 1917).

[30] Chisini (1917, p. 543).

[31] Ibid., p. 544.

[32] Ibid.

[33] Ibid., p. 547–548.

these calculations is to find numerical invariants of the surface $f_c = 0$,[34] what is new in the 1934 paper is that those calculations are re-embedded in a new configuration, which stresses how these degenerations, or "variations" (to use the term of the 1934 paper) can be presented on the level of the "characteristic bundle".

Indeed, the main result of the 1934 paper is to show how, after performing these "variations", the 'regenerated' branch curve satisfies the invariance conditions of Enriques, i.e. the existence of a map sending loops around the (regenerated) branch curve to the symmetric group—and this with the help of the "characteristic bundle". This is what Chisini meant, among others, by "translation" performed between epistemic configurations: the "variations" of the degenerated (double) curve and especially their presentation in form of a braid were facilitating the proof that the invariance conditions are satisfied. However, the role assigned to visualization and to material models was minimized compared to the former papers from the 1930s: Chisini draws only one figure (see Fig. 4.3), which illustrates how this variation looks—but only on the level of the associated braid, and only for a degenerated node.

The method of degenerations and "variations" was later generalized by d'Orgeval[35] and by Chisini himself. For example, in 1938, Chisini considered three generic lines, but showed that if their variation should be a branch curve of a generic projection of a triple plane (that is, a cover of degree 3), it cannot be a degeneration of a sextic with nine cusps (though this curve is also a branch curve, see below), but rather a sextic curve with 6 cusps, obtained first from a "variation" of the three generic lines into a conic and a line (so that one would have a variation of $n - 1 = 3 - 1 = 2$ curves, hence $n$, being the degree of the cover, is 3).[36] This shows that certain types of variations—if one follows only Chisini's method—are not possible, or that several types of surfaces cannot be obtained in that manner.

However, Chisini's degeneration method was rather limited. Chisini himself did not elaborate on this, but d'Orgeval, while trying to expand Chisini's method to other types surfaces, reached that conclusion. In his thesis from 1943, d'Orgeval praised Chisini for his "presentation of a plane curve by a bundle of real curves, giving the conditions of Enriques a more accessible form, and permitting to penetrate further into knowing the multiple $n$-planes".[37] He then continued to describe Chisini's method of the variation of a double-counted, degenerated curve, into a branch curve of certain type of surfaces, noting that "Chisini expressed the hope that his method can be generalized to all of the surfaces [...]

---

[34] Ibid., p. 544; explicitly, Chisini aims to find the class of $f_c = 0$ (i.e. the degree of the reciprocal surface) and the number of cuspidal points of the double curve. The formulas for these two invariants are given in: ibid., p. 548.

[35] See: (d'Orgeval 1938, 1942, 1945 [1943]).

[36] Chisini (1938, p. 537). Recall that Chisini's initial theorem dealt with a curve with $n - 1$ components, obtaining from it after its "variation" a branch curve of surface of degree $n$.

[37] d'Orgeval (1945 [1943], p. 73): "M. Chisini [...] a permis de donner aux conditions d'Enriques une forme plus accessible, permettant de pénétrer plus avant dans la connaissance des plans $n$-ples."

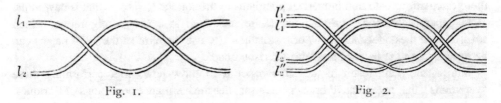

Fig. 1.                                      Fig. 2.

**Fig. 4.3** The braid diagrams corresponding to how a node of a degenerated branch curve (left) turns into, via "infinitesimal deformation", four nodes and two simple branch points (right). (Chisini 1934, p. 771)

[but] I showed that the view of Chisini was too optimist."[38] That is, d'Orgeval showed that not every branch curve of an algebraic surface can be obtained in this way. This is also to be noted in 1950, when d'Orgeval observed the following: "The question of characterizing the branch curve of a multiple plane [. . .] remains largely unresolved. We owe to M. Chisini, an interesting method to obtain such curves from degenerate models, but despite possible extensions of this method, it encounters, except for certain types of surfaces, great difficulties."[39] He then suggested examining another method—that of degenerating the surface itself—and not the branch curve—into a union of planes, a method which the mathematician Guido Zappa already employed successfully to research surfaces,[40] as d'Orgeval noted, and which I will survey later briefly.

This comment from 1950 shows already that Chisini's ideas regarding the degeneration of curves were not as productive as he hoped. However, there are other factors which might have prevented the dissemination of his theory. First and foremost, Chisini's insistence on the verification, or even proof, of several of his theorems with the usage of material models of strings was certainly not widely accepted anymore during the 1940s and the 1950s. While at the beginning of the twentieth century material models of mathematical objects were considered as pedagogic, and even sometimes as epistemic objects, the mathematical community of algebraic geometry outside, and partially inside of Italy, was either rejecting or disregarding these models as means of proof starting at the end of the 1930s. Second, and equally important, is Chisini's agreement with the isolation of the Italian mathematical community under fascism. Moreover, Chisini was part of the scientific committee, which decided on December 10, 1938, to expel all the Jewish members from the *Unione*

---

[38] Ibid., p. 74–75: "M. Chisini exprimait l'espoir que sa méthode puisse se généraliser à toutes les surfaces [. . .] [mais] j'ai été amené à montrer que la vue de M. Chisini était trop optimiste."

[39] d'Orgeval (1950, p. 351): "La question de caractériser la courbe de diramation d'un plan multiple [. . .] reste en gros irrésolue. On doit à M. Chisini, une intéressante méthode pour obtenir de telles courbes à partir de modèles dégénérés, mais malgré des extensions possibles de cette méthode, elle se heurte, sauf pour certains types de surfaces a de grosses difficultés."

[40] Zappa (1942).

*Matematica Italiana.*[41] One of the results of this decision was to deny Enriques (and other mathematicians) any access to the Roman Mathematical Institute; he was also removed from teaching. To recall, Enriques was Chisini's advisor and collaborator for years.[42] Chisini obviously was aware of this consequence (and others consequences which affected tremendously the life of Jewish mathematicians in Italy), but somewhat not surprisingly did not comment on it in his obituary to Enriques in 1947.[43] However the students of Chisini (Dedò or Tibiletti, for example) did not sign this decision (as only professors had to sign it, whereas in 1938 the students of Chisini were not in this status), and several of them (Dedò, for example) were in contact with anti-fascistic movements (see below). Chisini's agreement with this autarkic position, which encouraged in a certain sense the seclusion of the Italian mathematical community, might have led to inevitable consequence that his results on the characteristic braid, published during and after the 1930s, remained unknown to the international community during that decade and in the 1940s. This might also explain why d'Orgeval's methods remained also largely unknown, as he presented himself implicitly as a follower of Chisini. However, d'Orgeval's work also signifies the seclusion of the Italian mathematical community (and its results) in another way.

### 4.1.2.1 Bernard d'Orgeval in Oflag X B

Bernard d'Orgeval, who specialized in algebraic geometry, prepared his doctorate at the University of Rome, working there with Enriques, before being mobilized to participate in the Second World War. Between 1940 and 1945, d'Orgeval was successively transferred to three officers' camps (*Offizierslager*, abbreviated as *Oflag*): Oflag XIII A (1940–1941), Oflag XXI B (1941–1942) and finally Oflag X B (1942–1945).[44] While I discussed few of his mathematical works done at Oflag X B above, it is important to note that d'Orgeval was not only interested in mathematics during this time: during his imprisonment there he also prepared a doctor thesis in law, published in 1950 as *L'Empereur Hadrien*. After the war, he was a lecturer in Grenoble, then a professor in Algiers and finally in Dijon, where he remained until his retirement in 1979. Oflag X B, where d'Orgeval carried out most of his research during the war, was a camp for French officers, located at Nienburg am Weser, Germany, and was opened in May 1940.[45] The preface of d'Orgeval's thesis, which was written there, is telling in this respect:

---

[41] Guerraggio / Nastasi (2006, p. 263–264).

[42] Israel (2004, p. 42), Guerraggio / Nastasi (2006, p. 262).

[43] Although Chisini bothered to mention that Enriques loved his "Homeland" (meaning Italy), he also mentioned the fact that he was not "Arian", but rather a Jew (Chisini 1947a, p. 121–122); Chisini ends, nevertheless, the obituary, saying that the years 1908–1922 with Enriques "remain the most cherished of all my life" (ibid., p. 123). For forms of mathematicians' collaborations with and adaptations to fascist regimes, especially in Germany during the Third Reich, cf. (Mehrtens 1989, 1994, 1996).

[44] I thank Christophe Eckes for pointing this out.

[45] For the history of this camp, see (Sonnenberg 2005). See also: http://wir-wussten-nichts-davon.de/ (accessed on: 17.11.2021)

"This work, which began with the instructions of M. Enriques [. . .] was completed in April 1943, at Oflag X-B, at Nienburg/Weser, where I was imprisoned. The conditions of the captivity did not permit me to consult the abundant literature [. . .]. I am grateful to M. Chisini [. . .] for the precisions he gave me orally regarding his method of the construction of multiple planes, which I had to employ frequently. [. . .] I am beholden to MM. Enriques and Cartan, to whom I dedicate this modest work with a homage of gratitude and profound admiration.
      Nienburg am Weser, Oflag X-N, February-April 1943."[46]

The fact the d'Orgeval managed to write his thesis at this camp is remarkable—most likely due to the existence of a 'university' at Oflag X B—and is evidence that the prisoners there had relative freedom regarding their occupations in the camp (see Fig. 4.4). Prisoners of war, according to the Geneva Convention, were not required to work; as a result in numerous Oflags 'universities' or 'study centers'—also termed later as "barbed-wire universities"—were opened. These 'universities', besides having (sometimes extensive) libraries, also gave courses, lectures and exams, where degrees were also granted, which were to some extent recognized afterwards. To emphasize: d'Orgeval was not the only 'working' mathematician in the Oflags: another famous example is the development during the early 1940s of sheaf theory by Jean Leray, who was a prisoner of war in Oflag XVII in Edelbach, Austria.[47] Nevertheless, as it is also clear from d'Orgeval's preface, that even if there was a transfer of knowledge between the various mathematical communities (German, French and Italian) during the late 1930s and the Second World War, it was damaged during these years.[48]

### 4.1.2.2 Guido Zappa's degenerations

Another reason why Chisini's methods of the degeneration of the branch curve were not as productive as one could have hoped, lies not so much in historical and sociological factors, but is rather connected with the work of Guido Zappa. Zappa (1915–2015) was an Italian

---

[46]D'Orgeval 1945 [1943], preface: "Ce travail, commencé sur les indications de M. Enriques [. . .] a été achevé en avril 1943 à l'Oflag X-B, à Nienburg/Weser, où je me trouvais prisonnier. Les conditions de la captivité ne m'ont pas permis de consulter une abondante bibliographie [. . .]. Je suis reconnaissant à M. Chisini [. . .] des précisions qu'il a bien voulu me donner de vive voix, sur sa méthode de construction des plans multiples, que j'ai eue à employer fréquemment. [. . .] je suis redevable [. . .] à MM. Enriques et Cartan, que je dédie ce modeste travail, en hommage de reconnaissance et de profonde admiration."

[47]See: (Miller 2000; Eckes 2020). On the 'universities' at the Oflags, see: (Gayme 2015; Durand 1994, esp. p. 173).

[48]That mathematical research or ties between those communities were damaged was however not always the case for French mathematicians who were POW in the Oflags, as some sometimes did cooperate mathematically with German mathematicians and institutions while being prisoned; see: (Eckes 2020, 2021).

Regarding the transfer of mathematical knowledge during the 1930s and the 1940s between Italy and Germany, see (Remmert 2017). Remmert notes that despite willingness to cooperate, hardly any substantial cooperation between Italian and German mathematicians developed; this cooperation rather mostly remained on the level of episodic contacts and collaborations.

**Fig. 4.4** one of the prisoners' rooms in Oflag X B, as published in the report "Les camps de prisonniers de guerre en Allemagne" made in 1943 by the *Direction du service des prisonniers de guerre* (p. 177). The report notes, "the [prisoners'] rooms are bright, having large windows [. . .]. In addition, they each have a private library made up of books received individually by the prisoners." (ibid., p. 176)

mathematician, known mainly for his contribution to group theory. During the 1940s Zappa was highly productive in the field of algebraic geometry, which shows that mathematicians in Italy had to silently accept the fascistic regime (if they were non-Jewish and) if they wanted to continue working and researching in their universities. Zappa himself described this situation as adopting a system of a "double truth": on the one hand criticizing privately the actions of the fascistic regime, on the other hand, publicly agreeing with those actions.[49] Zappa published several papers on the topic of the degeneration of surfaces into a union of planes, inspecting also the changes occurring to the branch curve.[50] I would like to analyze one of the papers more thoroughly, as it highlights how Chisini's research on degeneration was slowly reaching its limits. The paper, published in 1945, deals with the fundamental group of the complement of the branch curve of a surface which can be deformed into a union of planes. The idea of investigating the fundamental group of the complement of the branch curve was of course not new, as we saw in Chap. 3. However, concentrating on the set of surfaces, which can be deformed into a specific set of simpler surfaces, helped to highlight the unique structure of this group.

---

[49] See: Zappa (1997, p. 39). For Zappa's research on algebraic surfaces, see: (Brigaglia / Ciliberto 1995, p. 47). As Zappa became assistant professor in Rome in 1940 and was close to Severi, who was known for being a staunch supporter of the Italian fascist regime, one could also read Zappa's account on mathematics in the time of fascism (Zappa 1997) as a retroactive attempt to explain this silent agreement.

[50] For example: Zappa (1942, 1943, 1945).

While Zappa surveyed the main contributions of Enriques, Zariski, B. Segre and Chisini, he noted two open research directions: the first, as Zariski suggested in *Algebraic Surfaces*,[51] is to find all normal subgroups of the fundamental group of finite index, in order to find "all types of surfaces which admit the corresponding curve of a branch curve [curve di diramazione]."[52] The second is to find a group theoretic description of Segre's results. Both lines of research underline a group theoretic investigation, which Zappa indeed followed, but in a different direction. Zappa, as mentioned above, concentrated on surfaces, which can be deformed into a union of $m$ planes $\lambda_1, \ldots, \lambda_m$. Taking $f$ to be the branch curve of such a surface, one sees that the branch curve of the deformation is a union of lines. Zappa denoted by $\rho_{ij}$ the line of intersection of the planes $\lambda_i$ and $\lambda_j$. He then proved that the fundamental group of the complement of $f$ is generated by the generators $g_{ij}$ such that:[53]

$$g_{ij}g_{iu}g_{ij} = g_{iu}g_{ij}g_{iu} \ (j \neq u), g_{ij}g_{uv} = g_{uv}g_{ij} \ (i,j \neq u,v), \Pi g_{ij}^2 = 1$$

Zappa noted that if the surface is a smooth surface in $\mathbb{CP}^3$ of degree $m$, then it can be deformed into $m$ distinct planes in $\mathbb{CP}^3$, and hence one obtains easily a presentation of the corresponding fundamental group, when one has $\binom{m}{2}$ generators $g_{ij}$, when $i,j = 1, \ldots, m$, $i < j$. Another example was given by Zappa,[54] for a rational curve of order $m$ without inflection points. Pointing out that this curve is also a branch curve, Zappa proved that the fundamental group of the complement of this curve is generated by $m - 1$ elements $\gamma_i$, such that

$$\gamma_i\gamma_{i+1}\gamma_i = \gamma_{i+1}\gamma_i\gamma_{i+1} \ (i = 1, \ldots, m - 2),$$

$$\gamma_i\gamma_j = \gamma_i\gamma_j \ (i = 1, \ldots, m - 3; j = i + 2, \ldots, m - 1),$$

$$\prod_{i=1}^{m-1} \gamma_i^2 = 1$$

Surprisingly, Zappa does not mention the braid group of Emil Artin; I discuss the braid group in the Appendix to this chapter, but for now it is enough to recall that Artin's presentation of the braid group of $m - 1$ strings, which was first presented in 1926, is the group generated by $m - 1$ elements $\gamma_i$, such that

---

[51] See: Zariski (1935, p. 163). Cf. also (Libgober 2014, p. 480).

[52] Zappa (1945, p. 144).

[53] Ibid., p. 150.

[54] Ibid., p. 151.

**Fig. 4.5** A drawing of the generator $\gamma_i$ (Artin 1926, p. 49)

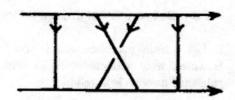

$$\gamma_i \gamma_{i+1} \gamma_i = \gamma_{i+1} \gamma_i \gamma_{i+1} \ (i = 1, \ldots, m - 2),$$

$$\gamma_i \gamma_j = \gamma_i \gamma_j \ (i \neq j, j + 1).$$

Here $\gamma_i$ represents a generator of braid group, according to Artin's presentation: it is the braid for which the $i^{\text{th}}$ string passes once above the $(i + 1)^{\text{th}}$ string, whereas the other strings are straight lines (see Fig. 4.5).[55]

While it is not known if during that time Zappa knew of Artin's presentation of the braid group, he certainly knew of Zariski's papers; Zariski published in 1936 a similar paper on the computation of the fundamental group of the complement of "rational maximal cuspidal curves of even order $2n - 2$ [...] [, which] practically coincides with the 'Zopfgruppe' of Artin."[56] While Zariski's methods are different, one has to wonder why Zappa did not mention this group. This might indicate how well (or poorly) disseminated the mathematical literature was on group theory in Italy.[57]

While Zappa did not continue working on this subject, his work attempted—in contrast to Chisini—to situate the research on degenerations in a broader configuration, that of computing the fundamental group. The computation of the fundamental group of the complement of the branch curve and finding its general properties was still an unsolved problem. While Chisini's method did not approach this problem, the degeneration of curves did not only focus on degeneration as such (hence focusing on degeneratable curves), but also on the "characteristic bundle"—an object much harder to deal with than the fundamental group. During the 1950s, several techniques were developed in order to deal with the "characteristic bundle", but, in a certain sense, they led to a stagnation of this research configuration. But before dealing with this research, I would like to survey another work of Chisini from 1944.

---

[55] For the history of the braid group between the 1920s and the 1950s, see: (Friedman 2019c).

[56] Zariski (1936, p. 607).

[57] As we will see in the next chapter, similar research—about the fundamental group of the complement of a branch curve of a surface, which can be deformed into a union of planes—will be considered by Moishezon during the 1980s.

### 4.1.3   Detour. 1944: Chisini's First 'Conjecture'

In 1944 Chisini published what is now known in the literature as 'Chisini's conjecture'. However, what was published, to emphasize, is a proof—though a wrong one. The published theorem is as follows:

> "(1) Given two algebraic surfaces $F$ and $\Phi$ which have the same branch curve $\varphi$ (under few restrictive hypotheses) there is a birational correspondence in the neighborhood of the [corresponding] ramification curves $C$ and $\Gamma$, whose projection is branch curve $\varphi$. (2) With some additional hypothesis, in particular that one of the two surfaces is given as a function $z(x\ y)$ of degree larger than 4, then the two surfaces are birationally equivalent."[58]

The restriction to surfaces of degree larger than 4 lies in the fact that Chisini noted a counter example: a sextic curve with 9 cusps is the branch curve of three non-strictly isomorphic degree 4 surfaces, and also the branch curve of a cover of degree 3.[59] Moreover, the proof of part (2) relies on two assumptions: first, the existence of a local isomorphism between two neighborhoods of $C$ and $\Gamma$ (being the content of part (1)); the second, that the branch curve can be degenerated in a certain way. However, what is the nature of this degeneration is not explicitly formulated, as Chisini only noted that one should "imagine [Si immagini] passing continuously from the curve $\varphi$ to its degenerated curve $\overline{\varphi}$."[60] What is actually meant by this act of "imagination", or how this deformation was done analytically—is not even explained.

While on the one hand this result can be considered as belonging to the research on degenerated curves, and on the other hand as opening a new research configuration, which situates the proof of the uniqueness of the branch curve as its aim, the above assumptions made by Chisini were problematic. This issue was already considered in 1947 by Chisini himself. He notes that "any indirect verification that tends to support the hypothesis of the general validity of the theorem would be very useful, also because it would serve as a spur for the arduous, and perhaps vain, undertaking to find a proof of the supposed general theorem."[61] However, as Chisini noted in a footnote to this statement, that without the

---

[58] Chisini (1944, p. 339).

[59] As Chisini (ibid., p. 349–350) observed, this implies that the condition that the degree of the surface is at least 5 is necessary: in modern terms, the generators of the fundamental group of the complement of this sextic curve can be sent to two different sets of three transpositions: to the three transpositions $(1, 2), (2, 3), (3, 1) \in Sym_3$ (in the symmetric group of three letters) and to the three transpositions $(1, 2), (1, 3), (1, 4) \in Sym_4$, both sets satisfy the same relationships (and the same invariance relations of Enriques); hence one obtains two coverings of the plane, one of degree 3 and one of degree 4, with the same branch curve. Chisini (ibid., p. 349, observation 3) only noted that for a set of three transpositions one may find two embeddings of this set, one into $Sym_3$, the other into $Sym_4$.

[60] Ibid., p. 348.

[61] Chisini (1947b, p. 6): "[. . .] qualunque verifica indiretta che tendesse ad avvalorare l'ipotesi della validità generale del teorema riuscirebbe molto utile, anche perché servirebbe di sprono per l'ardua, e forse vana, impresa di trovare una dimostrazione del supposto teorema generale."

degeneration hypothesis, "the supposed general theorem is probably not true [according to an] opinion of an authoritative master; and such an opinion would stop any enthusiasm from attempting [to find a proof], if some indirect verification of the supposed theorem has not been found before."[62] Who this 'master' is was not stated. But the fact that Chisini himself mentions that there is a 'higher' authority who doubts the validity of this general theorem, underlines that the research directions during the 1930s and the 1940s in some of the communities of Italian algebraic geometry were slowly criticized regarding their rigor—though, as we will see in the following section, this did not prevent Chisini from 'proving' a wrong theorem ten years later.

In 1979 Antonio Lanteri attempted to deal with this problem, and noted that the degeneration hypothesis should be removed,[63] and tried to re-prove the theorem, again under the assumption that a local isomorphism of the neighborhoods of the ramification curves $C$ and $\Gamma$ can be extended into a global one. However, also this assumption proved to be incorrect.[64] It was Fabrizio Catanese who noted in 1986 all of these wrong assumptions, and proposed a correct criterion to the problem, when two surfaces considered as covers "with the same branch curve would be strictly isomorphic 'under some suitable conditions of generality.'"[65] Catanese also termed Chisini's problem a conjecture (though for Chisini himself it was a theorem).[66] He proved that the two surfaces are indeed strictly isomorphic, if and only if a certain (marked) line bundle on $C$, which is of order 2, is in fact trivial. One must note here that the techniques of Chisini and of Catanese are very different, as the two are situated in different configurations: Catanese dealt with marked line bundles, doing calculation in the Picard groups of the ramification curve; Chisini attempted to construct the surface by examining to which transpositions loops encircling the branch curve should be sent (and this in order to verify Enriques's invariance rules), to obtain birational equivalence.

Surprisingly, while the criterion of Catanese looks more complicated than Chisini's 'proof', it turned out that Chisini's conjecture is indeed true and its proof is rather simple. At the turn of the twentieth century, in two papers written in 1999 and 2008, Viktor Kulikov proved that a covering of degree greater than 4 is uniquely determined by its branch curve. The Chisini 'conjecture' was first proven in 1999 for generic ramified coverings of degree bigger than 11 and for generic projections of any degree with one exception—the counter example of Chisini himself. In 2008 the other remaining cases were verified. However, how this conjecture was proven stands in direct contrast to how Chisini

---

[62] Ibid., p. 6, footnote 1: "Che il supposto teorema generale non sia probabilmente vero è l'opinione che mi scrive un autorevole maestro; e tale opinione fermerebbe ogni entusiasmo di tentare, se prima non si sia trovata una qualche verifica indiretta del supposta teorema."

[63] Lanteri (1979, p. 523).

[64] Cf. Lanteri (1987)

[65] Catanese (1986, p. 33).

[66] Ibid.

would have wished to see its proof: while Chisini promoted the approach that topology is a science "that instructs us to not make calculations",[67] Kulikov proved the conjecture via numerical calculations.[68] Kulikov assumed by negation that there are two non-isomorphic projective surfaces $S_1$ and $S_2$ with the same branch curve. Looking at a collection of curves on (the normalization of) the product $S_1 \times S_2 = \{(x, y) \mid x \in S_1, y \in S_2, f_1(x) = f_2(y)\}$, when $f_1$ and $f_2$ are the projections of $S_1$ and $S_2$ to the projective plane, Kulikov's crucial step is to use a variant of the Hodge index theorem[69] on the normalization of $S_1 \times S_2$, a variant which indicates that for every two divisors $D_1$ and $D_2$ on a nonsingular projective surface, one has $(D_1^2)(D_2^2) \leq (D_1 \cdot D_2)^2$. Since this inequality is eventually a numerical one, one obtains from it the necessary numerical restriction. This procedure was already well known during the 1970s and the 1980s,[70] but the way Kulikov approached the problem is very different from that of Chisini and Catanese. While both Chisini and Catanese tried to construct certain objects (Chisini—the surface itself from the branch curve, Catanese—a specific line bundle) and to prove that certain conditions hold with respect to those objects, Kulikov only concentrated on finding a numerical restriction, which would be not fulfilled if there were two non-isomorphic surfaces. In this respect, Kulikov's proof was non-constructive and was certainly positing Chisini's conjecture in another configuration.[71]

## 4.2    Chisini's Students: Isolation and Abandonment

As Giorgio Israel notes, only starting at the beginning of the 1960s did the study of modern algebra become well spread in the post-war curriculum for Masters in mathematics in Italy. During the 1950s, the "weight of tradition in Italian algebraic geometry seriously hindered the development of analysis".[72] We saw already how this "weight of tradition" led Chisini to find a very partial proof, with several assumptions, of a theorem which later became known as the first Chisini's 'conjecture'. However, Chisini's students, following in a sense their teacher, continued this 'traditional' research during the 1950s. The way the students of Chisini followed his research methods and topics was also noted by mathematicians not necessarily belonging to this group: in 1953 d'Orgeval called Chisini and his students "the school of Chisini", or the "geometers from Milan".[73] D'Orgeval designated hence the

---

[67] Manara (1968, p. 7).

[68] Kulikov (1999, 2008).

[69] For a modern proof of this variant, see: Badescu (2001, p. 20).

[70] . . . and the Hodge index theorem was already proved in the 1930s.

[71] Kulikov in 1999 notes that "if the degree of a generic morphism with given discriminant curve $B$ is sufficiently large, then this generic morphism is unique for $B$. Almost all generic morphisms interesting from algebraic geometric point of view satisfy this condition." (1999, p. 84).

[72] Israel (2004, p. 31).

[73] d'Orgeval (1953, p. 188).

entire group as sharing a common mathematical practice—researching branch curves with very specific methods (while not employing, for example, other methods). Maria Dedò recalls that for her father, Modesto Dedò, who was one of Chisini's students, higher algebra was not studied during the university year, but rather later, when one did research.[74] Following Israel's description, that the "recovery was slow and difficult",[75] I would like to consider two examples of how this recovery took place, by examining the works of two of Chisini's students: Dedò and Cesarina Tibiletti.

### 4.2.1  Dedò and the New Notation of Braids

Modesto Dedò (1914–1991) wrote in 1950 an influential paper, which was afterwards often cited by Chisini and his students. While the subject of the paper is only tangent to the investigation of branch curves, we will see it had a significant influence on how the "characteristic bundle" associated to the branch curve—or, as it was called starting in the 1950s, the "characteristic braid"[76]—was considered. Since I surveyed Dedò's work on braids extensively in a previous work,[77] I will concentrate here mainly on the aspects related to the branch curve.

After his military service during the early 1930s, Dedò obtained his PhD in mathematics in 1939 under the guidance of Chisini and became an assistant professor of Geometry at the Polytechnic of Milan. He was recalled to military service during the Second World War, and after the armistice treaty between Italy and the Allies in September 1943, he was stationed in Switzerland (in Münchenbuchsee). There he was in contact with members from Italian anti-fascist groups, among them with Ernesto Rossi, Filippo Sacchi and Luigi Einaudi. He delivered courses in mathematics, either at the University of Lausanne or University of Neuchatel.[78] Returning to Italy in 1945, this episode shows Dedò's disagreement with the fascist regime, in contrast to Chisini. Between 1953 and 1966 he was a professor at the *Accademia Aeronautica di Nisida* (near Naples), where he concentrated more on the didactics of mathematics;[79] this might be a reason why he wrote only one paper on the "characteristic braid". In 1966 he returned to the University of Milan. While in the earlier stages of his career he had investigated topics of classical algebraic geometry, he is now mainly known for his contributions to mathematics education.[80]

---

[74] Private communication with Maria Dedò (Email from 18.1.2020).

[75] Israel (2004, p. 31).

[76] Cf. Dedò (1950, p. 227).

[77] See Friedman (2019b).

[78] See: Manara (1993, p. 416). I thank Maria Dedò for highlighting this.

[79] His two volumes of *Matematiche Elementari dal punto di vista superior* are well known in this respect.

[80] For further and detailed biographical information, see (Manara 1993; Tibiletti 1999).

To recall, Chisini sought to associate to every (possibly singular) algebraic curve a 'characteristic bundle': collection of braids, which would describe what are the singularities and ramification points and how they are located. However Chisini did not consider any notation for these braids—and he hence did not use the notation that was used by Emil Artil at that time or Artin's presentation of the braid group (see the Appendix at Sect. 4.4). A possible notation, which Chisini did acknowledge, came with the article published in 1950 by Dedò, called "Algebra of Characteristic Braids" ["Algebra delle trecce caratteristiche"]. While Dedò presented Chisini's methods at the beginning of his paper, and mentioned explicitly his material models, he stated that the goal is "*to found an algebra which replaces these models*".[81] Given an algebraic curve and its 'characteristic braid', this algebra is enabled by notating the various braids obtained by following the various preimages along loops on the $x$-axis, which encircle branch points of the given curve. However, recalling that higher algebra was not a part of the Italian curriculum during the 1940s and 1950s, Dedò might have adopted Chisini's approach of not trusting it, and one may ask what is actually meant by "algebra" and which image did Dedò (or his colleagues) have of it. As we will see, algebraic proofs were hardly (or not at all) attempted in Dedò's paper, though one has to recall that it was Dedò who changed the terminology: from "bundle" to "braid"; this might indicate that he was aware of the acceptance of Artin's algebraic treatment of braids, hence wishing to show the relevance of his own research to this field of braid theory.

How this algebra is developed is presented in the first half of the paper. Dedò termed several braids as "canonical braids", denoting them as $i\,j$ or $j\,i$, for $i < j$; these braids are braids for which the $i^{th}$ string interchanges with the $j^{th}$ string, while going above it, and the other strings continue in straight lines, going above $i$ or $j$ if necessary. The notations are usually accompanied with diagrams—see for example the canonical braids $1\,2$ and $2\,4$ in Fig. 4.6.

Relations between these braids (for example, commutativity) are proven—notwithstanding Dedò's earlier statement—using a material model of strings. Moreover, it should be mentioned that it is clear that Dedò knew Artin's work. However, when Dedò referred to it, he also noted, that his braids are "more particular",[82] and hence one needs a special notation; a similar claim was expressed by Chisini in 1952, being the first time that Artin's name appears in Chisini's writings.[83] Moreover, Dedò emphasized, as we will see later, that his research concentrates on the factorizations of braids, an object which Artin did not consider. This shows that while Dedò was aware of Artin's work, he also wished to differentiate between his notation (and research) and Artin's.

---

[81] Dedò (1950, p. 228): "*Scopo di questa nota e quello di fondare un'algebra che sostituisca il maneggio di questi modelli.*"

[82] Dedò (1950, p. 228).

[83] Chisini (1952b, p. 19).

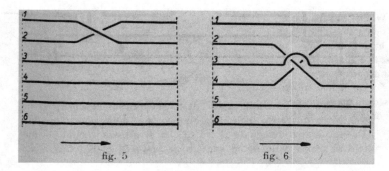

**Fig. 4.6** Dedò's notation of the "canonical braid" *i j*: on the left side the braid *1 2* is presented, on the right the braid *2 4* (Dedò 1950, p. 235)

**Fig. 4.7** An "elementary" braid
*i j(k)* (Dedò 1950, p. 245)

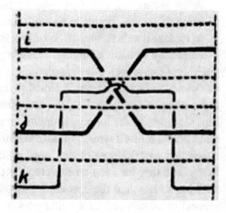

As became clear for Dedò, his initial notation was not enough to deal with all the different braids, which might appear when treating singular curves. Thus he introduced a new type of braid, notated as *i j(k)* and called an "elementary braid", to denote a braid whose strings *i* and *j* interchange, when another string *k* (when $k > j$ or $k < i$; recall that $i > j$ in any case) is intertwined between the two but returns to its original place afterwards;[84] see Fig. 4.7 for an example of such an "elementary" braid (Dedò presented several braids which would fulfill the conditions between *i, j* and *k*).

---

[84] If $i < k < j$, than the obtained braid would be a canonical one, denoted as *i j*.

**Fig. 4.8** While Chisini did not draw any figure illustrating as to how deformations of loops might look like, illustrations of such deformations were common. For example, Enriques drew the second transformation in 1923 (Enriques 1923, p. 191, Fig. 3)

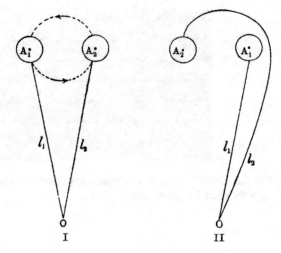

After presenting this new type of braid, the focus of Dedò's paper shifts from the notation of specific braids to an exploration of the "characteristic braid" as an object—that is, to look at it as a factorization of braids. In order to do that, Dedò returned to the setting presented by Chisini: a singular curve is given, projected to the $x$-axis, when a set of loops encircling the branch points is also given; thus obtaining with it the factorization of braids associated to this curve. Dedò aimed to find out what happens to this factorization once one deforms the set of loops on the $x$-axis (from which one obtains the various braids); see e.g. Fig. 4.1a, where the loop $\gamma_1$ is drawn on the $x$-axis.[85] Looking at two consecutive loops, it was Chisini himself who emphasized two types of basic deformations (see Fig. 4.8): "[I]t is noted that each change of the loop system is obtained by exchanges of two consecutive loops, and that for such an exchange two cases occur [. . .] according to whether the first remains as it is and the second is passed from left to right of this, or, vice versa, let the second stand still and let the first pass from the right to the left of it."[86] Fig. 4.8 illustrates the first type of this deformation.

However, Chisini was not interested in formulating an "algebra" of these deformations, a task that was fulfilled by Dedò, when he explored the algebraic relations between these transformations of the set of loops. Notating these two types of deformations presented by Chisini either by $P$ (the first case) or by $S$ (the second case), Dedò remarked that these notations are "symbolical descriptions".[87] The question arises, what happens to the induced braids as a result of these deformations. Dedò emphasized that the effect of these

---

[85] The question hence arises what happens when one deforms the loops $\gamma_1$, $\gamma_2$ etc.

[86] Chisini (1937, p. 60): "Qui si noti che ogni cambiamento del sistema di cappi si ottiene mediante scambi di due cappi consecutivi, e che per un tale scambio si presentano due casi (come è noto) secondo che si lasci fermo il primo e si faccia passare il secondo dalla sinistra alla destra di questo, o, viceversa, si lasci fermo il secondo e si faccia passare il primo dalla destra alla sinistra di questo."

[87] Dedò (1950, p. 232).

transformations on the resulting braids (see also below) are not only best seen with his notation, but actually are allowed by it:

"We will often need to work [operate] with $P$ and $S$ on the two consecutive sections [*tratti*; braids] $a$ and $b$: the operation $P$ will be indicated by

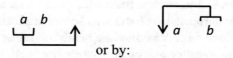

or by:

The previous notation will also allow us to indicate a sequence of operations $P$ and $S$. Thus, for example, the [following way of] writing [stands for consecutive operations of $P$ and $S$]";[88] Dedò drew immediately afterwards the following:

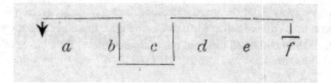

One should note here that Dedò's notation was also a notation which could be manipulated: the above drawing of a lettered sequence of braids, which are transferred via arrows from one side to the other, appears often in the work of Dedò (and also after, in the work of Tibiletti) and can be also considered as a dynamic diagram. In this sense, Dedò implicitly demanded from the reader to imagine a manipulation of the notation, hence charging it with a diagrammatic character.

What are therefore the operations $P$ and $S$ on the level of the factorization? the operation $P$ is considered by Dedò as an operation on a *factorization* of (two) braids, being the following: Given two braids, $a$ and $b$, one considers their product: $a \cdot b$. The operation $P$ transfers the braid $a$ to the right, and hence by operating $P$ on the product, one obtains the following product: $aba^{-1} \cdot a$ (whose end result is as the original product); one may then examine each of the factors by itself. In the same way, the operation $S$ sends the product $a \cdot b$ to $b \cdot b^{-1}ab$. Thus, the focus is now set on the operations $P$ and $S$ themselves and on *factorizations* of braids (and not on the braids as such).[89]

---

[88] Ibid.: "Ci occorrerà spesso di operare con le $P$ ed. $S$ sui due tratti consecutivi $a$ e $b$: l'operazione $P$ verrà indicata [...] La notazione precedente ci permetterà pure di indicare una successione di operazioni $P$ ed. $S$. Cosi ad esempio la scrittura [...]."

[89] Dedò, for example, proved the following theorem: assuming that the numeration of the strings $i, j, k$ is cyclical (e.g. when $i < j < k$), then "applying twice the operation $P$ on the pair on canonical [braids] $i j$ and $i k$ gives the same result as the operation S. With these hypotheses one has that $P^2 = S$, and since, in any case, $P = S^{-1}$, one obtains the relations $P^3 = S^3 = 1$ [and] $S^2 = P''$ (ibid., p. 239).

Concerning Chisini's results from 1937, Dedò remarked that one may carry them out "in a purely formal way, without resorting to the help of material models and without invoking the properties of the represented algebraic curve."[90] Several examples are given, showing how this may be done. The first example that is given is the deformation of a curve of degree 3. Looking at these curves, Dedò started from the characteristic braid corresponding to a smooth cubic curve, a characteristic braid which Chisini already calculated; then by using his theorems about the operations $P$ and $S$, and commutation relations between braids of the form $i\,j$, he obtained other 'characteristic braids' corresponding to *singular* cubic curves. This was done by operating *only* on the factorization itself, without proving that the resulting curve was actually analytically deformable into a smooth curve. In fact, it seems that the implicit assumption was that any sequence of algebraic operations on the factorization corresponds to an analytic deformation.

The second example concerns branch curves, which already delineates a trajectory for future research, which would be done later by Tibiletti. Dedò first of all noted that an exponent 2 (respectively 3) of the braid corresponds to a node (respectively to a cusp), as one has the same braid drawn twice (or thrice), one after the other. He then presented the following factorization and braid diagram (as in figure below) for a curve of degree 4 with 3 cusps:[91]

$$1\,2^3 \quad 2\,3(1) \quad 1\,3^3 \quad 3\,4(1) \quad 2\,3^3 \quad 2\,4.$$

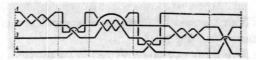

With this example, Dedò noted that it is easy to verify that this curve is a branch curve of a cubic surface (with a double line). Why is that? The explanation that is given is that one can associate to every string a transposition in $Sym_3$ (the symmetric group of three elements), which would describe the transposition of the sheets of the surface, such that one can verify, with the help of the diagram of the characteristic braid, that these transpositions would satisfy the "invariance conditions of Enriques".[92] What is implicitly described here is the map sending loops encircling the curve $C$ to the symmetric group, when $C$ is a curve of degree 4 with 3 cusps. This verification would be a proof that the curve is indeed a branch curve. Here it is essential to stress that Dedò was aware of Zariski's research on the fundamental group: at the end of his paper, Dedò cited Zariski's paper from

---

[90] Ibid., p. 240: "Come si vedrà è possibile effettuare questo passaggio in modo puramente formale, senza ricorrere all'aiuto di modelli materiali e senza invocare proprietà della, curva algebrica rappresentata."

[91] Ibid., p. 250.

[92] Ibid.

1929, underlining that Zariski's research on branch curve was done "through the analysis of the group of Poincaré."[93] Explicitly, Dedò was aware that one can present the above map as an epimorphism from "the group of Poincaré", i.e. the fundamental group $\pi_1(\mathbb{C}^2 - C)$ to the symmetric group $Sym_3$; however, any mention of such epimorphism was absent from Dedò's account, and group-theoretical considerations or any other mention of the fundamental group were not to be found. This shows that while Dedò aimed to found an algebra, the usage of group theory was not considered as a practice to be included in this image of algebra.

Another example is presented at the end of the paper:[94] Dedò computed the characteristic braid associated to the branch curve of a cubic surface under a generic projection. This branch curve is a sextic curve with six cusps, all of which lie on a conic; Dedò cited Zariski's conclusions (see Sect. 3.2.1) and noted "the equation of such a sextic is of the form $p^3 + \lambda q^2 = 0$, [. . .] when $p$, $q$ are two polynomials of the second and the third degree respectively".[95] Dedò computed the characteristic braid of the sextic curve with 6 cusps on a conic by performing the operations $P$ and $S$ on the second 'canonical form', which Chisini had computed in 1937, that is, on the characteristic braid associated to the intersection of 6 lines (intersecting at 15 nodes). The associated braid to this sextic is presented in Fig. 4.9. Also with this example one obtains the impression that with the help of a proper usage of the operations $P$ and $S$, one may obtain the characteristic braid associated to any singular curve, when operating on a well chosen, already known characteristic braid of another (perhaps simpler) curve with the same degree. Also here, Dedò noted that one can associate to the six strings six transpositions in the group $Sym_3$, and therefore "on a material model or with a drawing of braids one can verify that the conditions of invariance of Enriques are satisfied."[96] The question however remains, why operating the operations $P$ and $S$ is even allowed analytically—i.e. can an arrangement of 6 generic lines be deformed into a sextic curve with 6 cusps on a conic?

Before discussing the work of Tibiletti, it should be stressed that Dedò's paper aimed to found a new mathematical configuration for treating branch curves via factorizations of braids. His emphasis on symbolical reasoning was certainly a deviation from Chisini's mistrust in such reasoning, and one may claim that Dedò stressed the need to consider such 'characteristic bundles' or 'braids' algebraically and not via material models. Nevertheless, his image of algebra was somewhat limited, as any concrete usage of group theory was in practice inexistent.

---

[93] Ibid., p. 258, footnote 19.

[94] Ibid., p. 254–258.

[95] Ibid., p. 254.

[96] Ibid., p. 258: "Sul modello materiale o sul disegno della treccia si può verificare che sono soddisfatte *le condizioni di invarianza* di Enriques."

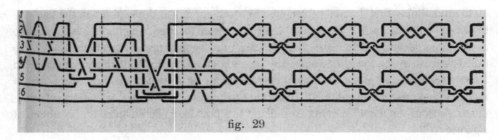

fig. 29

**Fig. 4.9** The braid diagram of the branch curve of degree 6 with 6 cusps, all of them lie on a conic (Dedò 1950, p. 257)

## 4.2.2  Tibiletti and the Second 'Theorem' of Chisini

How the work of Dedò influenced Chisini and his way of treating branch curves is to be seen with the work of his student Cesarina Tibiletti (1920–2005), who graduated in 1943. From 1950 till 1959 Tibiletti was a professor in Milan in charge of the course "Elementary mathematics from a superior point of view". Furthermore, in 1954 she obtained her teaching qualification in algebraic geometry. In 1955 she married her colleague Ermanno Marchionna (1921–1993), who was also a student of Chisini. Marchionna worked on similar subjects; during the early 1950s, he also researched branch curves; he graduated in 1944 and while in the post-war period he continued to work with Chisini on related problems arising from Chisini's research,[97] he turned during the 1950s to other subjects within algebraic geometry, such as linear systems of hypersurfaces, the Riemann-Roch theorem, or the arithmetic structures of the finite rings.[98]

In 1962 Tibiletti became a full professor in Milan. Apart from working on algebraic geometry and later on algebra (concentrating on this subject starting at the end of the 1950s), she also wrote several studies on the history of mathematics.[99] Just like the other students of Chisini, from the 1960s onwards she turned to research other subjects, different from ramified covers of the plane—another indication of the decline of the classical approach towards algebraic geometry. Notwithstanding, it is essential to examine her research, as it was leading to what may be called later the second Chisini's 'conjecture'.

In 1952 Tibiletti published two papers, which indicate how influential Dedò's and Chisini's methods were. The first paper, on "an a-priori"[100] construction of a sextic curve with nine cusps, presents two main results: the first, the construction of the characteristic braid of this sextic; the second, an explicit presentation of one of Chisini's

---

[97] Under the guidance of Chisini, and along the lines of research of B. Segre, Marchionna researched linear equivalence between different points of branch curves, for various surfaces and different projections (e.g. projection from an $n$-fold point on a surface) see: Marchionna (1950, 1951).

[98] See: Manara (1995).

[99] For a more detailed biography, see: Linguerri (2012).

[100] Tibiletti (1952a, p. 207).

**Fig. 4.10** Tibiletti's diagram of the "characteristic braid" of a regeneration of a conic with a tangent into a curve of degree 4 with three cusps and a simple branch point; only the diagram of the resulting curve is drawn (Tibiletti 1952a, p. 219, Fig. 8)

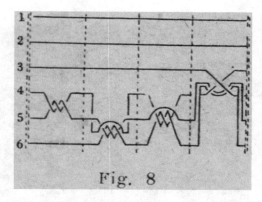

degeneration rules on the level of the characteristic braid diagram: how, *locally* a conic with a tangent is the limit form of a curve with three cusps and a simple branch point (see Fig. 4.10).

The paper begins with acknowledging the limitations of Chisini's method: "Given an algebraic curve it is particularly impossible to construct the characteristic braid of it [. . .]. Therefore, all the a priori (indirect) constructions of algebraic braids are of interest".[101] Hence, Tibiletti suggested, following Chisini, to work with degenerated curves. Moreover, concerning the first result, Tibiletti expressed explicitly what was implicit in Dedò's paper. She noted that before obtaining the characteristic braid of the sextic that "without under-taking a proof of this effective correspondence between topological and algebraic operations, we will ensure the result with an a-priori construction, starting from a limiting case [. . .]."[102] In a footnote to this paragraph, she remarked "that topological procedures [the deformations of algebraic curves] [. . .] have always led to braids that have been verified to actually exist as algebraic braids; so this procedure could be considered as a *constructive postulate.*"[103] It could easily be suggested that Tibiletti might have also hinted at the opposite "constructive postulate": that every operation on the factorization of braids corresponds to a deformation of an algebraic curve; this is also practically implied, since Tibiletti started with a limiting case—being a sextic curve with nine *nodes* (starting, in fact, with one of Chisini's canonical forms)—and after several operations on the characteristic braid (that is, on the corresponding factorization), she obtained a characteristic braid of a sextic curve with nine *cusps*. This curve, Tibiletti emphasized, is of a special importance, as

---

[101] Ibid.: "Data una curva algebrica è praticamente impossibile costruirne la treccia caratteristica [. . .]. Pertanto hanno interesse tutte le costruzioni a priori (indirette) di trecce algebriche [. . .]."

[102] Ibid., p. 213: "Senza intraprendere una dimostrazione su tale effettiva corrispondenza fra operazioni topologiche e algebriche, ci assicureremo del risultato con una costruzione apriori, a partire da un caso limite [...]."

[103] Ibid., p. 213, footnote 2: "Notiamo però che procedimenti topologici di questa natura hanno sempre condotto a trecce che sono state verificate effettivamente esistenti come trecce algebriche; cosi questo procedimento potrebbe essere considerato come un postulato costruttivo."

a

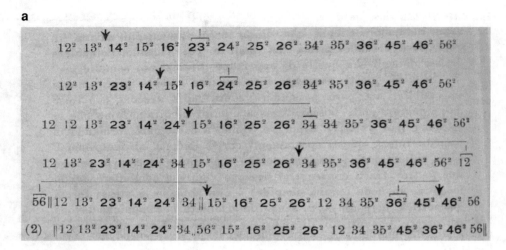

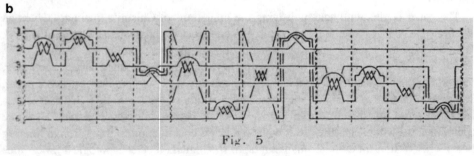

Fig. 5

**Fig. 4.11** (**a**) Several operations (*P* and *S*) done on the factorization, starting with factorization corresponding to a sextic with nine nodes (the bold letters "represent the actual nodes" (Tibiletti 1952a, p. 213) of the curve). Figure taken from (ibid, p. 214). (**b**) The characteristic braid diagram of a sextic with nine cusps (ibid., p. 215)

the sextic curve with 9 cusps is the branch curve of several coverings of degree 3,[104] as Chisini already indicated in 1944 (see Sect. 4.1.3). See Fig. 4.11a for several steps in Tibiletti's argument, being diagrammatic-symbolical, in the sense of Dedò (see above, Sect. 4.2.1), based on a moving, manipulable notation, whereas the end result (in Fig. 4.11b) is presented as a diagram.

The second paper of Tibiletti from 1952 deals with an explicit construction of the diagrams of the characteristic braids of several branch curves. Here Tibiletti used the method of Chisini: taking a degenerated curve—in this case, configurations of a conic with 1, 2 or 3 tangent lines—and using her former result, on how *locally* a conic with a tangent line is the limit form of a curve with three cusps (and a simple branch point; see

---

[104] Ibid., p. 208.

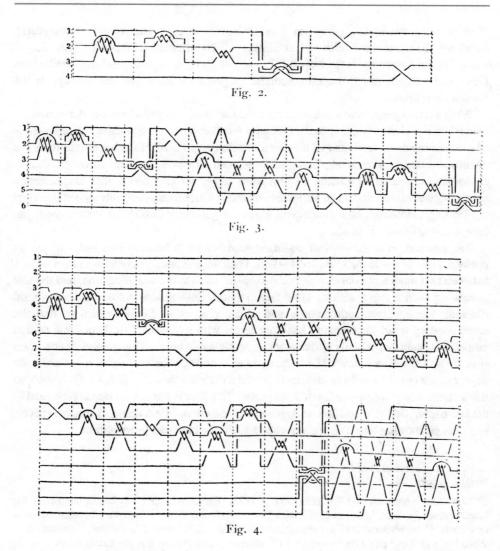

Fig. 2.

Fig. 3.

Fig. 4.

**Fig. 4.12** Three braid diagrams corresponding to three branch curves obtained from a limiting curve, being a conic with 1, 2 or 3 tangents (Tibiletti 1952b, p. 542)

Fig. 4.10), she depicted several characteristic braid diagrams. She obtained diagrams of the characteristic braids of the following curves: a curve of degree 4 with 3 cusps (which was already obtained by Dedò); a curve of degree 6 with 6 cusps and 4 nodes; and a curve of degree 8 with 9 cusps and 12 nodes (see Fig. 4.12). Tibiletti then noted explicitly that these curves are branch curves,[105] by associating to each string a transposition in the

---

[105]Tibiletti (1952b, p. 541–542).

corresponding symmetric group (in $Sym_n$, where $n = 3$, $4$ and $n = 5$ respectively), satisfying "Enriques's conditions of invariance."[106] To formulate it in modern terminology, what Tibiletti described, is the map $\pi_1(\mathbb{C}^2 - C) \to Sym_n$, though she did not, similarly as Dedò and Chisini, use this group-theoretic language or employed the concept of the fundamental group.

What the two papers show is not only a strong reliance on symbolical and diagrammatic reasoning, but also an implicit belief that with the diagrams, one can show the uniqueness of certain branch curves—that is, that for a certain (large) class of branch curves there are no other, non-isomorphic algebraic surfaces with the same branch curve.[107] This is obviously influenced by Chisini's claim from 1944, but also shows that diagrammatic reasoning was considered as an acceptable method of argumentation, although at that time the leading mathematicians developing algebraic geometry outside of Italy were using completely different methods.

The reliance on these ways of argumentation is also to be seen in a series of papers presented by Chisini in the years 1954–1955. These papers attempt to create an equivalence between two modes of considering nodal-cuspidal curves: the analytical one, and the one presenting such a curve with its braid factorization. Chisini conjectured that *every* braid diagram, locally representing cusps and nodes (i.e. using Dedò's notation, that the corresponding braid factorization has only terms with exponent 2 or 3) under a certain numerical condition, actually corresponds to an algebraic curve. This numerical condition is stated by Chisini as follows: If $d$ is the number of nodes, $c$ the number of cusps and $n$ the degree of the curve, then these invariants should satisfy $d + 2c + 3 < \frac{1}{2} n(n + 3)$. A proof to this statement is given in a series of three papers,[108] whereas the crucial step is presented in the last paper. This step consists of degenerating the curve into a union of simpler curves, and then performing a new operation, called $F$, on the obtained factorization:

---

[106] Ibid., p. 538.

[107] This belief is expressed by Tibiletti (1955, p. 183): "The knowledge of the braids [associated to] branch curves ([called:] braids of ramification [*trecce di diramazione*]) is important because the braid of a curve $C$, to whose strings are associated the transpositions related to the determinations [i.e. sheets] of a multiple plane (branched over $C$), represents in a unique way the family of surfaces $F$, birationally identical, which give the same multiple plane and clearly indicates for these $F$ many algebraic and topological properties. If the curve $C$ is a branch curve for a single multiple plane (what presumably happens for $n > 4$, and it is however easily verifiable with the braid), only the braid of $C$ is enough to represent the above mentioned family of surfaces $F$." ["La conoscenza delle trecce delle curve di diramazione (trecce di diramazione) risulta importante perché la treccia di una curva $C$, sui cui fili siano deposti gli scambi relativi alle determinazioni di un piano multiplo (diramato da $C$), rappresenta in modo univoco tutta la famiglia di superficie $F$, birazionalmente identiche, che danno lo stesso piano multiplo e indica chiaramente di queste $F$ molte proprietà algebriche e topologiche. Se poi la $C$ è di diramazione per un solo piano multiplo (ciò che presumibilmente accade per $n > 4$, ed. è comunque facilmente verificabile sulla treccia) la sola treccia di $C$ basta a rappresentare la suddetta famiglia di superficie $F$."]

[108] Chisini (1954–1955).

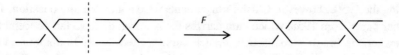

**Fig. 4.13** The operation $F$ transforms ("fuses", in Chisini's terminology) on the level of the braid diagram two branch points into a node. © Graphics: M.F

> "When we consider the braid as representative of a curve which can be varied, it happens that two consecutive braids merge, because the two corresponding branch points have come to coincide [...]. On the braid [diagram] this operation is obtained simply by removing the diaphragm that separates the two consecutive braids. This operation (extendable in the case of three consecutive [simple] braids, which occurs when a cusp originates), will be called *fusion* and will be symbolled by $F$."[109]

While Chisini did not depict this operation $F$ on any diagram of a factorization of braids, it is fairly easy to draw what $F$ performs, as depicted in Fig. 4.13: The operation consists of "removing the diaphragm" in the diagram itself, corresponding to the degeneration of the curve $xy = \varepsilon$ (having two branch points, when $\varepsilon > 0$) to the curve $xy = 0$ (having one node); but while the operation is local, Chisini did not note anything as to whether globally it was possible.

At this point Chisini considered a particular degeneration of the curve into simpler curves, concluding that "given the canonical braid of an algebraic plane curve of order $n$, we operate on this one by operations $P$, $S$, $F$ [...] [resulting in a] new braid with $d$ braids, which represent double points, and $k$ braids, which represent cusps; under the [above numerical] condition [...] the obtained braid is representative of a curve, which actually exists" of order $n$ with $d$ nodes and $k$ cusps.[110] However, two questions concerning the various analytical deformations of the curve arise: is the initial degeneration always possible? And is the operation $F$, that is, the 'fusion' always possible? Chisini assumes that these operations can always be done. However, as was seen already (Sect. 3.2.3), when in 1937 Zariski worked within a similar context—of 'smoothing' out a cusp of a curve—he emphasized that this can be done only when the various singular points are independent of each other.[111] Assuming that those arguments were known when Chisini employed his 'fusion' operation, one may wonder why Chisini did not pay attention to these restrictions.

---

[109] Ibid., p. 9–10: "Quando si considera la treccia rappresentativa di una curva variabile, accade che due tratti consecutivi vengano a fondersi, per il fatto che i due punti di diramazione corrispondenti sono venuti a coincidere, e cosi si sono riuniti in uno solo i due cappi relativi [...]. Sulla treccia tale operazione si ottiene semplicemente togliendo il diaframma che separa i due tratti consecutivi. Tale operazione (estendibile al caso di tre tratti consecutivi, cosa che avviene quando si origina una cuspide), verrà chiamata fusione e indicata col. simbolo $F$."

[110] Ibid., p. 13: "[...] la treccia cosi ottenuta è rappresentativa di una curva $Q_n$ (ancora d'ordine $n$) effettivamente esistente."

[111] Zariski (1937, p. 356). More precisely, Zariski checks that the induced relations in the fundamental group of the complement are group-theoretically independent.

During the 1950s, however, Chisini's arguments were not called into question. Tibiletti, in another paper from 1955, which summarizes the research done so far, concentrating on the characteristic braids associated to branch curves, cited the 1954 *theorem* of Chisini without questioning the validity of the action $F$.[112] Indeed, as noted above, from a local point of view (i.e. in a neighborhood of the singular point) the fusion was indeed possible, but from a global point of view the very same fusion was not always possible—and indeed, as we will see in Sect. 5.4.1, Boris Moishezon disproved this theorem in 1994.

## 4.3    Conclusion: Seclusion, Ignorance and Abandonment

While Chisini's results and the work of his students were known to a certain extent in Italy, they were barely known in the wider global context. The only manuscript to be found outside of Italy on the work of Chisini's group, by a member who was *not* Chisini's student, was a summary of two lectures given in Paris on November 26th and December third 1956 by Paulette Lévy-Bruhl in the *Séminaire Dubreil* titled "Algèbre et théorie des nombres." Lévy-Bruhl also used Dedò's notation and presented several of the diagrams used by Dedò and Tibiletti.[113] One of her talks was on "Plans multiples et tresses algébriques." However, in perusing the list of lectures given during this seminar throughout the 1956/57 academic year, one notes that Lévy-Bruhl's lecture was almost the only one focusing on classical algebraic geometry, whereas the other talks given at this seminar— by, among others, Wolfgang Krull and Zariski—concentrated mostly on algebra (most of the talks were on semi-groups, rings, ideals, modules), and several of the lectures dealt with algebraic curves and surfaces, fibered spaces or Chern classes. In contrast to Lévy-Bruhl's talks, which were held at the beginning of the seminar, Krull focused on a structural approach to algebra.[114] Zariski delivered his talk toward the end of the lecture series; he spoke about "Le problème des modèles minimum pour les surfaces algébriques," obviously representing a more modern approach to surfaces than that presented by Lévy-Bruhl. Moreover, apart from Lévy-Bruhl's talk, not a single mathematician used Dedò's notation during or after the 1960s.

---

[112]Tibiletti (1955, p. 199). It is interesting to note that in this paper Tibiletti focused mainly on notational reasoning, to prove, using the operations $P$ and $S$, equivalences of factorizations. However, there is a lack of a general (algebraic) *theory* of factorizations, beyond the various particular different case studies (cf. ibid., p. 187–189). What one finds is an emphasis on verification of simple procedures and claims with material models or with diagrams (ibid., p. 188, 205). For more elaborate complicated calculations, concerning branch curves of surfaces with isolated nodes, Tibiletti relies only on the factorization and the various operations ($S$, $P$ and $F$), and not on the resulting braid diagram (ibid., p. 210–213).

[113]Lévy-Bruhl (1956–1957a, 1956–1957b).

[114]Cf. Corry (2004, p. 263–268).

It is clear that the group concentrated around Chisini (Dedò and Tibiletti, as well as Manara and Marchionna) in the 1950s continued to use the practices that were customary in the 1930s: diagrammatic-material (and visual) reasoning, avoidance of algebra and a choice not to make use of other traditions or developments in algebraic geometry. One can suggest that the unique notation that they used was prompting a consolidation of their research practices. However, along with the fact that use of the notation itself was not evident to employ—at least when compared to Artin's notation of braids[115]—it was somewhat cumbersome. Moreover, Chisini's conjecture from 1944 and his 'theorem' from 1954 (which was eventually disproved in 1994) both employed fallible techniques, relying on assumptions and constructions whose correctness was unclear. Those factors led to the decline of this practice. Here we might also suggest that the self-isolation of the Italian mathematical community continued implicitly, at least during the first years after the war.

Nevertheless one cannot ignore the fact that Chisini and his students opened a new epistemic configuration with its own space of inscription. The novel notation emerged due to a conscious decision not to use Artin's notation,[116] and from a belief that one needs a special notation to describe the characteristic braid. Moreover, the braid diagrams of curves, and especially of branch curves, were considered the way to verify the invariance condition of Enriques; use of *only* Artin's notation (and no diagrams) would not have been suitable for that, as was argued by Dedò and Chisini.

The 'characteristic braid' of the branch curve also shows how it was used in the framework of new approaches to investigating surfaces—that is, the degeneration of curves. Once it was clear that one cannot obtain the 'characteristic braid' directly, one had to turn to "a-priori" methods, as described by Tibiletti. Investigating the branch curve with degenerations offered other configurations, and this approach was indeed was taken by d'Orgeval and Zappa; whereas Zappa (also) continued Zariski's line of investigation, d'Orgeval emphasized the limits of the method of degeneration—limits which Chisini himself eventually also acknowledged.

## 4.4   Appendix to Chap. 4: A Short Introduction to the Braid Group

This appendix aims to give a short introduction to the braid group, concentrating first on how Emil Artin researched braids during the 1920s.[117] The braid group—whose elements are possibly intertwined $n$-strings, and the defined action on them is concatenation—was officially considered a mathematical object to be investigated with the 1926 publication of

---

[115] Which was during the 1940s and the 1950s much more disseminated; see (Friedman 2019c).

[116] By "conscious decision" I mean that both Dedò and Chisini were aware of Artin's notation, but were at the same time critical of it; see: (Dedò 1950, p. 228; Chisini 1952b, p. 19).

[117] For a historical overview, see: (Friedman 2019c). However, note also that Hurwitz considered braids in 1891 as (in *modern terminology*) fundamental groups of configuration spaces of $n$ points in the plane (resp. on the complex line). See: (Epple 1999, p. 183–192).

**Fig. 4.14** Artin's drawing of a concatenation of two braids, depicting also what is a braid

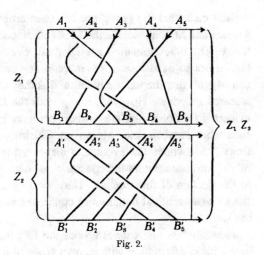

Fig. 2.

Emil Artin's paper "Theorie der Zöpfe". Artin's paper aimed at an algebraic treatment of this set of braids,[118] and his goal was to achieve a complete algebraic formalization of the set, giving it a structure of a group. In the course of the twentieth century, several definitions and representations were given to the braid group: as a group of $n$ strings together with their concatenation (as presented in 1926 by Artin), as the fundamental group of the unordered configuration space of the real plane, as homeomorphisms of the punctured disc up to isotopy, or Artin's representation of braids as automorphisms of the free group. In the following, for reasons of space, I will concentrate only on two definitions: (1) Artin's definition from 1926 and (2) braids as homeomorphisms of the punctured disc.[119] This Appendix hence does not aim under no circumstances to be exhaustive when presenting braid theory, and concentrates only on the themes and definitions appearing in this book.

**(1) Braids as strings:** If one follows Artin's paper from 1926, then braids as non-intersecting 'strings', connecting a set of $n$ points on a given segment with a another set of $n$ points on a parallel segment, "give rise to a group";[120] Artin noted that "schematically, a braid can [...] be represented by a drawing".[121] The drawings supplied, drawn by Artin, facilitate an understanding of what a braid is and what the composition of braids looks like in this group (see Fig. 4.14).

Artin defined the braid group of $n$-strings as generated by $n - 1$ generators, notated by $\sigma_i$, for $i = 1, \ldots, n - 1$, when $\sigma_i$ is the braid for which the $i^{\text{th}}$ string passes once *above* the $(i + 1)^{\text{th}}$ string, whereas the other strings are straight lines (see Fig. 4.15a).

---

[118] Artin also returned to work on the braid group between 1947 and 1950; see: Artin (1947, 1950).

[119] For an overview of all of the definitions, see e.g. (González-Meneses 2011).

[120] Artin (1926, p. 47).

[121] Ibid., p. 48.

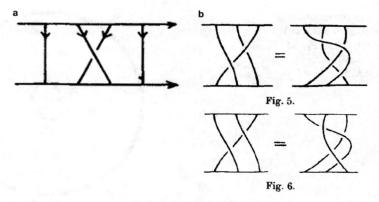

**Fig. 4.15** (**a**) A drawing of the generator $\sigma_i$ (Artin 1926, p. 49). (**b**) Inducing the relation $\sigma_i\sigma_{i+1}\sigma_i = \sigma_{i+1}\sigma_i\sigma_{i+1}$. (Artin 1926, 51)

The way Artin notated the generators of the braid group helped him to present the relations between them. Explicitly, the relations are:

$$\sigma_i\sigma_k = \sigma_k\sigma_i \text{ for } k \neq i, i+1 \text{ and } \sigma_i\sigma_{i+1}\sigma_i = \sigma_{i+1}\sigma_i\sigma_{i+1} \text{ for } i = 1, 2, \ldots n-2.$$

Artin's proof of these relations is visual. He, for example, explicated that one can "*extract from the [following] figures*" the relation $\sigma_{i+1}^{\pm1}\sigma_i = \sigma_i^{\mp1}\sigma_{i+1}\sigma_i^{\pm1}\sigma_{i+1}^{\pm1}$, from which he induced the relation $\sigma_i\sigma_{i+1}\sigma_i = \sigma_{i+1}\sigma_i\sigma_{i+1}$. The proof relies on the diagrams depicting (see Fig. 4.15b) what happens when one 'shifts' a crossing $\sigma_i$ from one side of the crossing $\sigma_{i+1}$ to the other. As we saw above, Chisini and Dedò did not use this presentation or these generators, and Dedò explicitly aimed to develop his own notation for braids. However, other mathematicians discussed in this book: Ron Livne and Boris Moishezon (see Chap. 5) did use Artin's definition when investigating braid factorizations associated to branch curves, and also made use of properties of the *center* of this group, which I will now present.

Explicitly, using Artin's presentation one can also find properties and generators of special subgroups of the braid group, such as the center. The *center* $Z(G)$ of a group $G$ is the set of elements that commute with every element of it, or: $Z(G) = \{x \in G : \forall g \in G, xg = gx\}$. It was proved by Frank Arnold Garside at the end of the 1960s that for the braid group $B_{n+1}$ (generated by the $n$ braids $\sigma_1, \ldots, \sigma_n$), the center of $B_{n+1}$, when $n > 1$, is generated by $\Delta^2$, when $\Delta = \Pi_n \cdot \ldots \cdot \Pi_1$, and when $\Pi_r \doteq \sigma_1 \cdot \ldots \cdot \sigma_r$. Hence, the elements in $Z(B_{n+1})$ are powers of $\Delta^2$. For example, for $n = 2$, i.e. for the braid group with three strings $B_3$, one obtains that $\Delta^2 = (x_1x_2 \cdot x_1)^2$.

**(2) Braids as homeomorphisms of the punctured disc:** Braids can also be defined as an automorphism of the punctured disc up to isotopy. This definition was employed by Moishezon in several of his papers, and hence it is presented here as well.

**Fig. 4.16** The automorphism $f_i$ of $D_n$, consisting of a clockwise 180-degree rotation of $P_i$ and $P_{i+1}$ inside a small disc. © Graphics: M.F.

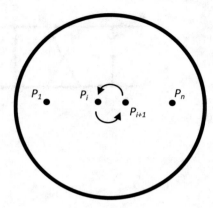

To be more precise, let $D^2$ be a disc in the real plane and $D_n = D^2 - \{P_1, \ldots, P_n\}$ be the punctured disc (i.e. $D^2$ minus $n$ distinct points). Consider now the set of orientation preserving homeomorphisms from $D_n$ to itself, fixing the boundary pointwise, when two homeomorphisms are equivalent if they can be transformed into the other by a continuous deformation. With this definition, the braid group is isomorphic to the group of such homeomorphisms (up to isotopy), with the action of composition of homeomorphisms. If one considers the points $P_1$, ..., $P_n$ arranged on a straight line, then according to this definition, the generator $\sigma_i$ (in Artin's definition) can be defined as the homeomorphism $f_i$ of $D_n$ to itself which consists of a clockwise 180-degree rotation of $P_i$ and $P_{i+1}$ inside a small disc surrounding only these two points, when $f_i$ outside this disc (and on its border) is the identity homeomorphism (see Fig. 4.16).

# From the 1970s Onward: The Rise of Braid Monodromy Factorization

**5**

As already indicated, from the late 1930s onward, algebraic geometry underwent a complete reformulation, when it was realized that the 'classical' language was inadequate and that a thorough rewriting in algebraic terms was needed. In Sect. 3.3, we saw how Zariski and Weil considered their own rewriting projects with commutative algebra.[1] However, after the reshaping of algebraic geometry in terms of the language of commutative algebra, in the 1950s and 1960s, algebraic geometry was reshaped again via the language of category theory, schemes and sheaves;[2] the classification of surfaces project was extended to surfaces over fields with characteristic $p > 0$, and a new classification of surfaces slowly emerged (that of Enriques–Kodaira), extending the existing classification of Castelnuovo and Enriques to non-algebraic compact complex surfaces, now classifying surfaces according to the Kodaira dimension with the help of Chern classes. The research on coverings was generalized: a definition of ramified covers was first given in algebraic terms (see Zariski, Sects. 1.2 and 3.3) and later formulated in terms of schemes (with

---

[1] Dieudonné, in *History of algebraic geometry* (1985, p. 59–90, esp. p. 64–67), describes this period as the epoch spanning from the 1920s till 1950s, which presented "new structures in algebraic geometry" (ibid., p. 59). On how geometry, algebra and algebraic geometry were transformed in the USA during this period, see: (Parshall 2022, p. 232–286). The need to rewrite the foundations of algebraic geometry can be seen, for example, in two monographs which addressed this question: André Weil's *Foundations of algebraic geometry* (1946) and Alexandre Grothendieck's *Éléments de géométrie algébrique* (from 1960 till 1967), when the latter work rewirtes algebraic geometry using schemes and sheaves. On Zariski's work, see (Parikh 2009); On Zariski's arithmetization of algebraic geometry, see (Slembek 2002).

[2] On how category theory was employed in algebraic geometry, especially in Grothendieck's work, see: Krömer (2007, p. 161–188), McLarty (2016). On Grothendieck's work on the fundamental group, see: Murre (2014); on how Grothedieck transformed the concept of topological space to topos, see: Bélanger (2010); on Grothendieck and his works, see the contributions in: Zalamea (2020), Zalamea (2019), Schneps (2014) and the review of Gray (2015a).

© The Author(s), under exclusive license to Springer Nature Switzerland AG 2022
M. Friedman, *Ramified Surfaces*, Frontiers in the History of Science,
https://doi.org/10.1007/978-3-031-05720-5_5

Grothendieck); the *algebraic* fundamental group was defined and the relations between this group and the *topological* fundamental group became clearer and clearer.[3] The ramification (and branch) varieties (or schemes) were defined with the terms of this new language,[4] which did not resemble Zariski's algebraic definition from his 1958 paper. But one cannot say that these new definitions, results and revolutions reshaped the research on branch curves. Indeed, in 1971, in a new appendix to the second edition of Zariski's book *Algebraic Surfaces*, David Mumford notes that the "classification of plane curves $C$ with $d$ nodes and $k$ cusps and the computation of the fundamental group $\pi_1(\mathbb{CP}^2 - C)$ has unfortunately not been pursued."[5] Recalling that the classification of these curves was one of Zariski's main research themes in the 1920s, one may be tempted to conclude that the new techniques of algebraic geometry prompted stagnation in the research on branch curves,[6] or simply, that the theme of branch curves was no longer considered epistemically, as something from which new research directions could emerge.

This chapter deals with the renewed interest that did take place, starting in the 1970s, thanks to Boris Moishezon. Section 5.1 reviews the developments during the 1960s and 1970s concerning the branch curve and how research on this subject almost disappeared, with practically the only counter-example to this situation being a paper by Jonathan M. Wahl from 1974 on the moduli space of nodal–cuspidal curves, which posits branch curves in a new epistemic configuration. I also discuss shortly in Sect. 5.1.1 the work of Shreeram S. Abhyankar on ramified covers with characteristic $p > 0$ in the late 1950s and 1960s, a research configuration that shows the similarities and differences of the employed methods and obtained results compared to those used for ramified covers with characteristic 0. Consequently, Sect. 5.1.2 surveys the second edition of Zariski's *Algebraic Surfaces*, analyzing how branch curves were treated in this new edition.

---

[3] See: Dieudonné (1985, p. 143): With the Italian school of algebraic geometry, "the concept of *ramified covering* [...] had a predominantly topological flavor, and when one associated a group to a covering, it was usually by the standard composition of loops and the use of homotopy. After 1950, and the introduction of the new concepts of 'abstract' algebraic geometry over an arbitrary field, it was necessary to give a purely algebraic foundation to these notions. The first steps in this direction were taken by Lang, Serre, and most actively by Abhyankar [...]." Under the notions given a "purely algebraic foundation", Dieudonné lists the notions of covering, branch locus, algebraic and tame fundamental group.

[4] This rewriting starts in the 1960s, with Grothendieck's SGA1. Explicitly, see the definition from 1984, presented at Sect. 1.1 from: (Barth/Peters/Van de Ven 1984, p. 41). To recall, for a covering $p : X \to Y$, the ramification divisor $R$ is the zero divisor of the canonical section in $\text{Hom}(p^*(\mathcal{K}_Y), \mathcal{K}_X)$, and since $\mathcal{K}_V = \bigwedge^n T_V^\vee$ (for a variety $V$), when a global section of this section is everywhere-regular differential form, one can prove that in the affine case, the definition can be reduced to Zariski's equivalent definition from 1958 of ramification variety (see Sects. 1.2 and 3.3).

[5] Zariski (1971, p. 231).

[6] This is to be seen also in (Popp 1970, p. 148), who asks for a further investigation of the structure of $\pi_1(\mathbb{CP}^2 - C)$ for $C$ a nodal-cuspidal curve.

Section 5.2, the main section of this chapter, deals with the work of Boris Moishezon and Ron Livne on factorizations during the 1970s, a work that signaled the opening of a new research configuration on branch curves with new tools. We already saw this with Chisini and Dedò and their own work on factorizations, but it was Livne and Moishezon who considered these factorizations as an object of research on their own; their work on factorizations was much more algebraic, and also did not only concentrate on complex plane curves or branch curves, but also on other types of factorizations. Section 5.3 concentrates on Moishezon, his time in the USSR and in Israel, and his research during the 1980s on 'braid monodromy factorization' associated to a branch curve. This section highlights the social conditions surrounding Moishezon, Shafarevich and the Jews emigrating from the USSR, concentrating on how scientists and mathematicians emigrated to Israel and how Moishezon was integrated at Tel Aviv University. My claim is that these social circumstances and factors are part of how the epistemic configuration was shaped and formed, and they are therefore elaborated in this section.

The last section, Sect. 5.4, reviews the work of Moishezon together with Mina Teicher in the 1980s and 1990s. It shows how renewed research on the structure of the fundamental group $\pi_1(\mathbb{CP}^2 - B)$, $B$ being a branch curve, prompted two new research configurations: first, finding the initial counter-examples of algebraic surfaces, whose non-existence was conjectured by what was known as the Bogomolov 'Watershed Conjecture'; second, finding new results concerning families of degeneratable surfaces. Moishezon and Teicher's work also showed that the braid monodromy factorization and the fundamental group functioned differently when situated in different epistemic configurations: sometimes as a tool, sometimes as an object of research. The chapter ends with a discussion on one result ensuing from the research on factorizations: Moishezon's refutation of Chisini's 'theorem' from 1954–55, and the split between the different types of factorizations that this refutation revealed.

## 5.1   The 1960s: Generalization and Stagnation or the "Rising Sea" and the Sunken Branch Curves

At the turn of the 1960s a series of papers and manuscripts was published, concentrating on more general results related to complex ramified coverings, which in turn also reframed the research on branch curves and coverings of the complex projective plane. However, one may ask whether this led to new results concerning branch curves, a question I will discuss shortly below. Zariski's famous result from 1958 on the purity of the branch locus (see Sect. 3.3) is certainly one of these results, whereas Hans Grauert's and Reinhold Remmert's article "Komplexe Räume" from the same year generalized the Riemann-Enriques existence theorem[7] of covering from the case of a covering of the projective line (resp. plane) to

---

[7] See Sect. 3.2.1: To recall, these two theorems (of Riemann and Enriques) present the necessary and sufficient conditions for the existence of a complex algebraic curve resp. surface, given a map associating loops around the branch points (resp. curve) to certain permutations.

an arbitrary complex variety.[8] The paper obviously mentions Riemann, and it is highly general in its settings, but notes the "classical algebraic geometry",[9] in order to exemplify, that while already this classical approach did handle ramification of algebraic varieties, in contrast to that, when considering complex manifolds,[10] ramification points were "excluded from any consideration".[11] Indeed, as Kiyoshi Oka commented in 1951: "We know almost nothing about the internally branched domains";[12] here Oka means singular varieties, and not smooth varieties, which are coverings of other smooth varieties. During the 1950s Grauert and Remmert, Heinrich Behnke, Karl Stein and Henri Cartan all dealt with the problem of finding a revised definition of the complex manifold, revising the definition of a Riemann surface for higher dimensions; however, they did not consider the case of complex varieties of dimension 2 as a special case;[13] indeed, "Komplexe Räume" does not mention these varieties, but rather deals with covering and their ramification for any dimension.

Alexander Grothendieck and Grauert met in 1957 and began to exchange their ideas about the necessity of permitting "nilpotent elements if one wants to utilize the full power of the methods of algebraic and analytic geometry",[14] at the same time the article "Komplexe Räume" was published. While Grauert worked on the analytic side, Grothendieck invented new techniques for algebraic geometry. Moreover, Grothendieck, during the years 1959–1961—with his novel concepts of *revêtements étales* and his categorical understanding[15]—suggested considering coverings over algebraically closed fields with characteristic $p > 0$ and their associated covering over an algebraically closed field with characteristic zero. This approach was also influenced by the work of Abhyankar

---

[8] Grauert / Remmert (1958). One of their results was that every finite, analytically ramified covering of a complex manifold is a normal complex space.

[9] Ibid., p. 246.

[10] That is, a manifold with an atlas of charts to the open unit disk in $\mathbb{C}^n$, when the transition maps are holomorphic.

[11] Ibid., p. 245: "Eine erste Beschreibung dieser sog. Riemannschen Gebiete wurde 1932 von H. Cartan und P. Thullen gegeben wurden; jedoch vorerst alle Verzweigungspunkte von der Betrachtung ausgeschlossen."

[12] Oka (1951, p. 204–205): "On ne sait presque rien sur les domaines intérieurement ramifiés." Forty years later after the publication of "Komplexe Räume", Remmert (1998, p. 227) explains this ignorance as follows: "A systematic study of singularities was started by Alfred Clebsch, Max Noether and Italian geometers in the last century. [...] When complex manifolds came into life it was clear from the very beginning that they were not general enough. The singularity of $w^2 - z_1 z_2 = 0$ at the origin shows that one has to admit spaces which locally are not even homeomorphic to an open set in $\mathbb{R}^n$. However singular points were not considered for a long time. When studying non-univalent domains over $\mathbb{C}^n$ in the [nineteen] thirties and forties, mathematicians excluded possible branching, because they were well aware of the mysteries lying hidden in the ramification points."

[13] For a historical overview, see Remmert (1998) or (Hartmann 2009, p. 151–155).

[14] Remmert (1995, p. 9).

[15] On these coverings, see: Zalamea (2020).

(see Sect. 5.1.1). It is in Grothedieck's *Séminaire de Géométrie Algébrique* 1 (SGA1), taking place in 1960–1961 that one notes another shift within the entire mathematical configuration of algebraic geometry, now using the language of schemes and categories.[16]

This highly abstract method is to be considered within the framework of Grothendieck's philosophy of mathematics presented in his work *Récoltes et semailles:* "the rising sea" in which the theorem is "submerged and dissolved by some more or less vast theory, going well beyond the results originally to be established".[17] Yet the question remains whether certain areas in this mathematical landscape—such as branch curves—were really ever fully submerged or if they actually did not return to the surface after having initially sunk.

Here I do not intend to survey in depth the transformations and revolutions that algebraic geometry went through during the 1960s. How this discipline was developed, especially under Grothendieck and his *Séminaire de Géométrie Algébrique du Bois Marie* is outside the scope of the book. This is the case for two reasons: first, already SGA1, published in 1971 under the title *Revêtements étales et groupe fundamental*, does not contain a single reference to branch curves or ramified surfaces—although the first pages of the book do deal with non-ramified covers.[18] This is not surprising, considering Grothendieck's style and his ambition to rewrite the foundations of algebraic geometry, in a language and setting as general as possible. That being said, this in no way implies that the members of the seminar were unaware of the results obtained regarding surfaces, branch curves or ramified covers. For example, Zariski's result regarding the purity of the branch locus is mentioned and the following comment is added: "Zariski's results [are] demonstrated topologically. [These] are far from having been assimilated by 'abstract' algebraic geometry and deserve new efforts."[19] The reference given to Zariski's results is chapter VIII of *Algebraic Surfaces*, which, to recall, deals with branch curves. Second, diving into the work of Grothendieck, even only in this context of ramified and non-ramified covers, would force us to make a detour, which has a far too large circumference for this book. It would force us to concentrate on techniques and concepts that, in the final analysis, were hardly used to investigate branch curves. I will however refer to several writings during the 1960s and the 1970s, which nevertheless referred to branch curves explicitly.

---

[16] In *SGA 1*, Grothendieck (1971, p. 2) defines the a non-ramified point using the language of schemes: "Given $X$ and $Y$ preschemes, $f : X \to Y$ a morphism of finite type, $x \in X$, $y = f(x)$. The following conditions are equivalent [and in this case the morphism $f$ is *non ramified* at $x$]:

(i) $\mathcal{O}_x / \mathfrak{m}_y \mathcal{O}_x$ is an finite, separable extension of $k(y)$.

(ii) $\Omega^1_{X/Y}$ is zero at $x$

(iii) The diagonal morphism $\Delta_{X/Y}$ is an open immersion at the neighborhood of $x$."

Compare also the definition in *EGA IV* (Grothedieck 1967, p. 65).

[17] Grothendieck (1985–1987, p. 552–3, p. 555). On Grothendieck's image of mathematics, see: McLarty (2007); the translation of the above citation is taken from: ibid., p. 302.

[18] See for example, Grothendieck (1971, p. 1–6).

[19] Ibid., p. 278: "[. . .] les résultats de Zariski [sont] démontrés par voie topologique. Ces derniers sont loin d'avoir été assimilés par la géométrie algébrique 'abstraite' et méritent de nouveaux efforts."

One of the talks at the *Séminaire Bourbaki*, which did indicate that the participants of the seminar were aware of former research concerning branch curves and ramified covers of the projective plane was the talk of Jean-Pierre Serre, held in May 1960, called "Ramified covers of the projective plane" ["Revêtements ramifiés du plan projectif"]. Serre begins the paper by recalling the "theory of existence of Riemann-Enriques-...".[20] He does so by indicating that this theory shows that "in a pictorial fashion, any topological covering of an algebraic variety $V$ is algebraic."[21] He then notes that one does not suppose that $V$ is projective, and remarks what Zariski already noted in *Algebraic Surfaces* regarding the projective plane: if one is interested in coverings of $W - F$, when $W$ is a projective variety and $F$ is a sub-variety, then "the subgroup of finite index of connected coverings $W' \to W$ which are non-ramified outside of $F$ correspond bijectively to subgroups of finite index of $\pi_1(W - F)$."[22] When $W$ is the projective plane, Serre notes, this is "the theory of existence [...] of Enriques".[23] For a general $V$, the theorem was proved by Grothendieck and his theory of descent, making use of the results of Grauert-Remmert for normal varieties. While situating Riemann and Enriques within the classical, and Grauert-Remmert and Grothendieck within the modern research on covers, several actors (e.g. Segre or Chisini's students) are forgotten.

Serre also notes that there is an "interest (also from the algebraic point of view) of determining of group $\pi_1(V)$". While this determination "is easy when $dim(V)$ is 1, it is not at all like that for dimension 2. The case most studied is when $V = \mathbb{CP}^2 - C$, $C$ being a plane curve."[24] Serre then refers to the general method of van Kampen and Zariski to find this group, and also mentions Chisini for finding "several particular case studies".[25] However, the focus of the talk is not an introduction of the results of Zariski or of some of the members of the Italian school, but rather a presentation of analog results for normal varieties over a closed field of any characteristics, following the results of Abhyankar (see below). As Serre underlines afterwards, at this setting one works with the algebraic fundamental group, and not with the topological one.[26]

---

[20] Serre (1960, p. 483): "commençons par rappeler l'énonce du 'théorème d'existence' de Riemann-Enriques..."

[21] Ibid.: "De façon plus imagée: tout revêtement topologique d'une variété algébrique est algébrique."

[22] Ibid. This is a reference to the theorem from the theory of topological coverings in algebraic topology. To present this theory in a modern language, given a path-connected, locally path-connected space $Y$, there is a one-to-one correspondence between (normal) subgroups $G'$ of $G = \pi_1(Y)$ and (normal) path-connected coverings $Y'$ of $Y$ such that $\pi_1(Y')$ is isomorphic to $G'$. Hence, normal subgroups of finite index correspond to finite coverings.

[23] While this is a reference to Enriques's invariance rules, it is essential to stress that Enriques did not deal with the fundamental group; this was, in fact, Zariski's critique on Enriques. See Sect. 3.2.1.

[24] Ibid.

[25] What these case studies are goes unmentioned.

[26] Ibid., p. 485.

### 5.1.1   Detour: End of the 1950s: Abhyankar's Conjecture

Abhyankar's work on ramified covers in characteristic $p > 0$ in the late 1950s and 1960s was fundamental for understating the similarities and the differences of ramified coverings over the complex numbers compared to the ramified coverings in characteristic $p > 0$, and his two conjectures, made in 1957, are in that sense a milestone in this research.[27] These differences can be already detected when considering covers of the affine line. Over the complex numbers, the line is simply connected, which means that it has no nontrivial unramified covers.[28] But over a closed field of characteristic $p > 0$ non-trivial non-ramified coverings do exist, as one can take for example the theory Artin-Schreier coverings, developed during the 1920s. One of the simplest examples is of the curve $y^p - y - x = 0$, being a cover of degree $p$ of the affine line, which is on the one hand not trivial, but on the other hand unramified: the derivative: $\frac{d}{dy}(y^p - y - x) = py^{p-1} - 1 = -1$ never vanishes.[29]

One of the questions that arose is hence the structure of the fundamental group of the affine line in characteristic $p$. Abhyankar's conjectures, solved by Raynaud and Harbater, describe the fundamental group of the affine line in positive characteristics. Here however one has to note, as Abhyankar did, that the notion of the fundamental group must be revised: Over closed fields of characteristic $p > 0$ one cannot use the same definition of the topological fundamental group (as the group of equivalent classes of loops). If, however, one notes that for covering spaces, the fundamental group is the group of deck transformations of the universal covering space,[30] then an analogue can be found. More precisely, since algebraic varieties often do not have a universal cover, "one works with the full inverse system of finite étale covers $Y \to X$ [...].[31] The automorphism group of this inverse system is then called the étale [or algebraic] fundamental group."[32] This group being the *algebraic* fundamental group, thus, according to Abhyankar this is a "good algebraic approximation to [the topological fundamental group] $\pi_1$".[33]

---

[27] Abhyankar (1957).

[28] This follows from the following theorem from algebraic topology (see footnote 21 above): as mentioned above, given a path-connected, locally path-connected topological space $Y$, then for every normal subgroup $G'$ of $G = \pi_1(Y)$ there corresponds a covering $Y'$ of $Y$ such that $\pi_1(Y')$ is isomorphic to $G'$. In our case, if $Y$ is the affine complex line, then $\pi_1(Y)$ is trivial; hence there are no nontrivial coverings, since there are no subgroups of $\pi_1(Y)$.

[29] Since the calculations are done in a field of characteristic $p$, $py = 0$ in this field.

[30] Given a covering $p: C \to X$, a *deck transformation* is an automorphism $f: C \to C$ such that $p \circ f = p$. A covering space is a *universal* cover if it is simply connected; this means that if $Y$ is the universal covering of $X$, then $\pi_1(X)$ is isomorphic to the group of the deck transformations of $Y$.

[31] The thorough investigation of étale covers was conducted in SGA 1, which was taught during 1960–1961 (Grothendieck 1971).

[32] Harbater et al. (2018, p. 244).

[33] Abhyankar (1957, p. 839).

While the precise formulation of Abhyankar's conjectures is outside the scope of this book,[34] Harbater et al. formulate these two conjectures as follows: "The first [conjecture], which limits the possible phenomena, is that the theory of covers of varieties should be the same in characteristic $p$ as in characteristic 0, after eliminating the prime $p$ from the situation. The second, which pulls in the opposite direction, is a kind of Murphy's Law: whatever can happen in characteristic $p$, will happen."[35] Already these conjectures, which dealt only with the case of curves, show the results regarding ramified and unramified coverings of the affine line in characteristic 0 cannot be automatically transferred to characteristic $p$. At the end of the 1950s, Abhyankar continued his investigation to the case of surfaces and branch curves, in a series of six papers called "Tame Coverings and Fundamental Groups of Algebraic Varieties", where he, among others, generalized some of Zariski's results on branch curves for fields of characteristic $p$.[36] An indication that Abhyankar's results were considered as essential concerning the state of the art with respect to branch curves of surfaces over fields with characteristic $p > 0$ can be seen in the new edition of Zariski's *Algebraic Surfaces* from 1971 (see Sect. 5.1.2), where Abhyankar wrote the first appendix to Chapter VIII: the chapter on branch curves.[37] The appendix is a summary of results from the end of the 1950s and beginning of the 1960s. Abhyankar introduced the notions of algebraic fundamental group and tame fundamental group, but he notes that a "reason why one would be particularly interested in the more restricted object $\pi$' [the tame fundamental group] is that in general, for $p \neq 0$, the structure of $\pi$ [the algebraic fundamental group] turns out to be very complex and quite unanalogous to that of $\pi_1$ [the topological fundamental group]."[38] The tame fundamental group, while presenting the similarities when one works with parts of coverings "prime to $p$",[39] hence shows the difficulties one encountered of investigating the structure of the *algebraic* fundamental group of the complement of branch curves for fields of characteristic $p > 0$.

---

[34] For the precise formulation, see (ibid., p. 840ff).

[35] Harbater et al. (2018, p. 240). A short historical survey of these conjectures is given in (ibid., p. 242–245)

[36] See for example Abhyankar (1959, 1960).

[37] Zariski (1971, p. 224–228). David Mumford wrote the second Appendix to this chapter (ibid., p. 229–231).

[38] Ibid., p. 225.

[39] For the exact definition of the tame fundamental group, see e.g.: Ibid., p. 224. Cf. also Harbater et al. (2018, p. 240): "Abhyankar's philosophy says that the Galois groups of order prime to $p$ over $U$ [$U$ is $X - B$, when $X$ is a $X$ is a smooth projective curve and $B$ a set of distinct $r$ points] should be the same as those that arise in the analogous situation for complex algebraic curves, and that this should be the only constraint on the types of Galois groups that arise."

## 5.1.2 1971: The New Edition of Zariski's *Algebraic Surfaces*

As was indicated above, Mumford still noted in 1971 that the classification of nodal-cuspidal curves and the investigation of the corresponding fundamental group of the complement are still not sufficiently understood. The same therefore applied concerning the research of branch curves, a research that was experiencing stagnation. Indeed, the research on branch curves over fields of characteristics 0 belonged, one might suggest, to classical algebraic geometry, whereas the research during the 1950s and the 1960s aimed to rewrite classical algebraic geometry: the configuration initiated mainly by Abhyankar concentrated on coverings over fields of characteristics $p > 0$, whereas Grothendieck examined schemes over commutative rings (and not necessarily over surfaces).

In 1971 a new edition of Zariski's *Algebraic Surfaces* was issued with numerous appendices. David Mumford, who was the "chief editor of the appendices",[40] apologized in his preface to the new edition for the "deficiencies of our contributions. Is any potential reader skilled enough to be familiar with all the diverse foundations and abstract tools referred to in these appendices, patient enough to unwind the tangled relationships between old and new lines of argument [...]?"[41] But while this apology highlights the "tangled relationships" between the various languages and rewritings of algebraic geometry, it also stresses that the "Italian school, judged by its own standard, had completed a mature theory of algebraic surfaces." This might imply that these standards were not only unclear within the new configurations of algebraic geometry, but also were not rigorous enough. Hence Mumford noted that one of his aims to "is to clarify the connection between the modern and the Italian terminology and between essentially equivalent modern and Italian theorems." According to Mumford, some chapters are hard for the modern reader to follow, since "to the Italians, a surface was essentially a birational equivalence class and the models used were almost always (non-normal) surfaces in $\mathbb{CP}^3$; whereas today two surfaces are thought to be 'the same' only if they are biregularly equivalent (i.e. isomorphic as schemes) [...]."[42] Hence, according to Mumford this was the motive for a new edition: before embarking on modern research of branch curves, one has to reformulate the classical results in modern language.

An example of how this could be done is to be found in the appendix to chapter V of Zariski's book, which deals with the problem of moduli of surfaces, whose formula, as we saw in Sect. 3.1.3, Enriques deduced using (also) invariants of the branch curve. Mumford

---

[40] Zariski (1971, p. vi). Besides Mumford, who wrote the appendices to chapters III, IV, V, VI, VII, Joseph Lipman and Abhyankar wrote appendices to chapter II resp. chapter VIII.

[41] Ibid.

[42] Ibid. See Dieudonné (1985, p. 52), who criticizes heavily the classical Italian geometers: "Italian geometers [were] [...] working only with positive divisors and most often with surfaces embedded in $\mathbb{CP}^3$, impose limitations on themselves that lead to considerable complications in their definitions and techniques. Unfortunately, the very widespread tendency of this school to lack precision in definitions and proofs [...]." Cf. the critics of Zariski and Weil in Sect. 3.3.

reformulates the formula of Enriques, now in terms of cohomology sequences and the dimensions of different cohomology groups,[43] but afterwards notes that while

> "the classical and modern theory tie up perfectly in regards to $1^{st}$ order deformations [...] Zariski points out [in the 1935 edition *Algebraic Surfaces*], the Italians never proved that $1^{st}$ order deformations either of non-singular curves on an irregular surface or of surfaces in $\mathbb{CP}^3$ with ordinary singularities, or of plane curves with nodes and cusps, could be realized as the $1^{st}$ order terms of actual deformations, i.e., that in each case 'the characteristic system is complete'. In Grothendieck's language, this is the question of whether the deformation problem is *unobstructed*, or whether the universal deformation space [...] is non-singular. [...] even in char $= 0$, obstructions can exist."[44]

Moreover, Mumford added later, regarding the question, whether the characteristic series of families of nodal-cuspidal curves is complete, that "[o]ne would expect that the branch curves to suitably generic projections onto $\mathbb{CP}^2$ of surfaces with obstructed moduli [...] would provide counter-examples. However, this should be looked into."[45] Similar to Zariski in 1935, the goal of Mumford was—using the re-writing of the question with the new language, with its resituating in another epistemic configuration (for example, using the language of cohomology, of scheme theory, or in "Grothendieck's language")—to "point out clearly the main gaps" in the former configuration.[46] Mumford's appendix to chapter VIII, in which he noted the stagnation of the research of nodal-cuspidal curves, also points out that branch curves may at least serve to disprove Enriques's assumption about the completeness of the characteristic series for nodal-cuspidal curves. And indeed, in 1974, Jonathan M. Wahl found a series of nodal-cuspidal curves, all of which are branch curves, whose characteristic series is not complete.[47] The first curve in this series is a branch curve of degree 104, with 900 cusps and 3636 nodes. In order to disprove Enriques's assumption, Wahl employed tools from scheme and category theory as well as infinitesimal deformation theory—obviously techniques that were not available during the first half of the twentieth century.[48]

---

[43] Ibid., p. 126–128.

[44] Ibid., p. 128.

[45] Ibid., p. 231. See footnote 48 below for Wahl's explanation, why one 'expected' this phenomenon from branch curves.

[46] Ibid., p. vii.

[47] Wahl (1974, p. 573): The examples Wahl found are "contrary to an assumption of Enriques."

[48] To see this, it is instructive to note how Wahl describes the procedure of finding his example, at the beginning of the paper. In 1962, in his paper "Further Pathologies in Algebraic Geometry", "Mumford found a family of non-singular space curves whose local moduli space [...] is generically nonreduced. Kodaira then suggested that if an irreducible surface in $\mathbb{CP}^3$ has ordinary singularities along one of these curves, and if a 'generic projection' of that surface were made onto $\mathbb{CP}^2$, then the branch curve should have only ordinary nodes and cusps as singularities, but a singular local moduli space. In this paper, we provide such an example [...]." (ibid., p. 530). Whether the moduli space is

However, also in this work of Wahl, branch curves were only serving as an example to prove another result regarding moduli spaces (or, alternatively, to disprove the assumption made by Enriques). These curves were not taken as an object of research by themselves. Moreover, "Chisini's school" (as d'Orgeval termed it) and its ideas regarding the investigation of factorization of braids as a way to 'represent' the branch curve, were either forgotten or not taken any more into account within newer epistemic configurations. However, starting 1975, another approach concerning branch curves, which did posit these curves at its center, began to emerge; its origins are to be found in the works of Boris Moishezon and Ron Livne.

## 5.2  The 1970s: Livne and Moishezon on Equivalence of Factorizations

As we will see in the following sections, Boris Moishezon revived the interest in factorizations of braids as a way to characterize complex plane curves, and especially branch curves. I will discuss Moishezon's work and life in more detail in Sect. 5.3, but the question that already arises concerns how he came to research this idea. Additionally, were complex plane curves, or more particularly, branch curves, the only motivation to consider factorizations as an algebro-geometric object? The answer to the latter question, as I will show in this section, goes beyond a simple 'yes' or 'no'. However, already here one must emphasize that 'revive' is used here retroactively: Moishezon and Livne were not aware in the 1970s of the research done by Chisini, Dedò and Tibiletti on these factorizations; hence one may consider the epistemic configuration established by Moishezon and Livne a new one. Only during the late 1980s did Moishzon become aware of Chisini's work (see Sect. 5.3.3.2). But before reviewing Moishezon's work, I would like to concentrate on the Master thesis o Livne.

In 1975 Ron Livne published his thesis "Rigidity of the centre of $B_3$", when $B_3$ denotes the braid group of three strings. Livne was at that time a MA student of Moishezon at Tel Aviv University, Israel. The thesis shows, as I will elaborate, that factorizations of elements in the braid group interested Moishezon in general, and not necessarily as a mere characterization of curves. Since this work marks the beginning of a renewed thinking on factorizations and on branch curves and the founding of a new epistemic configuration, a close examination of it is essential.

---

non-reduced or singular was either impossible to formulate (due to lack of algebraic techniques) or to prove with the techniques of the Italian school.

## 5.2.1   Livne's MA Thesis from 1975

As we saw in Sect. 3.2.1, Zariski proved in 1929, based on unproved assumptions, that the fundamental group of the complement of a nodal plane curve is abelian. Fulton and Deligne correctly proved the theorem in 1980, but obviously in 1975 these proofs were as yet unknown. Livne stated in the introduction to the 1975 thesis that Zariski's unproved theorem drove his motivation. What Livne proposed in the introduction is to "try a topological approach that leads, in turn, to algebraic problems concerning Artin's braid groups [. . .]. The work starts here with the first non-trivial case: a non-singular cubic, i.e. an elliptic curve."[49]

 The choice of a non-singular cubic was not explained in the thesis. However, a reconstruction of the explanation is possible. Let us recall the construction of Chisini presented in Sect. 4.1.1: a non-singular complex cubic curve in the complex projective plane $\mathbb{C}^2$ has, when projected generically to a projective line $l$, six branch points: $a_1, \ldots, a_6$.[50] Hence one can consider the following map (being in fact a homomorphism of groups):

$$\pi_1(l - \{a_1, \ldots, a_6\}) \rightarrow B_3,$$

sending a loop in the complex projective line, which encircles one (or several) of the branch points (being an element in $\pi_1(l - \{a_1, \ldots, a_6\})$), to the corresponding braid with three strings (i.e. an element in the braid group $B_3$), a braid which described the motion of those three preimages of the loop. This map was already presented by Chisini (see e.g. Fig. 4.1), but not in the context of group theory, and Chisini did not use the terms 'homomorphism', 'group' or 'fundamental group'. Nevertheless Chisini already noted, that a loop encircling all of the branch points can be decomposed as a concatenation of six loops (each loop encircling a single branch point), hence the corresponding element—denoted as $\Delta^2$—in the braid group of three strings $B_3$ can be decomposed into a factorization of six elements. It is on this backdrop that one can understand Livne's following remark: "The topological equivalent of the algebraic theorem [Zariski's theorem] here is that if a surface [curve] in $\mathbb{C}^2$ has a 'generic' projection (in the algebro geometric sense) then any such two are isotropic. This amounts to proving a theorem that states that all representations of $\Delta^2$ as a product of conjugates of the generator are equivalent with respect to some relation."[51] This means that if one wants to prove Zariski's theorem, one possible way would be to prove it via the equivalence of factorizations, since isotopic curves induce equivalent factorizations (and

---

[49] Livne (1975, Introduction)

[50] This was already known by Salmon (see Sect. 2.2): a non-singular complex (projective) plane curve of degree $n$ has $n(n-1)$ branch points when projected generically to the complex line; the branch points are projection of the intersection of the curve with its polar curve which is of degree $n-1$. If $n = 3$, as in the example in Livne's thesis, then $n(n-1) = 6$.

[51] Ibid. Though Livne wrote "surface" in the introduction, one may assume that he meant to write "curve", since no complex surface can be embedded in $\mathbb{C}^2$ except $\mathbb{C}^2$ itself.

vice versa). That is, once proving Zariski's theorem for one non-singular (or nodal cubic plane) curve, it would be then possible to prove it for all non-singular (or nodal cubic plane) curves; and from that, one obtains that once proving it for one factorization, it would be possible to prove Zariski's theorem for all equivalent factorizations.

To understand Livne's formulations and results, a short mathematical detour is needed. The *center* of a group is the set of elements that commute with every element of it. Frank Arnold Garside proved in 1965 that for the braid group $B_{n+1}$ of $n + 1$ strings, generated by the $n$ braids $x_1, \ldots, x_n,$[52] the center of $B_{n+1}$, when $n > 1$, is generated by $\Delta^2$, when $\Delta = \Pi_n \cdot \ldots \cdot \Pi_1$, when $\Pi_r \doteq x_1 \cdot \ldots \cdot x_r.$[53] Hence for $n = 2$, i.e. for the group $B_3$, one obtains that $\Delta^2 = (x_1 x_2 \cdot x_1)^2 = x_1 x_2 x_1 x_1 x_2 x_1 = x_1 x_2 x_1 x_2 x_1 x_2 = (x_1 x_2)^3.$[54] Denoting, as Livne does, $x_1 = x$, $x_2 = y$, the element $\Delta^2 = (xy)^3 = xyxyxy$ is the generator of the center of $B_3$. Livne noted that "this theorem"—and here Livne meant the theorem of Zariski for the cubic non-singular curve[55]—"can be expressed as a rigidity theorem on the generator of the centre of $B_3$, or [...] as rigidity of 1 in $\mathbb{Z}/2\mathbb{Z} * \mathbb{Z}/3\mathbb{Z}$."[56] I will return below to the notion of rigidity, but one may assume that when examining the factorization of $\Delta^2$ as $xyxyxy$, already in 1975 Livne and Moishezon may have considered the equivalence of factorizations of elements in the braid group as a way to examine isotopy of two curves. Livne's thesis approaches the problem not by investigating the curve with, for example, analytical methods, but rather in an algebraic way, i.e. by examining the factorization.

* * *

To see this, an inquiry into the concept of rigidity is needed, to which Livne referred to in the introduction. The rest of the section will be however quite technical and the reader may hence jump to beginning of Sect. 5.2.2. Given a group $G$ and a set of elements in it, $x_1, \ldots, x_n$, Livne denoted their product by $A = x_1 \cdot \ldots \cdot x_n$. He noted: "We apply on the $n$-tuple $x_1, \ldots, x_n$ the following transformations":[57]

$$\sigma_i : \ldots x_i, x_{i+1} \ldots \rightarrow \ldots x_{i+1}, x_{i+1}^{-1} x_i x_{i+1} \ldots$$
$$\sigma_i^{-1} : \ldots x_i, x_{i+1} \ldots \rightarrow \ldots x_i x_{i+1} x_i^{-1}, x_i \ldots$$

---

[52] Here Garside uses Artin's presentation of the braid group, when $x_i$ is the $i^{th}$ generator (being the braid for which the $i^{th}$ string passes over the $i + 1^{th}$ string). See the Appendix of Chap. 4 (Sect. 4.4).

[53] Frank Arnold Garside (1915–1988), who served as Lord Mayor of Oxford in 1984–1985, examined in his 1965 thesis the lattice structure of the braid group (Garside 1965). This enabled him to solve the conjugacy problem for this group. In this context, he investigated the properties of the element $\Delta$.

[54] Garside (1969, p. 246). In the third equality one uses the relation of the braid group: $x_1 x_2 x_1 = x_2 x_1 x_2$.

[55] I.e. that the fundamental group of the complement of a nodal curve is abelian.

[56] Livne (1975, introduction).

[57] Ibid., p. 1.

Livne noted implicitly that there are $n$ transformations $\sigma_1, \ldots, \sigma_n$ and explicitly that between these transformations the relations of the braid group hold (that is $\sigma_i\sigma_j = \sigma_j\sigma_i$ for $|i - j| > 1$ and $\sigma_i\sigma_{i+1}\sigma_i = \sigma_{i+1}\sigma_i\sigma_{i+1}$). Applying these transformations "one gets different $n$-tuples $[\ldots] y_1, \ldots, y_n$. [For] [t]hese new ones $[\ldots]$ their product is [also] $A$ and $[\ldots]$ each $[y_i]$ is conjugate to some $x_i$". However, Livne did not mention that these operations on the $n$-tuples were first introduced by Hurwitz in 1891, when $G = Sym_n$ (the symmetric group);[58] one may assume however that Moishezon (and, hence, one may assume, Livne also) did know of Hurwitz's results.[59]

After defining the transformations $\sigma_i$, Livne presents the concept of *rigidity*: "given an $n$-tuple $y_1, \ldots, y_n$, can we come from it to $x_1, \ldots, x_n$ by the $\sigma_i$ and their inverses?"[60] If the answer is yes, a group is said to be *rigid* if every two equal factorizations are equivalent under the transformations $\sigma_1, \ldots, \sigma_n$ and their inverses. Livne then noted that Artin proved the rigidity of two equal factorization for the free group of $n$ generators. The question that stands at the center of the thesis is whether this property holds also for other groups, and especially for the braid group. Concentrating on the group $B_3$, as noted in the introduction of his thesis, the question is whether every two different, six-element factorizations of $\Delta^2$ are equivalent.

Livne examined first the quotient group $B_3/\Delta^2$, which is isomorphic to the free product of two free groups: $\mathbb{Z}/2\mathbb{Z} * \mathbb{Z}/3\mathbb{Z}$, when $\mathbb{Z}/3\mathbb{Z}$ is generated by $b$ and $\mathbb{Z}/2\mathbb{Z}$ by $a$.[61] He then proved that given a product of six elements $y_1, \ldots, y_6$ in $\mathbb{Z}/2\mathbb{Z} * \mathbb{Z}/3\mathbb{Z}$ such that $y_1 \cdot \ldots \cdot y_6 = 1$, then "by successive applications of $\sigma_i$ and their inverses, we can transform the $y$'s to the $x$'s such that all the $x$'s are short",[62] which means that $x$ is one of the following elements: $s_0, s_1$, or $s_2$, when $s_0 = a^2b$, $s_1 = aba$, and $s_2 = ba^2$. He showed that one can—by suitable transformations—"make all the $y$'s $[\ldots] s_1$ or $s_2$". But since the generators of $B_3$, denoted by $x$ and $y$, are sent to $s_1$ and $s_2$ under the map $B_3 \to B_3/\Delta^2 \cong \mathbb{Z}/2\mathbb{Z} * \mathbb{Z}/3\mathbb{Z}$, Livne proved that every product of six elements in $B_3$ which is equal to $\Delta^2$ can be transformed to the "'canonical' form $xyxyxy$",[63] thus proving rigidity. This would imply that every cubic non-singular curve is isotopic to each other, as above—though this claim is not expressed explicitly. And since it was known that for a smooth cubic curve $C$, $\pi_1(\mathbb{C}^2 - C)$ is abelian, the claim is indeed proven.

---

[58] Hurwitz (1891, p. 31).

[59] See: Moishezon/Teicher (1988, p. 433).

[60] Ibid.

[61] The group $\mathbb{Z}/n\mathbb{Z}$ is the set $\{0, 1, 2, \ldots n-1\}$ with the action + (mod $n$); For example, the group $\mathbb{Z}/3\mathbb{Z}$ is isomorphic to the set $\{0, 1, 2\}$ with the action + (mod 3), $\mathbb{Z}/2\mathbb{Z}$ is $\{0, 1\}$ with the action + (mod 2).

[62] Ibid., p. 4

[63] Ibid., p. 8.

## 5.2.2   Separations of Configurations and Shifts of Contexts

Livne got the problem from Boris Moishezon in the summer of 1973, but the Yom Kippur War (from October 6 to 25, 1973) caused the universities in Israel to be closed during the fall semester of 1973–74, and Livne, who entered military service in February 1974, could return to research only several months later. After submitting the thesis in June 1975, he hardly saw Moishezon anymore, and did not return to work these ideas.[64]

Therefore, one may assume that already in 1973 Moishezon and Livne started researching equivalence of curves (i.e. being isotopic) by means of factorizations. However, as the thesis shows, the proof of being isotopic was purely algebraic. Bringing any factorization of $\Delta^2$ of six elements in $B_3$ to the canonical form is not proved by analytical or topological means or with diagrammatical reasoning, as Dedò and Tibiletti did; moreover, how this proof of rigidity has consequences regarding the fundamental group of the complement of any cubic non-singular curve is not elaborated in the thesis. Indeed, in order to prove completely what was stated in the introduction, one had to state (or prove), that for at least one, generic non-singular cubic curve $C$, the fundamental group $\pi_1(\mathbb{C}^2 - C)$ is abelian, and also elaborate what is the topological meaning of the transformations $\sigma_i$ with respect to this group.

Livne also remarks[65] that he proved in an unpublished manuscript that for the braid group with 4 strings $B_4$, $B_4$/Center($B_4$) is isomorphic to a semi-direct product of $B_3$ and the free group $F_2$;[66] hence "the result for $B_3$ gives automatically something [. . .][i.e. a similar result] for $B_4$."[67] This implies that Livne, and hence probably Moishezon as well, may already have considered generalizing the result of braid group with *any* number of strings, and hence proving Zariski's theorem for *any* nodal plane curve, i.e. of any degree; Livne noted however in his thesis: "the difficulty [in this method of investigating factorizations] lies in the relations holding in the braid group modulo its center when expressed as a free product $\mathbb{Z}/n\mathbb{Z} * \mathbb{Z}/(n + 1)\mathbb{Z}$ with some relations."[68]

Both the (published) result on $B_3$ and the (unpublished) one on $B_4$ use in course of their proof *algebraic* properties of the group $B_3/\Delta^2$ (resp. $B_4$/Center($B_4$)), but employ neither the geometric-visual definition of the braid group (as strings) nor properties of curves of the third (or the fourth) degree. I claim that with the newly offered research configuration, a separation, or rather a shift of research techniques and contexts occurred: whereas the problem originated from a study of curves and their associated fundamental group of their

---

[64] Private communication with Ron Livne (Email from 13.7.2018). Indeed, starting the end of 1976, Moishezon was already at Columbia University.

[65] Private communication with Ron Livne (Email from 15.7.2018).

[66] In his MA thesis Livne proved a similar result that $B_4$/Center($B_4$) is isomorphic to $(\mathbb{Z}/3\mathbb{Z} * \mathbb{Z}/4\mathbb{Z})/\langle b^2 aba = abab^2 \rangle$. (Livne 1975, p. 3)

[67] Private communication with Ron Livne (Email from 15.7.2018).

[68] Livne (1975, introduction).

complement, the research that followed is purely algebraic, concentrating on proving a certain property of factorizations of elements in the braid group of three strings—i.e. bringing any six-element factorization to the *canonical form*, this form being an algebraic expression. One may suggest that a similar separation happened in Dedò's configuration (see Sect. 4.2.1), but here, to stress only one difference, the use of algebraic techniques was much more significant, as Livne underlined the group theoretical properties which are employed.

* * *

Two years after the publication of Livne's thesis, in 1977 Moishezon published his book *Complex Surfaces and Connected Sums of Complex Projective Planes.*[69] As we will see, in the second part of the second chapter Moishezon used Livne's theorem of factorization to prove an algebro-geometric result, which posits the usage of factorizations in yet another context, having nothing to do with proving Zariski's statement or with branch curves, but rather with maps $p : X \to B$, when $dim(X) = 4$, $dim (B) = 2$, called *Lefshetz fibration.*[70]

Moishezon was interested in those fibrations, for which the genus of a generic fiber is 1, i.e. when the generic fiber is a torus, and when $B$ is a two-dimensional sphere (a generic fiber of $p : X \to B$, given a generic point $b \in B$, not being one of the critical values $a_1, \ldots, a_n$ of the map $p$, is defined to be the manifold $p^{-1}(b)$).[71] He employed the notion of the *mapping class group* associated to a manifold $M$: this is the group (of isotopy-classes) of automorphisms of $M$. Moishezon used the known fact that the mapping class group of a torus is the group $SL_2(\mathbb{Z})$.[72] Since the quotient $SL_2(\mathbb{Z})/\mathrm{Center}(SL_2(\mathbb{Z}))$ is isomorphic to $\mathbb{Z}/2\mathbb{Z} * \mathbb{Z}/3\mathbb{Z}$, Moishezon used Livne's result on the rigidity of factorizations in $\mathbb{Z}/2\mathbb{Z} * \mathbb{Z}/3\mathbb{Z}$ to prove a result on the uniqueness of Lefshetz fibrations.

Explicitly, Moishezon noticed that a loop in $B$ around one critical value $a_i$—an element in $\pi_1(B - \{a_1, \ldots, a_n\})$—induces an automorphism of the fiber, that is, an element in the mapping class group of the torus, i.e. in $SL_2(\mathbb{Z})$ (see Fig. 5.1). Hence, since the loop in $B$ that encircles all the critical values is, on the one hand, a concatenation of all the $n$ loops, each encircles only one critical value, and on the other hand, corresponds to the identity element in the mapping class group (since over this loop the fibration is trivial), one obtains

---

[69] I will discuss this book later in Sect. 5.3.2.

[70] I cite a modern definition of Lefshetz fibration from (Degtyarev 2012, p. 101): Let $X$ be a compact connected oriented smooth 4-dimensional manifold and $B$ a compact connected smooth oriented surface. A *Lefschetz fibration* is a surjective smooth map $p : X \to B$ with the following properties:

(1) $p(\partial X) = \partial B$ and the restriction $p : \partial X \to \partial B$ is a submersion;

(2) $p$ has but finitely many critical points, which are all in the interior of $X$, and all critical values are pairwise distinct; denote these critical values in $B$ by $a_1, \ldots, a_n$.

(3) about each critical point $x$, there are charts $(U, x) \cong (\mathbb{C}^2, 0)$ and $(V, b)) \cong (\mathbb{C}^1, 0)$, $b = p(x)$ in which $p$ is given by $(z_1, z_2) \to z_1^2 + z_2^2$.

[71] That is, the fiber $p^{-1}(b)$, when $b \neq a_1, \ldots, a_n$ (see the former footnote).

[72] This group is the set of $n \times n$ matrices with integer entries and determinant 1, together with the action of matrix multiplication.

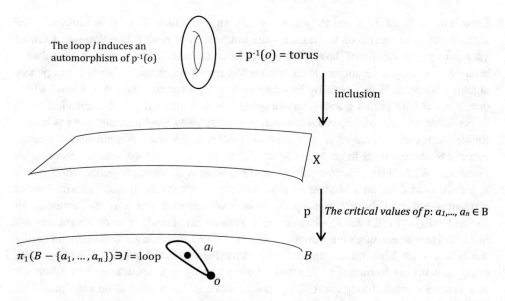

The loop $l$ induces an
automorphism of $p^{-1}(o)$

$= p^{-1}(o) =$ torus

inclusion

X

p   *The critical values of $p$: $a_1,..., a_n \in B$*

$\pi_1(B - \{a_1, ..., a_n\}) \ni l = $ loop

$a_i$

$B$

$o$

**Fig. 5.1**  A depiction of Lefschetz fibration, when a loop in $\pi_1(B - \{a_1, ..., a_n\})$ based at the point $o$ induces an automorphism of the fiber $p^{-1}(o)$, which can be also represented as an element of $SL_2(\mathbb{Z})$. © Graphics: M.F.

a factorization of elements in $SL_2(\mathbb{Z})$ which is equal to the identity element. This enables Moishezon to use Livne's results.

With respect to Livne's work, there is certainly a development and generalization in Moishezon's book. First, in an appendix to the second chapter, written by Livne, the theorem from 1975 proven for a factorization of six elements is proven now to *any number* of elements; explicitly, Livne proved that given a product $y_1 \cdot \ldots \cdot y_n = 1$ such that $y_i \in \mathbb{Z}/2\mathbb{Z} * \mathbb{Z}/3\mathbb{Z}$, and all the $y_i$ are conjugates of $s_1$,[73] then "by successive applications of elementary transformations the $n$-tuple $(y_1, \ldots, y_n)$ can be transformed to an $n$-tuple $(h_1, \ldots, h_n)$ with each $h_i$"[74] of the form $s_0$, $s_1$, or $s_2$. Second, Moishezon generalizes Livne's theorem, which appeared in the appendix, proving that the product $y_1 \cdot \ldots \cdot y_n$ (when the $y_i$ are as above) is equivalent to the product $(s_1 \cdot s_2)^{n_2}$.[75] Noting that

$$SL_2(\mathbb{Z}) \rightarrow SL_2(\mathbb{Z})/\text{Center}(SL_2(\mathbb{Z})) \cong \mathbb{Z}/2\mathbb{Z} * \mathbb{Z}/3\mathbb{Z},$$

Moishezon generalized the result from products in $\mathbb{Z}/2\mathbb{Z} * \mathbb{Z}/3\mathbb{Z}$ to products in $SL_2(\mathbb{Z})$. This enabled him to characterize this class of Lefschetz fibrations by looking only at

---

[73] See Sect. 5.2.1 for the definition of $s_1$. The elements $s_2$ and $s_0$ are conjugates of $s_1$.
[74] Livne (1977, p. 225).
[75] Moishezon (1977, p. 180). Moishezon proves that in this setting, $n$ is always even.

factorizations in $SL_2(\mathbb{Z})$, and to prove that any such fibration $X \to B$ is unique.[76] The algebro-geometric result of Moishezon—the uniqueness of such a fibration—is obtained via a purely algebraic proof. To emphasize, the fact that certain factorizations can be always brought to a *unique* 'canonical form' via certain transformations, which is a completely algebraic theorem, is then used by Moishezon to prove a certain result in a domain which does not—at least at first glance—have anything to do with results on factorizations.

As noted above, while Livne's research was completely algebraic, and the motivation for his work was the computation of the fundamental group of the complement of a nodal curve, Moishezon used these algebraic results to prove algebro-geometric theorems on fibrations. While there is here a shift of context, what is common, together with Livne's appendix in Moishezon's book and his unpublished work on $B_4$, is that factorizations of elements—either in the braid group $B_n$ or in quotients of it (such as $B_n/\mathrm{Center}(B_n)$ or $B_3/\mathrm{Center}(B_3) \cong \mathbb{Z}/2\mathbb{Z} * \mathbb{Z}/3\mathbb{Z}$)—is considered as an object of study in its own right, and not just a tool to research other mathematical objects. This implicit insight is to be found during the 1980s, when Moishezon, together with Mina Teicher, developed his ideas regarding braid monodromy factorization. However, before turning to a discussion of Moishezon's and Teicher's work during the 1980s, another work of Livne should be examined—his PhD thesis "On certain covers of the universal elliptic curve".

### 5.2.3  On Surfaces with $c_1^2 = 3c_2$ and Livne's 1981 PhD Thesis

As was noted briefly in the introduction to this chapter, a new classification of surfaces was carried out during the 1950s and the 1960s, using (also) Chern classes.[77] Hassler Whitney and Eduard Stiefel initiated in the same year, 1935, research into those classes. While Stiefel introduced 'characteristic' homology classes determined by the tangent bundle of a smooth manifold, Whitney researched the case of an arbitrary sphere bundle, using later cohomology theory, from which the concept of a characteristic cohomology class emerged.[78] Shiing-Shen Chern introduced in 1946 what was later called the Chern classes, by attaching to each complex vector bundle $E$ on a differential manifold $M$ of (real) dimension $n$ an element $c_j(E)$ of the cohomology group $H^{2j}(M, \mathbb{Z})$, when $2j \leq n$.[79] While I will not discuss the history of how Chern classes were defined and treated,[80] of particular interest for the classification of algebraic surfaces are the Chern classes associated to the

---

[76] Ibid., p. 175, Theorem 9. In fact, Moishezon proves that this factorization is determined by the Euler characteristics $e(X)$ of $X$, and that $e(X) > 0$, $e(X) \equiv 0 \ (mod \ 12)$.

[77] Cf. Gray (1999, p. 65).

[78] See: Milnor / Stasheff (1974, preface).

[79] See: Dieudonné (1985, p. 93); Chern (1946). Note also the shift in the definition of Chern classes of Grothendieck (1958), using category theory.

[80] See e.g. Brasselet (1998).

tangent bundle $\mathcal{T}_X$ of an algebraic complex surface $X$; those Chern classes are denoted by $c_j(X)$. Since $H^n(M, \mathbb{Z})$ is isomorphic to the group of integers $\mathbb{Z}$, one obtains for a complex algebraic surface $X$ (whose real dimension is 4) that $H^4(X, \mathbb{Z})$ is isomorphic to $\mathbb{Z}$. Hence one can associate to $c_2(X) \in H^4(X, \mathbb{Z})$ and also to $c_1^2(X)$ positive integers, called the Chern numbers.[81] While several invariants associated to algebraic surfaces were re-written in terms of the Chern classes,[82] one of the research directions, led by Fedor Bogomolov, Shing-Tung Yau and Yoichi Miyaoka, among others, was to find restrictions on the Chern classes of algebraic surfaces—and this in order to restrict the possible topological types of the underlying real 4-dimensional manifold. It was proven in the late 1970s first by Bogomolov that $c_1^2 \leq 4c_2$ and afterwards that $c_1^2 \leq 3c_2$; this, together with other restrictions, such as $5c_1^2 + 36 \geq c_2$ or the divisibility of $c_1^2 + c_2$ by 12, led, starting at the end of 1970s, to the investigation of the "geography of surfaces",[83] dealing with the question of the classification of surfaces, employing, in a similar manner to Enriques and Castelnuovo, a geographical metaphor. We already encountered this classification project of algebraic surfaces according to their genera, done by Enriques and Castelnuovo, which was completed in 1914. The current classification dealt with a special class of surfaces called minimal surfaces of general type;[84] one hence asked: Which pairs of integers are the invariants $(c_1^2, c_2)$ of those surfaces? As the boundaries of the region of the possible pairs $(c_1^2, c_2)$ were known, due to the above mentioned inequalities, the employed geographical metaphor was somewhat different than how it was used by Enriques and Castelnuovo, as we saw in Sect. 3.1.2 (cf. also Sect. 3.3): no longer an exploration of the complete unknown, but rather a search within a confined region. Regarding the boundaries of this region, it was also proven by Hirzebruch-Miyaoka-Yau that surfaces of general type satisfy $c_1^2 = 3c_2$ if and only if the surface is isomorphic to a quotient of the unit ball of $\mathbb{C}^2$ (i.e. of $|z_1|^2 + |z_2|^2 = 1$, when $z_1$ and $z_2$ are the complex coordinates) by an infinite discrete group.[85]

---

[81] Note that on the cohomology groups $H^j(M, \mathbb{Z})$ one can define multiplication, such that if $x \in H^p(M, \mathbb{Z})$, $y \in H^q(M, \mathbb{Z})$ then $x * y \in H^{p + q}(M, \mathbb{Z})$. Hence, since $c_1(X) \in H^2(X, \mathbb{Z})$, $c_1^2(X) = c_1 \cdot c_1 \in H^4(X, \mathbb{Z}) \cong \mathbb{Z}$.

[82] For example, the arithmetic genus $p_a$ of an algebraic surface is $\frac{1}{12} \left( c_1^2 + c_2 \right)$, and the signature of the intersection form on the second cohomology is $\frac{1}{3} \left( c_1^2 - c_2 \right)$.

[83] The expression appears for example in: Holzapfel (1980, p. 230). See also (Chen 1987), whose title is "On the Geography of Surfaces".

[84] For the definition of surfaces of general type—being surfaces whose Kodaira dimension is 2—see below (Sect. 5.3.2). Given a surface the *minimal model* of it (which always exists and is unique if the Kodaira dimension is greater than 0) is a (possibly another) surface $X$ birational to the original surface, such that any birational morphism $f: X \to X'$ to another surface $X'$ is necessarily an isomorphism. If $X$ is the unique relatively minimal model in its birational equivalence class, then we say that $X$ is a minimal model.

[85] Hirzebruch (1958); Yau (1977).

Algebraic surfaces as branched covers also played a role in this research. Two research directions should be mentioned, before further discussing Livne's and Moishezon's works. First, it is important to emphasize the work of Friedrich Hirzebruch (1927–2012) during the 1980s. Hirzebruch, in his paper from 1983 "Arrangements of Lines and Algebraic Surfaces", associated to any arrangement of lines on the projective plane an algebraic surface, branched over this arrangement. Taking a minimal model of this surface, he applied the Miyaoka-Yau inequality $c_1^2 \leq 3c_2$ on this surface to obtain new restrictions on line arrangements, restrictions which were not known till then and were not at all obvious.[86] In the book from 1987 *Geradenkonfigurationen und algebraische Flächen*, the proof of the Miyaoka-Yau inequality is described by Hirzebruch as a great achievement for the research of surfaces, prompting new research directions (among others) in the field of algebraic surfaces and configurations of lines.[87] This book itself can be described as "exemplary of his [Hirzebruch's] 'philosophy' of mathematics: a profound theorem is illustrated by explicit examples".[88] Indeed, finding restrictions on complex plane curves by considering them as branch curves of surfaces with normal singularities was not a new technique; recall that Enriques already used this technique to construct double covers with specific genera. Here Hirzebruch used branch curves (being in this case line arrangements) as a tool in order to discover numerical restrictions of line arrangements.

While Hizebruch's research described above dealt specifically with line arrangements, the second research direction dealt explicitly with algebraic surfaces. Around 1977 Fedor Bogomolov, after proving that $c_1^2 \leq 4c_2$, conjectured the following: "All surfaces $X$ for which [...] $c_1^2 - 2c_2$ is positive have infinite fundamental group $\pi_1(X)$".[89] The conjecture was based on several examples,[90] and especially on the Hirzebruch-Miyaoka-Yau theorem that all surfaces with $c_1^2 = 3c_2$ are a quotient of the unit ball, hence having an infinite fundamental group. In 1983 Feustel and Holzapfel called this conjecture the "watershed conjecture",[91]—using yet another geographical metaphor—but added that "we are not convinced of the effectiveness of the watershed conjecture", pointing to several attempts to disprove this conjecture.[92] As we will see in Sect. 5.4, Moishezon and Teicher eventually disproved the conjecture around 1985—starting with an investigation of the fundamental group of a branch curve of certain surfaces.

---

[86] Hirzebruch (1983, p. 132–133). See also: Hirzebruch (1985) and (Barthel/Hirzebruch/Höfer 1987). For example, one of the restrictions discovered is the following (Hirzebruch 1983, p. 132): Given an arrangement of $k$ lines in the complex projective plane, let $t_r$ ($r > 1$) be the number of $r$-fold points of the line arrangement. If $t_k = t_{k-1} = 0$, then $t_2 + t_3 \geq k + t_5 + 2t_6 + 3t_7 + \ldots$

[87] Ibid., p. V.

[88] Scharlau (2017, p. 235)

[89] Reid (1977, p. 626); See also: Van de Ven (1978, p. 164).

[90] See: Ibid., p. 163–164.

[91] Feustel/Holzapfel (1983, p. 10).

[92] Ibid., p. 11, 37.

To return to Livne's work, in 1981 he published his PhD thesis "On certain covers of the universal elliptic curve", with the aim to "analyze a certain new class of surfaces of general type".[93] One of the main results of the thesis was the construction of new surfaces with $c_1^2 = 3c_2$. Livne noted that "Hirzebruch showed, using his proportionality principle that co-compact quotients of the ball by discrete torsion-free subgroups of $PU(2, 1)$ satisfy $c_1^2 = 3c_2$". Hence, "[m]ost of the constructions of surfaces of general type with $c_1^2 = 3c_2$ are done by defining a discrete subgroup $\Gamma \subset PU(2, 1)$."[94]

What is unique in Livne's thesis is his construction of complex surfaces with $c_1^2 = 3c_2$, which are not constructed explicitly as the above quotient, providing the first "'algebro-geometric' examples of [such] surfaces". Livne noted that a "natural question arises: to determine the fundamental group [of the surface] as a discrete, co-compact, torsion-free subgroup of $PU(2, 1)$."[95] Here Livne used the theorem that if a surface $X$ with $c_1^2 = 3c_2$ is a quotient of $|z_1|^2 + |z_2|^2 = 1$ by a discrete subgroup $\Gamma \subset PU(2, 1)$, the fundamental group of this surface is $\Gamma$. One of the steps in the determination of this group consists of computing a fundamental group of a complement of certain curve, using the methods of Zariski-van Kampen implicitly. While the following presentation of Livne's treatment is somewhat technical, it is important to stress that the computation of this fundamental group is done within the framework of a larger research configuration (i.e. construction of surfaces with $c_1^2 = 3c_2$), and the work with the method of Zariski-van Kampen is not an end for itself. Nevertheless, Livne did coin terminology that would become central in the 1980s when researching branch curves: the "braid monodromy".

* * *

Explicitly, Livne aimed to compute the fundamental group of the complement of the set

$$V' = \left\{ (t, u) \in \mathbb{C}^2 : t \neq 0, 27; 4u^3 - tu - t \neq 0 \right\}$$

In order to do that, he "draw[s] the picture of [. . .] $V'$ in the real $(t, u)$–plane [when] $V'$ is the complement of the solid lines."[96] (see Fig. 5.2) In order to compute $\pi_1(V')$ one has to find the branch points of the curve $4u^3 - tu - t = 0$ with respect to the projection $p : (t, u) \mapsto t \in \mathbb{CP}^1$; these are $t = 0, 27$ and $t = \infty$. Denoting by $T$ the complex projective line, whose (complex) coordinate is $t$, the usage of the Zariski-van Kampen method begins

---

[93] Livne (1981, p. i).

[94] Ibid., p. ii. $U(n, 1)$ is a subgroup of $(n + 1) \times (n + 1)$ matrices over $\mathbb{C}$, which preserve the quadratic form $|z_1|^2 + \ldots + |z_n|^2 - |z_{n+1}|^2$. Let $PU(n, 1)$ its projectivization. This means that the action of $PU(n, 1)$ on $\mathbb{CP}^n$ leaves invariant the set of vectors whose norm is negative (when the norm of $z$ is defined as its quadratic form). Taking $n = 2$, $PU(2, 1)$ acts on the ball in $\mathbb{C}^2$, since it leaves invariant the set $|z_1|^2 + |z_2|^2 - |z_3|^2 < 0$ or $|z_1|^2 + |z_2|^2 < |z_3|^2$, or in the neighborhood of the complex plane $\mathbb{C}^2$ when $z_3 \neq 0$, $|z_1/z_3|^2 + |z_2/z_3|^2 < 1$, being the unit ball in $\mathbb{C}^2$.

[95] Ibid., p. ii-iii.

[96] Ibid., p. 86–87.

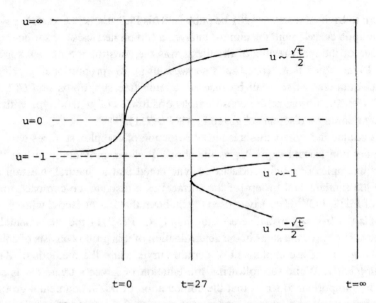

**Fig. 5.2** The curve $4u^3 - tu - t = 0$ and the lines $t = 0, 27$ (Livne 1981, p. 87). © Ron Livne

**Fig. 5.3** The loops $y$ and $z$, being the homotopy classes of $\tilde{y}$ and $\tilde{z}$ (Livne 1981, p. 88). © Ron Livne

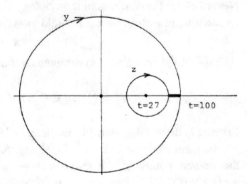

with determining a basis for $\pi_1(T - \{0, 27, \infty\})$, when $T$ is also considered as a real 2-dimensional plane (see Sect. 3.2.3). To find this basis Livne drew two loops $\tilde{y}, \tilde{z}$ in $T$, both exiting from $t = 100$ (as is shown in Fig. 5.3).

Defining these two loops, another loop $\tilde{x}$ is defined in the following way, such that the loop $\tilde{y}$ is "homotopic to $\tilde{z}\tilde{x}$". This means that the loop $\tilde{x}$ exits also from $t = 100$ and encircles the point $t = 0$.[97] Livne denoted a basis $x, y, z$ for the group $\pi_1(T - \{t = 0, 27, \infty\})$, where $x, y, z$ are the homotopy classes of $\tilde{x}, \tilde{y}, \tilde{z}$. Considering the map $p : (t, u) \mapsto t$, he

---

[97] Ibid., p. 88.

**Fig. 5.4** The loops $A$, $B$ and
$C$ at the fiber $t = 100$ (Livne
1981, p. 88). © Ron Livne

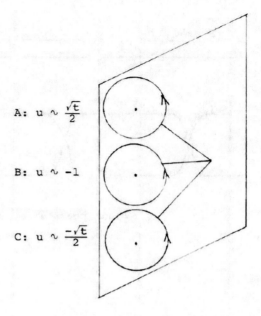

$$A\!:\ u \sim \frac{\sqrt{t}}{2}$$

$$B\!:\ u \sim -1$$

$$C\!:\ u \sim \frac{-\sqrt{t}}{2}$$

instructed "in the fiber over $t = 100$ to choose loops $\widetilde{A}, \widetilde{B}, \widetilde{C}$ [as in Fig. 5.4]",[98] when $A$, $B$ and
$C$ are their homotopy classes. More explicitly, denoting by $u_1$, $u_2$, $u_3$ the solutions of
$4u^3 - 100u - 100 = 0$, the loops $\widetilde{A}, \widetilde{B}, \widetilde{C}$ encircle in the fiber $p^{-1}(t = 100)$ the points
$\tilde{u}_1 = (u_1, 100), \tilde{u}_2 = (u_2, 100), \tilde{u}_3 = (u_3, 100)$ , and $A$, $B$ and $C$ are generators of
$\pi_1(p^{-1}(t = 100) - \{\tilde{u}_1, \tilde{u}_2, \tilde{u}_3\})$.

This construction enables Livne to consider $V'$ as a fibration over $T - \{0, 27, \infty\}$.
Hence, by denoting $F_2$ the free group with the generators $x$ and $y$, and $F_3$ the free group with
the generators $A$, $B$ and $C$, one obtains from the map (called "the homotopy sequence of a
fibration"):

$$V' \to T - \{0, 27, \infty\},$$

the following sequence of groups:[99]

$$1 \to F_3 \to \pi_1(V') \to F_2 = \pi_1(T - \{0, 27, \infty\}) \to 1.$$

It was Zariski and van Kampen who formulated how to determine the fundamental group
$\pi_1(V')$, when one knows the induced action of the generators of $T - \{0, 27, \infty\}$ on the
generators of the fiber.[100] But neither Zariski nor van Kampen employed terms such as

---

[98] Ibid.

[99] Ibid., p. 89.

[100] See Sect. 3.2.3 for how van Kampen described his method.

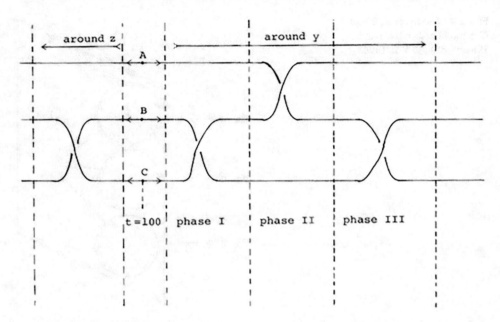

**Fig. 5.5** The drawing of braids as preimages of the loops around the various branch points (Livne 1981, p. 90). © Ron Livne

"homotopy sequence of a fibration" due to the obvious reason that this vocabulary did not yet exist. One may claim that, on the one hand, the method introduced by Livne is a modification of Zariski-van Kampen's method; Livne did not mention the names Zariski or van Kampen explicitly,[101] and the modification is to be noticed in the following remark of Livne: "To determine the action we consider the braiding [as presented in Fig. 5.5]".[102] But this modification is, in fact, an embedding of the computation of the fundamental group in a new configuration.

Returning to Livne's construction, what did he mean by braiding? What Livne drew is the movement of the points of the fiber, when one goes around the loops $z$ and $y$. This kind of drawing and construction already appeared in Chisini's writings,[103] as we saw in Chap. 4, and was certainly not new. This however is not to claim that Chisini and Livne were working in the same mathematical configuration, or that Livne knew of Chisini's work. Moreover, Livne's terminology is new, as he introduces the term "braid monodromy".[104] Even if the term "monodromy" was employed during the twentieth

---

[101] Though obviously he knew his work, as is clear from Livne's MA thesis.

[102] Ibid.

[103] and also in the writings of many other mathematicians (see Sect. 4.1.1).

[104] Ibid., p. v.

century often to describe motion of points in a fiber of a covering, the term "braid monodromy" was new.

With the help of the figure of the braid (see Fig. 5.5), Livne computes the action of $z$ and $y$ on the generators $A$, $B$ and $C$. He notes "going around $z$ we get the simple braid

$$\begin{pmatrix} A \\ B \\ C \end{pmatrix} \xrightarrow{z} \begin{pmatrix} A \\ BCB^{-1} \\ B \end{pmatrix}$$

and around $y$ we have a composition of three simple braids",[105] resulting in the action:

$$\begin{pmatrix} A \\ B \\ C \end{pmatrix} \xrightarrow{y} \begin{pmatrix} ABCB^{-1}A^{-1} \\ ABA^{-1} \\ A \end{pmatrix}$$

By noting that $y = xz$, one can deduce the action of $x$ on $A$, $B$ and $C$. Hence, Livne proves that the fundamental group $\pi_1(V')$ is a semi-direct product $F_2 \times F_3$, when the action of $F_2$ (generated by $x$ and $y$) on $F_3$ is as described above.[106]

* * *

Taking a step back from the technical aspects of Livne's construction, although the calculation of the fundamental group of the complement of complex plane curves was not new during the 1970s and the 1980s, it is clear that the techniques and the terms that Livne used were. Therefore he clearly was working in a new mathematical configuration. Emphasizing that the action on the generators of the fibers is achieved by considering the "braid monodromy"[107] was not present in Zariski's or van Kampen's writings; moreover, although Chisini and his students already took notice that the motion of the points in the fibers along a loop gives rise to a braid, their work, as we saw, was hardly known outside of Italy till the late 1980s. The connection to the research on the braid group is also not a direct one: on the one hand, when Livne noticed that the deduced braid (either from $y$ or from $z$) gives rise to an action on the free group, he mentioned explicitly Emil Artin's paper "Theory of Braids" from 1947.[108] On the other hand, there was no usage or mention of any notation of the braid group or its generators, as was developed by Artin.

---

[105] Ibid., p. 91.

[106] A *semi-product* group can be constructed as the Cartesian product of two other groups, when the first group operates on the second group (for example, by conjugation, but not necessarily), obtaining as a result a particular multiplication.

[107] Ibid., p. v.

[108] Ibid., p. 89.

What is also essential to note are the visual steps and reasoning that Livne employed. The actions induced by $y$ and $z$ are deduced only from looking at Fig. 5.5. No analytic computation is given as to how the points in the fiber change position when one moves along the loop $z$, for example, and only a drawing of a simple crossing, which switches between the strings $B$ and $C$, is drawn. This visual-diagrammatical step helps considerably in the understanding of the induced action of the semi-product; i.e. how the element $z$ here operates on the elements $A$, $B$ and $C$. In that sense, the diagram serves as a key step in algebraic reasoning and justifies the algebraic formulation of how the elements ($y$ and $z$) of the group $F_2$ operate on the elements ($A$, $B$ and $C$) of the group $F_3$. It replaces in fact any algebraic computation, though one may claim that the different steps describing what happens when one goes around the loop $y$, as presented in Fig. 5.5, are not as intuitive as in the case of the loop $z$. This reliance on diagrammatic reasoning echoes former epistemic configurations: also Chisini and Dedò during their own work on braids, induced from algebraic curves, relied on diagrammatic reasoning.

In this work of Livne, braid monodromy is considered as a tool to compute the fundamental group; it is not an object of research, considered, for example, as a factorization of elements. Moreover, as Livne has stated, "I met Moishezon and told him about the result [of the PhD thesis] before I wrote my thesis, but I believe that his work was not dependent on it in any way."[109] Hence, though Moishezon was not directly influenced from Livne's PhD thesis, one may assume that these ideas regarding the braid monodromy were in a way present in Moishezon's work, as they were discussed between Moishezon and Livne already during the 1970s. How Moishezon reshaped the concept of braid monodromy during the 1980s is the subject of following section.

## 5.3    Moishezon's Program

Boris Moishezon (1937–1993) was one of the mathematicians who, during the 1980s, renewed interest in branch curves and the associated braid factorization, called "braid monodromy factorization". His interests and mathematical influence were nevertheless much wider, and I would like to describe shortly his biographical background, as his biography reflects also how Jewish scientists and mathematicians emigrated outside the USSR during the 1970s and the 1980s.

The question, which should be addressed before continuing, concerns the relevance of bringing Moishezon's biographical details to a book dealing with the history of the various configurations of branch curves. To state the obvious: the history unfolded here should not be viewed as 'internal' one; that is, the material, social and political 'external' conditions and events must also be taken into account when considering how those epistemic configurations transform and emerge. These aspects were already seen with the Italian

---

[109] Private communication with Ron Livne (Email from 15.10.2018).

school of algebraic geometry and its growing seclusion, also due to the rise of fascism; or with d'Orgeval's work in captivity in Oflag X B. Hence, the emigration story of Moishezon as well as the political views of Igor Shafarevich and his colleagues are certainly relevant to how new mathematical configurations were developed.

### 5.3.1 From the USSR to Israel and to the USA

Moishezon, born in Odessa, Ukraine (then a part of USSR), defended his Master thesis in 1962 at the Institute of Applied Mathematics in Moscow. From 1964 to 1967 he taught mathematics as a senior lecturer at the Pedagogical Institute in Orekhovo-Zuyevo and in 1967 he obtained his PhD. During his studies in Moscow in the 1960s, Moishezon was researching at the Steklov Institute of Mathematics. As he recalled later, "though [Israel] Gelfand and [Ilya] Pyatetskii-Shapiro were my advisors, my real teacher throughout those years was Shafarevich."[110] During the years 1961–1963 Shafarevich organized a seminar at Moscow University, in which the classical works of the Italian mathematicians on the theory of algebraic surfaces were studied. I will return below to Shafarevich's conception of the Italian school of algebraic geometry. What is important to note here, however, is that Moishezon was one of the participants at these seminars. The result of this two-year seminar was the publication of the book *Algebraic Surfaces*, published in Russian in 1965 and translated to English in 1967. The name of the book was a reference to Zariski's 1935 book, though the subjects and the methods used were different. Each chapter was written by a different mathematician; the sixth chapter "Surfaces of Fundamental Type" was written by Moishezon.

As Viktor Bukhstaber and Sergeĭ Novikov underline, Moishezon introduced into this seminar ideas from Kählerian geometry, whereas the other participants were more "algebraically inclined".[111] Thus, for example, in 1966 Moishezon proved one of his famous theorems, that a compact complex manifold, possessing a Kählerian metric and a field of meromorphic functions of transcendence degree equal to the dimension of this manifold, is a projective manifold.[112] The manifolds having this property are today called *Moishezon manifolds*; see Sect. 5.3.2 below.

During the years 1967–1969 Moishezon and Novikov organized a seminar on algebraic and Kählerian geometry in Moscow, which was to prove influential for numerous students.[113] Moishezon introduced in this seminar not only the methods of Kählerian geometry, but also the works of Heisuke Hironaka on the resolution of singularities. Moreover, Moishezon was supported by Shafarevich during his career in Moscow, at

---

[110] Moishezon (1992, p. 61).

[111] Bukhshtaber / Novikov (1995, p. 613).

[112] See: Moishezon (1966).

[113] Bukhshtaber / Novikov (1995, p. 614).

least till he left USSR. As Aron Katsenelinboigen recalls, "he [Shafarevich] helped Boris gain admittance to the graduate school at the Institute of Applied Mathematics, the former USSR Academy of Sciences, and he continued supporting him for many years."[114] Moishezon also recalled warmly how Shafarevich supported him and his work: "Shafarevich was considered one of the most brilliant mathematicians in the country. In my wildest dreams I had never imagined that such a person would be my thesis director . . . I literally worshipped [him]."[115] However, at the beginning of the 1970s, Moishezon had to leave the USSR. This was due to the worsening conditions for Jews in the country and the rise of institutionalized anti-Semitism, a subject I will discuss below. He submitted an application for permission to leave for permanent residence abroad, thus becoming a 'refusenik';[116] as an obituary from 1993 notes, "[h]e and nine peers signed a public statement in 1972 decrying the high fees for exit visas imposed on holders of doctoral degrees. The fees, as much as $25,000, were intended by the Soviet Government as repayment for the émigrés' education."[117] When he took part in a demonstration that condemned the murder of 11 Israeli athletes at a terrorist hostage taking at the 1972 Olympics in Munich, he was detained. It remains unclear whether, as a consequence, he lost his job.[118] Eventually, at the end of 1972 he obtained the permission needed and immigrated to Israel (see also below), working as a professor in the Tel-Aviv University between 1973 and 1977.[119] Vitali D. Milman recalls that "[t]he emigration of the mid-1970s had already brought mathematicians of the highest caliber and of all ages to Israel [among them] [. . .] Moishezon". Tel Aviv University hired most of the mathematicians. This was principally due to the efforts of the president of Tel Aviv University, Yuval Ne'eman, who was acting president between 1971 and 1975. Moreover,

---

[114] Moishezon (2001, p. 217).

[115] Moishezon (1992, p. 61).

[116] The term *refusenik*, derived from "refusal", denoted individuals whose emigration was refused by the authorities of the USSR. See also below.

[117] Saxon (1993).

[118] See: Columbia University Record Archives (1993). The obituary from Columbia University notes that according to his wife, Moishezon did lose his job, while Moishezon himself, in an interview done with him in November 1972, says that "[s]urprisingly, he had not been dismissed from his job after applying for an exit visa, but continued to work until their [his wife and his] departure." (Moishezon / N.N. 1972, p. 3)

[119] A document of the faculty of exact sciences, titled "Social integration of Immigrants-teachers" from sixth July 1973, lists Moishezon in a roster of "Immigrants-teachers" (in: Historical Archives of Tel Aviv University, "Immigrant teachers" ["Morim Olim"], 1971–1973, 901.1061/021, Exact Sciences fond); a list titled "USSR emigrants, who emigrated during the years 1973/74 and their salary was financed by the Emigration office", lists Moishezon as a member in the mathematics department, who started working there on the first February 1973 (in: Historical Archives of Tel Aviv University, "USSR scientists", 900.0429/89, Rector chamber fond).

"until the arrival of Pyatetskii-Shapiro [in 1976], Israel had no specialists in representation theory; until Moishezon came [in 1972], there was no one in algebraic geometry."[120] Indeed, among his students, of whom Moishezon was the supervisor (for Master or PhD in mathematics) in Israel during the 1970s, one can find Anatoly Libgober, Ron Livne, Mina Teicher and Dov Wajnryb. Indeed, as Mina Teicher notes, the first course on algebraic geometry to be given in the year 1973–74 at Tel Aviv University was given by Moishezon, a course which included only four students, among them Livne, Ofer Gaber and herself.[121]

Livne and Teicher are professors for mathematics at the Hebrew and Bar-Ilan universities in Israel, and Wajnryb was until his retirement a professor at the Technion. After participating in the movement to change immigration policies in USSR, Libgober immigrated to Israel in 1973. In the late 1970s he moved to Chicago, where he is professor Emeritus. It is clear therefore that Moishezon, with his other colleagues, who emigrated from the USSR at the same time and after him, helped building an Israeli community of algebraic geometry. But while Milman presents Moishezon as the founder of this community in Israel, one has to wonder how much this account reflects the involvement of Moishezon in Tel Aviv University, as the annual syllabi of the faculty of exact sciences tells a slightly different story. Moishezon indeed taught a two-semester long course on "Algebraic Geometry" during the year 1973–74.[122] However, as the Yom Kippur war ended on the 25 October 1973, and since the first semester of the year 1973–74 was supposed to begin on the 28 October 1973, it was decided to postpone this semester to the next one, and to hold a summer semester during the summer of 1974.[123] These events might have influenced how many students participated in Moishezon's course. Moreover, during the following year, 1974–75, he was teaching other courses: "Introduction to set theory", "topology" and "functions of several variables".[124] Only during the following year, 1975–76, did he again teach a course in "Algebraic Geometry", though now the course spanned over just one semester.[125] However, starting the year 1976–77 Moishezon was already on leave without pay, as is indicated in the syllabus of this year as well as in the syllabus of the following year.[126] The syllabus of 1978–79 does not list Moishezon anymore as a member of the department of mathematics.[127] Indeed, in 1977 Moishezon

---

[120] Milman (2006, p. 217–218).

[121] Private communication with Mina Teicher (Email from 23.1.2019).

[122] Tel Aviv University Curriculum (1973, p. 73–74, p. 67).

[123] Private communication with Mina Teicher (Email from 30.1.2020). See also the letter from Shlomo Simonsohn, the Rector of Tel Aviv University between 1971 and 1977, from 1 January 1974 titled "Teaching arrangements in the emergency year 1974" (in: Historical Archives of Tel Aviv University, "War 1973–1974", 901.0451/04, Rector chamber fond).

[124] Tel Aviv University Curriculum (1974, p. 85, 93).

[125] Tel Aviv University Curriculum (1975, p. 120).

[126] Tel Aviv University Curriculum (1976, p. 89; 1977, p. 93).

[127] Moishezon was already at Columbia University starting the academic year 1976–77. The protocol of the meeting of mathematics department at Tel Aviv University from 27.10.1978 confirms that

moved to Columbia University and in the same year he published his book *Complex surfaces and connected sums of complex projective planes*. At Columbia he was a professor of mathematics until his death, in 1993, there he also took interest also in archeology and anthropology, as I will elaborate in Sect. 5.3.2.

### 5.3.1.1 Moishezon's Emigration and Jewish Mathematicians in the USSR

The emigration story of Moishezon takes place, as was already noted above, within the larger context of Jewish mathematicians emigrating away from the USSR. Often this emigration was denied or postponed for several years, which was compounded with a worsening of political and academic conditions of those concerned.[128] Indeed, following the doctrinaire standpoint in the USSR at that time that any religious or ethnic affiliation is void of belonging to the communist nation, the Soviet authorities refused to recognize the national aspirations of the Jews, including what was thought of as a possible connection to Israel. After the Second World War, Jewish survivors of concentration and refugee camps began to immigrate to Israel. This movement was anathema to the communist authorities, and the more they tightened their rule over Eastern Europe, the more they tightened their hold on the Jews of these countries until the borders were completely blocked.

In the 1960s and 1970s the Soviet Union had a procedure according to which potential emigrants must obtain a permit from a government office called OVIR (abbreviated for the "Office of Visas and Registration"). Many potential emigrants received a negative answer or waited for approval for a long time; Eugnee Levich, Alexander Voronel, or Alexander Lerner were famous for being refused to emigrate outside of USSR.[129] Some of those who were refused emigration received the desired permit only after several years. In reaction to this crisis, Tel Aviv University decided to create dozens of "non-standard" positions for scientists from "states in distress". The mathematician Saul Abarbanel, who was in 1972 the vice rector of the Tel Aviv University, sent a letter on 28.8.1972 to all the other

---

Moishezon "left the department" (see: Historical Archives of Tel Aviv University, Mathematics 1978–1980, 901.1201/59, Exact sciences fond).

[128] The literature on Soviet Jews and their attempts to emigrate from USSR after the Second World War, and the role that USA and Israel played in these attempts, is vast. For example, see: (Peretz 2015; Feingold 2007; Friedman / Chernin 1999; Pinkus 1989).

[129] On the *refusenik* movement during these years, see e.g. (Kosharovsky 2017); see also: (Feingold 2007, esp. p. 79). Eugnee (Yevgeny) Levich applied in 1972 for an exit visa to leave to Israel, was denied and as a result he joined the refusenik movement. After spending a year in a labor camp in the Siberian Arctic (Bulletin of the Atomic Scientists 1973, p. 2), he was released in 1974 and was a professor at the Weizmann Institute, Israel between 1975 and 1978, before leaving to the US. His recollections of this period are described in (Levich 1976). Alexander Voronel was dismissed from his job and stripped of his academic titles after applying for an exit visa in 1972; he immigrated to Israel in 1974 and from that time onwards became a professor at Tel Aviv University; see: (Voronel 1991). Alexander Lerner, one of the leading scientists in USSR, applied for an exit visa already in 1971 to immigrate to Israel, but was denied, and lost all of his positions. Only in 1988 was he granted such a visa, and he was appointed a professor at the Weizmann Institute (Saxon 2004).

presidents of Israeli universities, informing them that Yuval Ne'eman encourages these universities to create similar positions.[130] Those "non-standard" positions were temporary, in order to decide whether the guest doctors or guest professors could be accepted later as regular department members in the corresponding departments.[131] At the same time, Ne'eman, together with Dan Amir and David Prital, among others, founded in 1971 the "Israel Public Council for Soviet Jewry", which was the national organization to support for the right of Jews to emigrate from the Soviet Union to Israel. Along with this organization, the "Scientists Committee of the Israel Public Council for Soviet Jewry" was founded in August 1972, of which Ne'eman was the first chairman.[132] The "Scientists Committee" issued numerous bulletins, reporting on struggles of Jewish scientists in USSR and in other countries and recounting their personal stories.

One of the personal accounts published was an interview with Moishezon, done in November 1972, two weeks after he arrived to Israel. The interview delineates a less severe image of the state of Jewish scientists in USSR. Moishezon noted in this interview that

> "I applied for my 'characteristika' [the immigration visa] at work in May, 1972 because I wanted the OVIR to have the documents before Nixon's visit [in USSR, on May 1972]. We felt that if there were many Jews who applied, the voices would be louder and the chances greater that they would be heard. [. . .] I formally applied to OVIR on July 5th. In September, I received a verbal refusal on the grounds that I was a valuable specialist, according to the OVIR inspector. It is always given verbally, never in writing. Unexpectedly, one month later, I was called to the OVIR and given permission to depart. I left on November 3rd and arrived in Israel on November 6th through Vienna."[133]

Moishezon did acknowledge that he was politically active, being a "part of Levich-Voronel-Lerner group" but had not lost his job as a result of that.[134] He called then to legislate an "open law openly declaring permission to emigrate" but ended the interview with an optimistic, but not so realistic tone, saying that the USSR wants good relations with the US, and the "outcry of world Jewry [. . .] was undoubtedly of great help in influencing

---

[130] Letter from S. Abarbanel to the presidents of the Universities, 28.8.1972, in: Historical Archives of Tel Aviv University, USSR scientists, 900.0429/89, Rector chamber fond.

[131] Letter from S. Abarbanel to the deans, heads of schools and faculties of Tel Aviv University, 27.8.1972, in: Historical Archives of Tel Aviv University, USSR scientists, 900.0429/89, Rector chamber fond.

[132] The first bulletin of the *Scientists Committee of the Israel Public Council for Soviet Jewry* was published on 1 September, 1972, and starts with the following declaration: "On Sunday, August 20th, an emergency meeting [of the 'Scientists Committee'] was convened to protest the latest steps taken by the Soviet authorities against Jews possessing academic education. The meeting was organized by the heads of Israel's institutions of higher learning and leading scientists."

[133] Moishezon / N.N. (1972, p. 2).

[134] Ibid., p. 3.

the Kremlin [regarding this.] Jews have been allowed freedoms in Moscow—writing open letters, demonstrating [. . .]."[135]

Why did Moishezon put such an effort to present this optimistic tone at the end? Indeed, Moishezon was quite lucky receiving his permission so fast. Moreover, it is worth recalling that the interview was taking place only two weeks after Moishezon arrived to Israel, hence he was still very much influenced by the political climate in USSR.

However, even Moishezon himself noted, in a somewhat sarcastic tone, "[t]here is no way of proving that one is not a 'valuable specialist' or has no security connections".[136] Hence, Moishezon's description is rather an exception to the situation of Jewish scientists in the USSR during that time. In a statement published at the beginning of June 1973, several scientists, among them Mark Azbel, Anatoly Libgober, Alexander Luntz and Alexander Voronel, decided to start a hunger strike. The statement given below, written originally in English, was done over the telephone to Yuval Ne'eman; it was requested that it be delivered to their colleagues:

"For some time past officials of the soviet ministry of the interior have established a practice of frankly explaining to scientists and specialist trying to leave for Israel that the reason for their detainment in the USSR are their professional qualifications. Verbiage about 'secrecy' is more and more often spread nowadays, and similarly are term 'state interest' and one one's 'value as a specialist' prevail. Thus simultaneously with suspending the use of the education tax law it has become more and more apparent that we are enslaved without any right of redemption as most of the specialists lose their jobs when they apply for exit visas. The 'State interest' involved in out detainment is simply the right of the state to dispose of our fate at will. The fact that one of our colleagues, Eugene Levich, has recently been drafted arbitrarily as a soldier shows that the scope of our possible 'use' to the state is quite unlimited. For all we know, this new kind of serfdom may be sanctioned silently by many political leaders in the world.

Relying therefore on ourselves we intend to prove that we belong to nobody but ourselves. Being considered 'state property' we prefer as a matter of principle and provided all other ways are exhausted, to destroy this 'property' rather than recognize that anybody is entitled to our souls and bodies.

As a first step we declare, beginning 10 June [1973], a hunger strike of many days in protest against the principle of 'property' in human being as manifest in our forced detainment in the USSR."[137]

The above description reflects more faithfully the situation of Jewish mathematicians—students as well as professors. A well-known account is the personal essay of Grigori Freiman *It Seems I Am a Jew*, which was translated into English and published in the USA in 1980. The samizdat (in Russian, "self-publishing" or an underground work) of Freiman

---

[135] Ibid.

[136] Ibid.

[137] "Statement by 7 Jewish scientists in Moscow (received by Prof. Y. Ne'eman telephonically—10/ 6/73)", In: Historical Archives of Tel Aviv University, Soviet Union 9556, 1972–1973, 901.1251/11, Exact sciences fond.

shows also how widespread the problem was. Melvyn B. Nathanson, who translated Freiman's work, writes in the introduction: "Ten years ago [in 1970], by doing extremely well on the entrance examination, brilliant young Jewish students might be admitted to Moscow State University. Now it is almost impossible, not only for Jews, but also for others whose background makes them politically suspect. [. . .] The result [. . .] is the emigration of large numbers of mathematicians out of the Soviet Union to the United States, Western Europe, and Israel".[138] Israel Gohberg recalls a similar account.[139] Anatoly Vershik paints a bleak picture of the almost impossible situation of Jewish mathematics students and of the admission to the Mathematics faculties in USSR in the 1970s and 1980s.[140] Misha Shifman summarizes the situation as follows:

> "[I]t is a well-known fact that the Russian mathematical establishment was pathologically anti-Semitic. Such outstanding mathematicians as [Lev] Pontryagin, Shafarevich and [Ivan] Vinogradov, who had enormous administrative power in their hands, were ferocious anti-Semites. The tactics used for cutting off Jewish students were very simple. At the entrance examination, special groups of 'undesirable applicants' were organized. They were then offered killer problems which were among the hardest from the set circulated in mathematical circles [. . .]."[141]

While I will return to Shafarevich's anti-Semitic views in the next section, which were more theoretical than practical in reality (at least during the 1960s and the 1970s), it is indeed well known that Pontryagin and Vinogradov were anti-Semites and did exercise their power to obstruct the careers and research of Jewish mathematicians—accounts on Pontryagin's views were published as the end of the 1970s.[142] Whereas Freiman gives a vivid description of Vinogradov's anti-Semitism,[143] Pontryagin is known that in "only a few years [. . .] [as] the new editor-in-chief of a leading mathematics journal, [he] has lowered the proportion of articles authored by Jews from one third of the total to practically zero."[144] How and whether this situation affected the formation algebraic geometry school of Shafarevich is to be seen in the next section.

---

[138] Freiman (1980, p. xii–xiii).

[139] Gohberg (1989, p. 21–24, 45–46, 53–54).

[140] Vershik (1994). Cf. also Fuchs (2007, p. 215): "As a rule in the years 1970–1988 there were some three to five Jews or half Jews among the 500 students of Mekh-Mat [the department of mechanics and mathematics of the Moscow university] each year."

[141] Shifman (2005, p. 8).

[142] See: Bari Kolata (1978) and Pontryagin's response, in: Pontryagin (1979).

[143] Freiman (1980, p. 43–49). Edward Frenkel (2013, p. 68) notes also that "the Mathematical Branch of the Academy was for decades controlled by the director of the Steklov Mathematical Institute in Moscow, [. . .] Vinogradov, nicknamed the 'Anti-Semite-in-Chief of the USSR.' Vinogradov had put in place draconian anti-Semitic policies at the Academy and the Steklov Institute, which was in his grip for almost fifty years."

[144] Zaslavsky / Brym (1983, p. 112).

## 5.3.2    Before Braid Monodromy: The Shafarevich School, Moishezon and the Decomposition of Algebraic Surfaces

"Around 1960 there was a renewed interest in these questions [of classification of surfaces]. [...] Shafarevich and his school put the arguments of Castelnuovo and Enriques on more secure foundations than their 'intuitive' proofs."[145]

Dieudonné's succinct description above summarizes Shafarevich's seminar, emphasizing, using an architectural metaphor, the need for "secure foundations". That being said, the description hardly notes what these "secure foundations" were, neither their motivation nor the mathematical and sociological background of this seminar. As was mentioned above, during the years 1961–1963 a seminar at Moscow University was organized by Shafarevich, in which the theory of algebraic surfaces was studied and reformulated from a more modern point of view. Moishezon, as noted above, was an active participant at the seminar.[146] Shafarevich noted, in the introduction to his book *Algebraic Surfaces* that it became clear for the Italian school "that the results making up the 'classical' theory of algebraic surfaces fall into two fundamentally different classes."[147] The first class includes all results, which are special cases of theorems on algebraic varieties of arbitrary dimension. At this point a reference to Zariski's 1935 book *Algebraic Surfaces* pops up: "It is interesting to note that almost all the results in the classical survey of Zariski are exactly of this kind,"[148] but the explicit title of his book is only given at the end in the bibliography. The second class includes "the great complex of results which are grouped together by the Italian algebraic geometers under the heading of 'classification of algebraic surfaces'. It seems that none of these results can be extended to varieties of higher dimension [...]."[149] The aim of the seminar, and of the resulting book, was to present these results with modern techniques. Indeed, Shafarevich noted that the

"results presented in this book may almost all be found in the survey of Enriques' [*Le superficie algebriche*] [...] The proofs of a large part of the theorems are based on [the] ideas of Enriques. At the same time, it would hardly be possible to carry out the details of Enriques' proofs, for example, following the customs of the time and his school, he frequently limited himself to the consideration of a 'general' case, not choosing the most unpleasant cases that might be examined. On the other hand, for certain questions we can supplement the classical results with new ones."[150]

---

[145] Dieudonné (1985, p. 121).

[146] The other participants of the seminar who contributed to the book *Algebraic surfaces* were, besides Shafarevich, Yurii Ivanovich Manin, Ju. R. Vainberg, Aleksei Borisovich Zhizhchenko, Andrei Nikolaevich Tyurin, and Boris Gershonovich Averbukh.

[147] Shafarevich (1967, p. iii).

[148] Ibid. The text refers here to Zariski's *Algebraic Surfaces*. However, there is only one reference to a result from the book (see: ibid., p. 4).

[149] Ibid., p. iii.

[150] Ibid., p. v–vi.

Indeed, in 1989 Shafarevich described his feelings while preparing for the seminar as frustrating yet fascinating: "you read Enriques's book and it is totally unclear whether you will be able to extract something from it to give your talk. But it was very interesting. We wrote the book *Algebraic Surfaces* from it."[151]

Shafarevich noted that the basic method used in the book "is the method of coherent sheaves",[152] referring to two papers: Serre's 1955 "Faisceaux algébriques cohérents" and Zariski's 1956 "Algebraic sheaf theory". Making also use of the analytic methods of Kodaira and Spencer,[153] Shafarevich and his students not only made Enriques's results more rigorous, but also gave new proofs[154] or corrected mistakes and cases which Enriques ignored.[155] Moreover, as Aleksei Nikolaevich Parshin notes, "this seminar and the book served as the main impetus for the further development of algebraic geometry in Moscow."[156] Taking this into account, it is essential to note the obvious: the name of Shafarevich's book refers to Zariski's 1935 book *Algebraic Surfaces*, and hence posits itself implicitly as having the same 'classical' status as his book. However, the subjects of the chapters of the book have a substantially different focus than Zariski's: the book begins with chapters on birational transformation, minimal models and rationality, and afterwards discusses ruled and rational surfaces, "surfaces of Fundamental Type" (called today surfaces "of general type"), algebraic surfaces with Kodaira dimension 0 (see below) as well as Enriques and Kummer surfaces. Any equivalent treatment of the topics of chapter VIII of Zariski's book on branch curves is not to be found.[157] If branch curves were a somewhat marginal topic in this book (and probably of the seminar), how did this influence Moishezon's work? Before examining this topic, I would like to remark on Shafarevich's anti-Semitism, which may have played a role in Moishezon's emigration.

* * *

The subject of Shafarevich's anti-Semitism and political views has been discussed thoroughly in several works.[158] To recall, it was Shafarevich's samizdat "Russophobia", published in the western world in 1988 where he expressed his anti-Semitism. However,

---

[151] Zdravkovska / Shafarevich (1989, p. 26).

[152] Shafarevich (1967, p. 2).

[153] See: Parshin (2006, p. 313).

[154] Shafarevich (1967, p. 193).

[155] Ibid., p. 182.

[156] Parshin (2006, p. 314).

[157] Also, Zariski's book did not deal with the classification of surfaces, but Shafarevich's book did.

[158] See for example (Berglund 2012), which rejects the view of Shafarevich as anti-Semitic. Borenstein, on the other hand, criticizes heavily Berglund's defense of Shafarevich and does consider Shafarevich as anti-Semitic (2019, p. 122–132). He notes: "Berglund rejects the label, both because Shafarevich himself claims not to be antisemitic, and because she does not find explicit antisemitism in Shafarevich's writings. While it is certainly true that [Shafarevisch's] *Russophobia* is exceedingly careful in its terminology, it is nonetheless rather clear in its intent." (ibid., p. 124) Moreover, Svetlikova (2013, p. 5) notes that the "combination of mathematics, politics, ultra-nationalism, and

Serge Lang argues that Shafarevich's views and actions do not admit a simplistic formulation. During the 1960s, Shafarevich did accept many Jewish students, and never considered
it as important, according to Igor Dolgachev, to check the religion of his students.[159]
Moishezon was one of Shafarevisch's associates, as well as Piatetski-Shapiro. While one
may claim that Shafarevich was not hostile to any Jew individually, perhaps considering
highly the mathematical prowess of an individual, appreciating it enough to disregard her
or his religion, it seems, on the other hand, according to Dolgachev, that cultural prejudices
toward Jews "may apply to Shafarevich [...] defined on the basis of accusing Jews of
corrupting a given culture by supplanting it with a uniform, crude 'Jewish culture'. The
main purpose of ["Russophobia"] [...], and of his whole life outside mathematics, was not
to express hatred of Jewish people and Jewish culture, as it was claimed, but rather to
defend Russian people [...]."[160] However, Borenstein claims that "Shafarevich's own
preoccupations, when it comes to both Jews and Russophobia, demonstrate how a basic
conspiratorial structure can simultaneously reinforce an existing prejudice (antisemitism)
and become so portable that it no longer relies on the initial target of its hatred as the source
of its power. Shafarevich's critique of the Jewish people is based on a set of misguided
notions [...]."[161]

While it is clear that during the 1960s and 1970s Shafarevich did not express openly any
anti-Semitic opinion towards individual Jews (in contrast to his colleagues, such as
Vinograd) and that he collaborated with and supported his Jewish students, he did not
support the emigration movement—neither immigration to Israel nor to the USA. Piatetski-
Shapiro noted that Shafarevich's reaction was very negative when he heard that he wanted
to leave USSR, and tried to persuade him not to do that. On the same note, Piatetski-
Shapiro recalled that Shafarevich also tried to convince Moishezon not to emigrate, which
might indicate a certain ignorance to the situation of Jewish mathematicians in USSR
during the 1970s.[162]

Moishezon    described    his    own    shock    when    reading    Shafarevich's    ideas    in
"Russophobia", and though Moishezon refused to write to him a letter of protest,[163] one
may suggest that Moishezon expressed his disappointment from and rejection of
Shafarevich's ideas in two books he wrote. The first book, published in 1990, is called
*Creation of the myth: notes on the identification of "Yurovsky"* [Сотворение мифа:
заметки об идентификации "Юровского"].[164] The book deals with the execution of

---

anti-Semitism was present in Moscow as early as the beginning of the twentieth century", and
underlines that the research of Shafarevich's views must take this into account.

[159] Lang (1998, p. 738).

[160] Dolgachev (2018).

[161] Borenstein (2019, p. 124).

[162] Piatetski-Shapiro (1993, p. 208).

[163] Moishezon (1992, p. 62).

[164] Moishezon (1990).

the Russian Czar, Nicolas II, and his family in 1918. In the USSR, for Moishezon, "there was an officially-accepted version that the head of the execution team was Jacob Yurovsky. The Soviet propaganda exploited the Jewish origin of Yurovsky, implying the crucial role of Jews in the tragedy."[165] In his book Moishezon challenged this "official version", and reached the conclusion that the real organizer of the murder was a well-known polish communist. Moishezon's second book, published posthumously in 2001, is called *Armenoids in Prehistory*. The book "attempts to reconstruct the oldest periods of the prehistory of the Ancient Near East and the neighboring areas by combining recent discoveries in several fields", as the philologist Vyacheslav V. Ivanov notes in the foreword, calling Moishezon a "brilliant scholar."[166] As Moishezon noted, the books ends with "the problem of Israel's emergence" and how to show its historical accuracy: "The attempt to understand the *Mystery of Israel* brings us back to the Biblical concepts of Holy Land [...]."[167] These two books, while showing that Moishezon's interests and horizons were far wider than only mathematics, indicate also how he devoted a lot of energy during the last years of his life to present a more positive image of Jews and the Jewish people; this, one might suggest, came also on the background to Shafarevich's anti-Semitic views.

* * *

Returning to discuss the mathematical research of Shafarevich and Moishezon, as was noticed above, Shafarevich and his students were considering again, now with new techniques, the classification of surfaces. As described in Sect. 3.1.3, Enriques and Castelnuovo already obtained a classification of surfaces in 1914, according to the plurigenera $P_i$, the geometric and the arithmetic genera ($p_a$ and $p_g$), and $p^{(1)}$. Shafarevich introduced in his book another classification: "In order to examine all these algebraic surfaces, we divide them into four groups on the basis of the value of an important invariant, which we denote by $\kappa$. The symbol $\kappa$ stands for the maximal dimension of the image of the surface under rational mappings corresponding to different multiplicities of the canonical class."[168] This symbol is later called the *Kodaira dimension*, which I will explain below; the dimension is an invariant introduced in Kodaira's 1960 paper: "On compact complex analytic surfaces".[169] Not only did Kodaira extend the classical results of Enriques and Castelnuovo concerning algebraic surfaces, but also treated non-algebraic surfaces, that is, all compact complex surfaces. The "Table of Types of Algebraic Surfaces",[170] to be found at the introduction of Shafarevich's *Algebraic Surfaces*, introduces the classification of surfaces according to this dimension (which can be $-1$,

---

[165] Moishezon (2001, p. 216).

[166] Ibid., p. xi.

[167] Ibid., p. 4.

[168] Shafarevich (1967, p. iv).

[169] Kodaira (1960).

[170] Shafarevich (1967, p. 6).

0, 1 or 2),[171] though the usage of the plurigenera is also to be found in this table. This certainly underlines the shift not only in the language of algebraic geometry but also in the research configuration one worked with, and hence also in the classification project; when compared to Enriques and Castelnuovo's classification table, Enriques and Castelnuovo's classified surfaces according to the plurigenera $P_{12} = 0$, $P_{12} = 1$, and $P_{12} > 1$, and then checking the subcases according to the other different genera.[172]

As the Kodaira dimension $\kappa$ is defined as the "maximal dimension of the image of the surface [under various mappings]",[173] its maximal value is 2 (as the dimension under a mapping can either stay the same or decrease). The class of surfaces for which $\kappa = 2$ is called surfaces of a 'general type'. It was Moishezon who wrote the chapter on these surfaces in the book *Algebraic surfaces*, concentrating on their possible birational embedding via the pluricanonical maps.[174] However, besides his work on the embedding of surfaces of a general type, one should mention two other important pieces during the 1960s. The first is his work from 1964 on a necessary and sufficient condition for a divisor to be ample.[175] The second important result concerns what is now known as *Moishezon manifolds*, being compact complex manifolds of dimension $n$ with $n$ algebraically independent meromorphic functions. Moishezon proved that such a manifold is a projective algebraic variety if and only if it admits a Kähler metric;[176] this result emphasizes Moishezon's research of Kählerian manifolds. However, none of the results during the 1960s suggest Moishezon showed a renewed interest in branch curves.

One of the main works of Moishezon during the 1970s was his book *Complex Surfaces and Connected Sums of Complex Projective Planes*. Already discussed in Sect. 5.2.1, in this book, to recall, one can find the beginnings of Moishezon's work on braid monodromy factorization. To present Moishezon's results regarding the braid monodromy factorization in a larger context, one of the main theorems presented in the book is that for any algebraic surface $S$ in a large class of algebraic surfaces (containing all simply connected elliptic surfaces and complete intersections) the connected sum $S \# \mathbb{CP}^2$ decomposes as a connected

---

[171] Note that if "the linear systems corresponding to all the multiplicities of the canonical class are empty, then we set $\kappa = -1$." (ibid., p. iv)

[172] Enriques (1949, p. 463–464).

[173] Shafarevich (1967, p. iv).

[174] Ibid., p. 113–161. One of the important results of Moishezon in this chapter was to prove that "for any surface of fundamental type [i.e. of general type] the system $|9K|$ gives a birational imbedding" (ibid., p. 113); see also: ibid., p. 136–137.

[175] Moishezon (1964). For the case of algebraic, smooth projective surface $S$, the Moishezon-Nakai criterion states that the divisor $D$ is ample if and only if its self-intersection number $D \cdot D$ is strictly positive, and for any irreducible curve $C$ on $S$ we have $D \cdot C > 0$.

[176] Moishezon (1966). Note that complex algebraic varieties always have this property, but the converse is not true (the counter example is known as Hironaka's example from 1962). For compact manifolds one can show that the field of all meromorphic functions on the manifold has *at most n* algebraically independent meromorphic functions.

sum of copies of $\mathbb{CP}^2$ and $\overline{\mathbb{CP}}^2$ (where the bar denotes reversed orientation).[177] This theorem already indicates that Moishezon was interested in obtaining a simpler presentation of algebraic surfaces; this goal—achieving such a presentation, or even achieving the simplest one—emerges also, as we will see, when Moishezon will discuss normal forms of braid monodromy of plane curves.

Looking closer at how Moishezon formulated his results, one can note that he called a surface $S$ "almost completely decomposable"[178] when the connected sum $S\#\mathbb{CP}^2$ decomposes as a connected sum of copies of the complex projective plane $\mathbb{CP}^2$ and complex projective plane with inversed orientation $\overline{\mathbb{CP}}^2$. A surface is *completely decomposable* when it decomposes as a connected sum of copies of $\mathbb{CP}^2$ and $\overline{\mathbb{CP}}^2$. Theorem $A$, presented in Moishezon's book, deals with any manifold $V$, which is "compact simply-connected 4-manifold which admits a complex structure"; in particular, all the simply connected algebraic complex surfaces are in this set. The theorem finds explicitly the numbers $k_1$ and $k_2$, such that $V\#k_1\mathbb{CP}^2\#k_2\overline{\mathbb{CP}}^2$ is completely decomposable, when the numbers depend on the intersection form of $V$, being a topological invariant.[179] However, theorem $B$ states that "any simply-connected elliptic surface $V$ is almost completely decomposable",[180] which means that in this case $k_1 = 1$, $k_2 = 0$. While theorem $A$ is, according to Moishezon, "theoretical", theorem $B$ is "empirical". Explicitly, Moishezon declared:

> "Note that Theorem $B$ together with [other results] [...] shows that all big explicit classes of simply-connected algebraic surfaces considered until now have the property that their elements are almost completely decomposable 4-manifolds. That is, the 'theoretical' Theorem $A$ gives much weaker results than our 'empirical knowledge'. The interesting question is, how far we can move with such 'empirical achievements' in more general classes of simply-connected algebraic surfaces."[181]

What do the terms "empirical" vs. "theoretical" mean in this framework? I will discuss these terms later in Sect. 5.3.3.2, once I have examined Moishezon's works on surfaces and branch curves, which is the subject of the next section.

### 5.3.3  From 1981 to 1985: (Re)introducing Braid Monodromy

> "Moishezon [in 1981] [...] attempted to [...] obtain invariants of an embedded projective surface $X \subset \mathbb{P}^N$ using the branching curve $C$ of a generic linear projection onto $\mathbb{P}^2$, and he

---

[177] See: Moishezon (1977, p. 2–4) for the summery of the results.

[178] Ibid., p. 3.

[179] Ibid.

[180] Ibid.

[181] Ibid., p. 4.

viewed $\pi_1(\mathbb{P}^2 \backslash C)$ as an invariant of deformation type of the pair $(X, \mathbb{P}^N)$. [. . .]. In a series of papers [. . .] explicit calculations for many concrete pairs $(X, \mathbb{P}^N)$ were carried out. Moishezon used the interpretation of automorphisms [. . .] in the Zariski–van Kampen presentation as *braids* (as was done earlier by Chisini), viewing the latter as the isotopy classes of diffeomorphisms of a disk fixing a finite subset of points which are identity on the boundary. [. . .] A special type of combinatorial analysis, partially motivated by Hurwitz's analysis of monodromy with values in a symmetric group, was developed for finding 'normal forms' for the braid monodromy of singular curves."[182]

As Libgober notes in the above citation, the novelty of Moishezon's introduction of braid monodromy during the 1980s was the understanding—in a similar manner as Chisini understood before—that the "crucial piece of data associated with the curve"[183] is the factorization, or the tuple, of elements from the braid group, a factorization which characterizes the branch curve and eventually determines the surface.[184] For Libgober, "Moishezon used the interpretation" of automorphisms "as braids". This implies that the epistemic configuration in which Moishezon worked was different from that of Zariski and, more decisively, from that of Chisini. One may however claim that both Chisini and Moishezon dealt with the factorizations of braids. Hence, how was Moishezon's approach essentially different from that of Chisini? An indication of this difference was already seen in Sect. 5.2, when during the 1970s Moishezon considered, on the one hand, factorizations of elements in the braid group of three strings associated to curves, and on the other hand, factorizations of elements $SL_2(\mathbb{Z})$ related to Lefshetz fibrations. In order to fully answer this question, I will examine in depth Moishezon's 1981 paper "Stable Branch Curves and Braid Monodromies", followed by an analysis of his two papers "Algebraic Surfaces and the Arithmetic of Braids I" resp. "II", published in 1983 and 1985.

### 5.3.3.1 1981: The Search for Normal Forms

Moishezon's 1981 paper begins with comparing between complex curves and complex surfaces. Given a smooth complex curve $X$ and a finite morphism $g : X \to \mathbb{CP}^1$, let $n$ be the degree of $g$ and $M$ the set of branch points on $\mathbb{CP}^1$. Moishezon noted "[t]here is a natural surjection $s : \pi_1(\mathbb{CP}^1 - M) \to Sym_n$ ([the] symmetric group of degree $n$), which is called 'the monodromy of $g$'." In other words, this is the epimorphism sending a loop encircling a branch point to the induced permutation describing the exchange between the sheets of the Riemann surface. Moishezon added that it is a classical theorem that a "normal form for [this] monodromy" exists, associating to every loop in $\pi_1(\mathbb{CP}^1 - M)$ a permutation in a unique way; and that one can always choose a certain basis for $\pi_1(\mathbb{CP}^1 - M)$, denoted by $\gamma_1, \ldots, \gamma_n$, such that $\gamma_1 \cdot \ldots \cdot \gamma_n = 1$ and that $s(\gamma_k) = (k, k + 1)$ for $k = 1, 2, \ldots, n - 1$;

---

[182] Libgober (2014, p. 483).

[183] Ibid.

[184] Moishezon was not only working on this subject during the 1980s. For example, the first examples of differentiably distinct but oriented homeomorphic surfaces of the general type are due to Moishezon and can be found in (Friedman / Moishezon / Morgan 1987).

$s(\gamma_k) = (2n - k - 1, 2n - k)$ or $k = n, n + 1, \ldots, 2n - 2$; and $s(\gamma_k) = (1, 2)$ for all the other $k$'s. For Moishezon, this "theorem provides a clear way for construction and classification of [. . .] Riemann surfaces".[185]

The question that arises is whether an analogous "normal form" exists for surfaces. As was already seen, given a complex (algebraic) surface $Y$ and a finite morphism $f : Y \to \mathbb{CP}^2$ of degree $n$, $\Sigma$ the branch curve of $f$ in $\mathbb{CP}^2$, there is, so Moishezon, a "natural surjection" $s : \pi_1(\mathbb{CP}^2 - \Sigma) \to Sym_n$. This is the epimorphism, already described by Enriques, sending a loop encircling the branch curve to the induced permutation, describing the exchange of sheets of $Y$ (when following the preimages of the loop). The explicit problem, according to Moishezon, "is that we know little about [the] groups $\pi_1(\mathbb{CP}^2 - \Sigma)$", indicating—without proof or reference, as many before him—that the branch curve "is a singular plane curve with only nodes and cusps".[186] The implicit problem is that knowing only the surjection $s$ does not yield the same results as in the case of the complex curve. In order to characterize the branch curve more accurately, Moishezon turns to "studying singular plane curves based on the notion of the so called braid monodromy."[187]

In order to do that, Moishezon takes the definition of the braid group $B_n$ as the "group of homotopy equivalence classes of homeomorphisms" of a disk $P$ fixing a finite subset of $n$ points, when these homeomorphisms are identity on the boundary.[188] Taking $\overline{P}$ as the projective complex line, $B'_n$ is defined in a similar way to $B_n$ (as the group of classes of homeomorphisms of $\overline{P}$), with the additional condition that the point at infinity remains fixed. Moishezon then proves $B'_n \cong B_n / \mathrm{Center}(B_n)$.

The second step in Moishezon's configuration is looking at a complex curve $C$ of degree $n$ in $\mathbb{CP}^2$, and a projection $p$ of $C$ from a point $O$ not on $C$ onto $\mathbb{CP}^1$. Moishezon denoted by $M$ the set of branch points in $\mathbb{CP}^1$ with respect to $p$. Choosing a point $u$ in $\mathbb{CP}^1 - M$, Moisezon defined "the braid monodromy [. . .][as] the following naturally defined homomorphism:"[189]

$$\overline{\theta} : \pi_1\left(\mathbb{CP}^1 - M, u\right) \to B'_n$$

which sends a loop to the braid induced from the motion of the preimages of $p^{-1}(u)$, in a manner similar to Chisini's construction (see Fig. 4.1). Here Moishezon used the definition of $B'_n$: the points of $p^{-1}(u)$ are considered as situated in a disc $\overline{P}$, and the loop induces a homeomorphism of $\overline{P}$ which fixes those points.

---

[185] Moishezon (1981, p. 107–108).

[186] Ibid., p. 109.

[187] Ibid.

[188] Ibid., p. 111. At: Ibid., p. 112, he proved the equivalence of this definition to the algebraic one. For the various definitions of the braid group, see the Appendix to Chap. 4 (Sect. 4.4).

[189] Ibid., p. 114.

For the same curve $C$ in the affine space $\mathbb{C}^2$, he defines the homomorphism into the braid group, replacing $B'_n$ with $B_n$:

$$\theta : \pi_1\left(\mathbb{C}^1 - M, u\right) \to B_n$$

Moishezon's declaration that he works with "braid monodromy" may lead one to think that Moishezon's configuration has parallel traits to Chisini's; however, Moishezon's configuration is different: not only is the formulation different, but he also uses a group-theoretic language, which Chisini, as we saw, rejects; the above definition of the braid group was not known to Chisini.

Moreover, one might say that these "naturally defined" homomorphisms are what Chisini considered visually with his material models and diagrams: the homomorphism $\theta$ (or $\bar{\theta}$) sends a loop based at $u$ to the (homotopy equivalence class of the) motion of the points of the fiber $p^{-1}(u)$, while moving along the loop, hence inducing a braid (compare Chisini's 'braid' in Fig. 4.1). But it would be erroneous to assume that Chisini and Moishezon researched the *same* object, as already the initial setting is very different: While Moishezon emphasized group-theoretic language, Chisini stressed the usage of material models. Also, in contrast to Chisini, Moishezon did not draw a single image in the 1981 paper.

Returning to Moishezon's construction, he chose a set of generators for $\pi_1(\mathbb{C}^1 - M, u)$ (resp. $\pi_1(\mathbb{CP}^1 - M, u)$), denoted as $\gamma_1, \ldots, \gamma_N$ (resp. $\bar{\gamma}_1, \ldots, \bar{\gamma}_N$), and a set of generators to the braid group, denoted as $X_1, \ldots, X_{n-1}$. He then noted that $\prod_{i=1}^N \bar{\theta}(\bar{\gamma}_i) = 1$ and $\prod_{i=1}^N \theta(\gamma_i) = \Delta^2$, where $\Delta^2$ is the generator of Center($B_n$) (see Sects. 4.4 and 5.2). At this point Moishezon concentrated on the case of branch curves: "From the point of view of the theory of algebraic surfaces [...] the most interesting case is when all singularities of $C$ are only nodes and cusps."[190] Noting that as group elements, $X_2, \ldots, X_{n-1}$ are all conjugate to $X_1$,[191] Moishezon formulated in a group theoretic language what should be considered as the generic form of the homomorphism $\theta$: "For a plane curve $C$ with only nodes and cusps [...] the braid monodromy $\theta$ is given by $\theta(\gamma_i) = Q_i^{-1} X_1^{\rho_i} Q_i$ where $\rho_i = 1$ for ordinary branch points, $\rho_i = 2$ for nodes and $\rho_i = 3$ for cusps", and $Q_i$ are elements of the braid group $B_n$.[192]

In contrast to Chisini's configuration, the space of inscription of Moishezon's configuration consists of "symbolically" formalized expressions, as he explicitly indicated: "The braid monodromy $\theta$ [...] is *symbolically* given by a formula $\Delta^2 = \prod_{i=1}^N Q_i^{-1} X_1^{\rho_i} Q_i$", and

[190] Ibid., p. 116.

[191] Ibid., p. 116. An element $a$ is conjugate to an element $b$ in a group $G$, if there exists an element $g$ such that $gag^{-1} = b$.

[192] Ibid., p. 117.

while choosing another set of generators to $\pi_1(\mathbb{C}^1 - M, u)$ "then our braid monodromy will by symbolically given by an 'equivalent' formula [. . .]."[193] The main task is therefore to choose the proper basis for $\pi_1(\mathbb{C}^1 - M, u)$, so that the formula of $\Delta^2$ "has a nice and simple form. Then (at least, on [an] *intuitive* level) we can say that we found a 'normal form' for the braid monodromy."[194] Immediately afterwards, Moishezon noted that choosing another basis (instead of $\gamma_1, \ldots, \gamma_N$) is equivalent to performing a sequence of elementary transformations on the corresponding tuples of $\gamma_1, \ldots, \gamma_N$, as described by Livne (see Sect. 5.2). Operating $\theta$, it is possible to consider these elementary operations on the $N$-tuple of $Q_i^{-1} X_1^{\rho_i} Q_i$. Hence Moishezon concluded by saying that "two formulae [for $\Delta^2$] [. . .] are equivalent [. . .] if the sequence [corresponding to the second tuple] [. . .] could be obtained from the sequence [corresponding to the first tuple] [. . .] by a finite sequence of elementary transformations." Therefore, "[w]e could ask also a purely algebraic question: Find some 'natural' expressions for $\Delta^2$ as a product $\prod_{j=1}^{s} Y_j$, where each $Y_j$ is a conjugate of some $X_1^{\rho_i}$, $\rho_i \in \{1, 2, 3\}$."[195]

It is clear that the emphasis of Moishezon is algebraic, and that the "symbolic" normal form is considered as a "natural" expression, as the "intuitive" one. The topological setting of determining the homomorphism $\theta$, in which the definition of the braid monodromy was situated before—that is, considering the induced homeomorphism of the disc (or plane) fixing the fiber points—is now replaced with an algebraic one, when the question is formulated in terms of finding "nice and simple form[s]" of products of $\Delta^2$. Methods of verifying various operations on the braid monodromy factorization are done only algebraically, without even imagining using a material model, as Chisini did. Moreover, it is not the (drawing of the) motion of the points in the fiber or its material model that is intuitive, but rather finding the algebraic symbolic "normal form". In this sense, one can see a major difference in how Moishezon considered "intuition" and "experiment" in mathematics, in comparison to Enriques and Chisini.

* * *

After declaring one of the main questions of the paper, Moishezon presented several examples of finding these normal forms. For a non-singular plane curve $C$ of degree $n$, it is well known (already with Monge and Salmon, for example, see Chap. 2), that it has $n(n - 1)$ branch points; Moishezon proves that "braid monodromy [of the curve] [. . .] is symbolically given by the formula"[196]

---

[193] Ibid.

[194] Ibid.

[195] Ibid., p. 119.

[196] Ibid., p. 120. Note that Chisini proved a similar result in 1937 (see Sect. 4.1.1).

$$\Delta^2 = \left(\prod_{i=1}^{n-1} X_i\right)^n.$$

This formula is a generalization of the result of Livne in his thesis, that every product of six elements in $B_3$ which is equal to $\Delta^2$ can be transformed to the "'canonical' form $xyxyxy$";[197] since Livne clearly referred in the introduction of his MA Thesis to a smooth cubic curve, it is clear that the formula above, when substituting $n = 3$, gives Livne's result. Moishezon then supplied the formulas for $\Delta^2$ for a union of $n$ lines in general position[198] and for rational plane curves (of degree $n$) with only nodes as singularities. The first part of the paper ends with what one may call a summary as well as a reformulation of Zariski's and van Kampen's results in the language of the braid group: the presentation of the fundamental group of the complement of a curve, as well as the relations between the generators in terms of the conjugating *braids* $Q_i$. The fact that Moishezon presented this group as derived from the braid monodromy may indicate that he perceived his epistemic configuration—the research on braid monodromy factorization—as more basic than the one only focusing on the fundamental group of the complement: indeed, it is the factorization that is described as "natural" and "intuitive", and not the fundamental group.

In the second and the third parts of the paper Moishezon turned to a more algebro-geometric approach, concentrating only on branch curves; he began by calculating the braid monodromy of a branch curve of a specific surface in $\mathbb{C}^3$, $z^n + xz + y = 0$, and then of a deformation of this surface, given by the equation $z^n + \varepsilon z^{n-1} + xz + y = 0$. In the third part Moishezon computed the factorization of $\Delta^2$ associated to the branch curve $S$ of a smooth surface $V$ of degree $n$ in $\mathbb{CP}^3$, proving that:[199]

$$\Delta^2 = \prod_{j=n-2}^{0} \left(\prod_{k=0}^{n-2} Z_{j,k;j+1,k}\right) \cdot \prod_{j=0}^{n-1} \left[Z^3_{j,0;j,1} \cdot \prod_{k=2}^{n-2}\left(\left(\prod_{k'=0}^{k-2} Z^2_{j,k';j,k}\right)Z^3_{j,k-1;j,k}\right)\right]$$
$$\cdot \prod_{j=n-2}^{0}\left(\prod_{k=0}^{n-2} Z_{j,k;j+1,k}\right).$$

He then used this presentation of $\Delta^2$ to prove that $\pi_1(\mathbb{CP}^2 - S)$ is isomorphic to $B'_n$. The way to prove it was Moishezon's realization, that from the factors of $\Delta^2$ one can induce all of the relations in $\pi_1(\mathbb{CP}^2 - S)$. This result exemplifies Moishezon's method: one can derive from an algebraic expression a topological invariant. This result is also a generalization of Zariski's result for cubic surfaces, as Moishezon remarked:[200] indeed, for $n = 3$,

---

[197] Livne (1975, p. 8).

[198] Note that a similar result was obtained in Chisini in 1937, though presented in Chisini's notation and configuration.

[199] Moishezon (1981, p. 161). The terms of the form $Z_{j, k; i, l}$ are elements in the braid group which are defined previously in the same paper.

[200] Ibid., p. 176, 178; see also Sects. 3.2.1 and 3.2.3 on Zariski's result.

$\pi_1(\mathbb{CP}^2 - S) \cong B_3' = B_3/\Delta^2 \cong \mathbb{Z}/2\mathbb{Z} * \mathbb{Z}/3\mathbb{Z}$, which is exactly what Zariski proved in 1929, though using completely different methods.

The fourth and the final part of the 1981 paper presents "braid monodromies as a way for understanding algebraic surfaces".[201] As presented at the beginning of the 1981 paper, the "general ideology"[202] of Moishezon is to study (smooth) complex surfaces embedded in a projective space $\mathbb{CP}^r$ as a covering of the projective complex plane. "In this case", noted Moishezon, "we can first project [the surface] to $\mathbb{CP}^3$ to get a surface $V_n$ ($n = $ degree $f$) with ordinary singularities $V_n$ could be considered as a degeneration of a nonsingular hypersurface $W_n$ in $\mathbb{CP}^3$."[203] Moishezon's hence aimed to investigate the effect of this degeneration on the corresponding branch curves—when the final degeneration is a union of $n$ planes. Indeed, already in 1977 Moishezon (in the book *Complex Surfaces and Connected Sums*) dealt with projections of projective algebraic surfaces into $\mathbb{CP}^3$, examining their double curve $D_0$ and its singularities.[204] Given $Y$ a degeneration of a surface $W$, if $\Sigma$ is the branch curve of the covering $f: Y \to \mathbb{CP}^2$, $D$ is the image of $D_0$ under $f$, and $S$ the branch curve of a generic projection $W \to \mathbb{CP}^2$, then—"$D \cup \Sigma$ will be a 'degeneration' of $S$, and the braid monodromy of $D \cup \Sigma \subset \mathbb{CP}^2$ will be a 'degeneration' of the braid monodromy of $S \subset \mathbb{CP}^2$. The last 'degeneration' could be understood as a 'degeneration' of a finite presentation of $B_n' \cong \pi_1(\mathbb{CP}^2 - S)$ to a finite presentation of $\pi_1(\mathbb{CP}^2 - D - \Sigma)$ [...]. We hope that this approach could be a certain clue to the 'mystery of surfaces of general type'."[205] As we saw in Chap. 4, degenerations of surfaces and curves were a common way for several mathematicians in the Italian school of algebraic geometry to investigate algebraic surfaces (between the 1930s and the 1950s), though Moishezon did not mention this tradition and the associated mathematicians, and probably was not aware of their results.

The motivation for presenting his "general ideology" lies not only in unraveling the "mysteries of surfaces of the general type", which was Moishezon's specialty, as he wrote the corresponding chapter on these surfaces in Shafarevich's book. In the larger context, it should be noted that this is yet *another program of classification of surfaces*, though now concentrating on braid factorizations associated to the branch curve of their projective models (and not on their Chern classes, the Kodaira dimension or the plurigenera, as other research configurations did). But another motivation should be made explicit: the fact the branch curve $\Sigma$ is not a typical nodal-cuspidal plane curve. This was, to recall, already noted by Enriques, Zariski and Wahl; Moishezon gave another reason why the branch curve is special: this is since for a branch curve, being a nodal-cuspidal curve, if one

---

[201] Ibid., p. 180.

[202] Ibid., p. 109.

[203] Ibid., p. 110.

[204] See "Appendix 1: Generic projections of algebraic surfaces into $\mathbb{CP}^3$" in: (Moishezon 1977, p. 72ff)

[205] Moishezon (1981, p. 110).

denotes by "$\rho$ the number of cusps of $\Sigma$ and by $d$ the number of nodes of $\Sigma$, [then] it is not difficult to prove (using, for instance, $c_1^2 + c_2 = 0$ ([*mod*] 12)) that $\rho = 0$ (*mod* 3) and $d = 0$ (*mod* 4). Both these facts, together with the fact of existence of an epimorphism $\pi_1(\mathbb{CP}^2 - \Sigma) \to S_n$, show that $\Sigma$ must be a curve of a very special type".[206] To stress: the statement that the number of cusps (resp. nodes) is divisible by 3 (resp. 4) appears here for the first time in the literature, though without any explicit proof.[207]

How special the situation is from an algebraic point of view is emphasized in the following "commutative diagram of epimorphisms",[208] and S and $\Sigma$ are, as denoted above, the branch curve of a smooth surface and the branch curve of a degeneration of this smooth surface respectively:

$$\begin{array}{ccc} \pi_1(\mathbb{CP}^2 - S) & \to & B'_n \\ \downarrow & & \downarrow \\ \pi_1(\mathbb{CP}^2 - \Sigma) & \to & Sym_n \end{array}$$

Here Moishezon noted that "[t]his diagram gives [an] explanation of the [...] 'experimental evidence' that $\pi_1(\mathbb{CP}^2 - \Sigma)$ 'closely related' to $B'_n$ and that $\pi_1(\mathbb{CP}^2 - \Sigma) \to Sym_n$ is 'close to being unique.'"[209] The uniqueness of this epimorphism and hence of the branch curve, that is, the fact that a branch curve is a branch curve of *only one* surface, is the content of Chisini's 'theorem' from 1944 (see Sect. 4.1.3), but it is not clear whether Moishezon was aware of the 'theorem' of Chisini at that time.[210] Moishezon hence emphasized his program to prove this uniqueness with braid monodromy and possible degenerations: by employing "symbolical" expressions (to use Moishezon's terms). This program—and a relativization of it—are expressed in the following: "The point of view formulated above on using a regeneration of $V_n$ to a nonsingular $W_n$ [embedded in] $\mathbb{CP}^3$ and the braid monodromies [...] sets a *program* which will possibly demand a lot of time and energy to be carried out."[211] The starting point of this program is to consider "from the experimental point of view" a degeneration of $Y$ into an arrangement of hyperplanes in $\mathbb{CP}^r$, being the complete degeneration of a surface.

---

[206] Ibid., p. 180.

[207] The proof that the number of cusps is divisible by 3 is given in (Moishezon/Teicher 1990, p. 132–133); the proof that the number of nodes is divisible by 4 is given neither in (Moishezon/ Teicher 1990) nor in any other paper, although a simple computation derived from the computation for the number of cusps confirms that this is indeed the case.

[208] Moishezon (1981, p. 189). Recall that $S$ is the branch curve of a smooth surface of degree $n$ in $\mathbb{CP}^3$, $\Sigma$ is the branch curve of another (generic) surface of degree $n$ in $\mathbb{CP}^3$, being a projection of a smooth surface in $\mathbb{CP}^r$.

[209] Ibid., p. 190.

[210] Indeed, Moishezon became aware of Chisini's papers on branch curves only at the end of the 1980s (see Sect. 5.3.3.2).

[211] Ibid. (cursive by M.F.)

Explicitly, Moishezon presented in the last pages of his paper his epistemic configuration for the following years. Considering a degeneration of a surface, Moishezon was interested in the effect that this degeneration has on the branch curve and its associated braid monodromy, and in particular he aimed to investigate the inverse direction—i.e. investigating first the degenerated curve and then drawing conclusions about the original one, considering it *locally* and not globally. Explicitly, Moishezon's idea is to investigate the braid monodromy in the *neighborhood* of singular points of the degenerated branch curve, and examining—in a similar fashion to Chisini—what happens to these singular points (and the associated braid monodromy) when examining the passage back to the smooth surface, with which one begins. This process is termed "regeneration", and Moishezon emphasized the local point of view as follows: "We can consider $S_W$ [the branch curve of a smooth surface] as a *regeneration* of [. . .] $D \cup \Sigma$ [the branch curve of a degenerated surface] (where, (1) each cusp, node or branch point of $\Sigma$ produces (respectively), a node or a branch point $S_W$; (2) triple point of produces six cusps and a number of branch points of $S_W$; [. . .] (7) each tangency point of $\Sigma \cap D$ produces three cusps of $S_W$)".[212] Moishezon listed in 1981 seven scenarios of these regenerations, without, however, supplying any proof, why certain regenerations occur as stated, or which effect do they have on the braid monodromy. Nevertheless, several of these regeneration processes are treated explicitly in the papers of 1983 and 1985.

However, not a single mention of Chisini's work on this topic is to be found. This indicates that even though Shafarevich emphasized the achievements of the Italian school, one may assume that the works read during the 1960s in Shafarevich's seminars were of the 'main players' (e.g. Castelnuovo, Enriques, Zariski) and not the 'minor' groups, which could be explained by the fact that the results of Chisini and his students were not well known outside of Italy. It also shows that the transfer of mathematical knowledge between the various mathematical configurations and communities was not complete: Chisini's work was at that time completely forgotten.

### 5.3.3.2 1983/1985: The Arithmetic of Braids and the Language of Factorizations

The shift from investigating branch curves and their associated fundamental group to investigating first their braid monodromy factorization as *algebraic objects* and only afterwards deducing results regarding the corresponding surfaces, is to be seen much more clearly in the two papers of Moishezon, published in 1983 and 1985.

The 1983 paper, dedicated to Shafarevich, repeating shortly the construction of the braid group presented in the 1981 paper, opens with the definition of the braid group $B_n$ as the group of diffeomorphisms of the Euclidean plane, generated by $X_1, \ldots, X_{n-1}$. The group has a "(normal) semi-group $B_n^+$ [. . .] which is generated by all $QX_1Q^{-1}$",[213] when $Q \in B_n$.

---

[212] Ibid., p. 186–187.

[213] Moishezon (1983, p. 201).

Moishezon aimed to develop an "arithmetic of braids": "A study of factorization properties of elements of the semigroup $[B_n^+]$ we call [...] 'arithmetic of braids.'"[214] Indicating that one can associate to a singular curve a factorization of $\Delta^2$, Moishezon noted that:

> "There is an empirical evidence that very often we can find for a singular curve a factorization of $\Delta^2$ (corresponding to braid monodromy) which is remarkably short and natural. We call such factorizations 'normal forms of braid monodromies.' Existence of 'normal forms' gives a strong indication that an essential part of the (yet to be discovered) theory of singular plane curves is 'written' in the language of (positive) factorizations of $\Delta^2$."[215]

Emphasizing again the need to find "normal forms", Moishezon here justified the shift toward re-writing plane singular curves "in the language of (positive) factorizations" with the "empirical evidence" found; this empirical evidence apparently points to the paper of 1981, but also to the works of Moishezon and Livne during the 1970s. The metaphor of re-writing and the usage of a new "language" is certainly not new: as we saw, it was already expressed by Weil and by Klein, both evoking the metaphor of the 'Tower of Babel', as a call for the foundation of a clearer language. This call appears also later in the introduction of the 1983 paper of Moishezon. When discussing the case of branch curves, Moishezon remarked that "we expect existence of a reach [sic!] braid-arithmetic *language* carrying essential parts of the theory of algebraic surfaces. It is important already to *translate* to a braid-arithmetic language known descriptions of different classes of algebraic surfaces."[216] What is presented here is a shift between the different mathematical configurations as a translation between languages, when it is obviously meant that this translation is bilateral, in the sense that theorems proved with the "braid-arithmetic language" might have an equivalent result with the "theory of algebraic surfaces" and vice versa.

   In order to fulfill this program, Moishezon followed the plan sketched at the end of the 1981 paper: one considers "degenerations of nonsingular hypersurfaces in $\mathbb{C}^3$ to union of a nonsingular hypersurface and planes in general position [...]. We will consider here only the case of splitting just one plane from generic Fermat surface $X_n$ [...]."[217] The rest of the paper deals with the results of the regeneration of the braid monodromy, and it is clear that Moishezon put a great emphasis on mastering the techniques of equivalence between factorizations. Only the later sections deal with what happens to the braid monodromy of the branch curve when adding a plane $W$ to a generic Fermat surface $X_{n-1}$. This leads to one of the main results: the section dealing with the "regeneration of $X_{n-1} \cup W$."[218]

---

[214] Ibid.

[215] Ibid., p. 202.

[216] Ibid., p. 204 (cursive by M.F.)

[217] Ibid., p. 205. The generic Fermat surface $X_n$ is defined as being "very close" to the Fermat surface $F_n$, "defined by $z^n - y^n + x^n = -1$ [...] [we] let $X_n$ be a generic Fermat surface close to $F_n$. It is convenient to take $X_n$ very close to an 'intermediate' surface $V_n$, defined by $z^n + \epsilon(1 + \rho x^{n-1})z - y^n + x^n = -1$, where $0 < \rho \ll \epsilon \ll 1$." (Ibid., p. 219).

[218] Ibid., p. 238.

The title of the section is however misleading: the section does not present how the surface regenerates into $X_n$, but rather how the algebraic expression of the braid monodromy factorization of the branch curve of $X_{n-1} \cup W$ is modified into braid monodromy factorization of the branch curve of $X_n$. To recall, in 1981 Moishezon noted, "each tangency point of $\Sigma \cap D$ [the degenerated branch curve] produces three cusps of $S_W$ [the desired branch curve]".[219] In 1983 he showed how this regeneration is expressed at the level of factorizations: by "direct computations it is possible to show that each of these tangency points produces three cusps of the branch curve of $X(\epsilon_1)$ and that the factor $Y^4_{n-3;k,k}$ [the element in the factorization that corresponds to a simple tangency point] has to be replaced by [three factors corresponding to the cusps] $\left(Y^{(1)}_{n-3}\right)^3 \cdot \left(Y^{(2)}_{n-3}\right)^3 \cdot \left(\alpha_k Y^{(2)}_{n-3} \alpha_k^{-1}\right)^3$."[220] As in the 1981 paper, no proof is given to why this would be the correct transformation at the level of the factorizations.[221] Other transformations that are presented correspond to other intersections of $\Sigma \cap D$, as presented in the 1981 list. All of these transformations are only on the level of the algebraic presentation of factorizations, giving prominence to the "braid-arithmetic language".

The 1985 paper follows the line sketched in the 1983 and 1981 papers, now bringing two concrete examples of a complete degeneration of surfaces and the calculation of the braid monodromy factorization, both of the degenerated branch curve and the regenerated branch curve.[222] The aim of Moishezon is explicitly stated, along with two possible results: "Our aim is to write explicit formulae describing the braid monodromy of $S$ [$S$ being the branch curve of these surfaces] [...]. Such formulae immediately give finite presentations for $\pi_1(\mathbb{CP}^2 - S, *)$ [...]. Another application of our results is the construction of the first examples of simply-connected algebraic surfaces of positive index (joint paper with M. Teicher [...])."[223] While the first result was a well-known computational result, as was discussed already by Zariski and van Kampen, the second result was a counter-example to Bogomolov's watershed conjecture (see Sect. 5.2.2). This was the first time this counter-example was announced in print, although, as we will see in Sect. 5.4, it was published only in 1987.

The paper begins with the degeneration of two surfaces—the Veronese embedding of $\mathbb{CP}^2$ and an embedding of $\mathbb{CP}^1 \times \mathbb{CP}^1$—into a union of planes (see Fig. 5.6a for the degeneration of the Veronese surface of degree $n^2$), computing the resulting braid monodromy factorization. As was already noted, Moishezon's investigation of the branch

---

[219] Moishezon (1981, p. 187).

[220] Moishezon (1983, p. 239).

[221] The proof is given only in (Moishezon/Teicher 1990, p. 173–179).

[222] The two surfaces are: (1) the Veronese embedding of degree $n$ of the projective complex plane $\mathbb{CP}^2$, thus obtaining a projective complex surface of degree $n^2$; and (2) the projective embedding of $F_0 = \mathbb{CP}^1 \times \mathbb{CP}^1$, "corresponding to the linear system $|m\ell_1 + n\ell_2|$, where $\ell_1, \ell_2$ are generators of $F_0$." (Moishezon 1985, p. 311)

[223] Ibid. The index can be given in terms of the Chern classes of the surface: $\frac{1}{3}\left(c_1^2 - 2c_2\right)$.

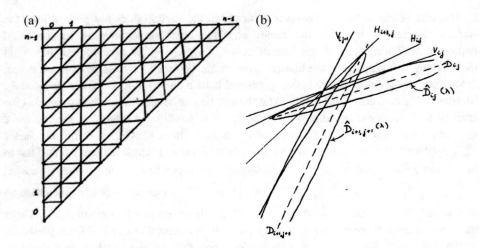

**Fig. 5.6** (a) The degeneration of the Veronese surface of degree $n^2$ embedded in $\mathbb{CP}^{n^2+1}$ into a union of planes; each triangle represents a plane (Moishezon 1985, p. 316) (b) Depiction of a regeneration of six intersecting lines into an intersection of four lines and two conics, tangent to the 4 lines (ibid., p. 329)

curve and its regeneration is only *local*, considering only what happens to the factorization at the neighborhood of the singular points of the branch curve as a result of the regeneration of the surfaces. It is essential to note that during this investigation Moishezon used numerous drawings, in order to visually present the local regeneration of several singular points. Thus, for example, Fig. 5.6b depicts the result of a regeneration of a singular point called "6-point" (being the intersection of six lines),[224] when two lines of the degenerated branch curve are regenerated into a conic. Although these drawings are not an essential part of the process of finding the regenerated braid monodromy factorization, they certainly serve as a heuristic tool of visualizing the steps of the regeneration.

To conclude, Moishezon's procedure is as follows: computing the local changes of the regeneration at the level of the symbolic factorization and then performing locally another regeneration of the surface. After performing the last regeneration, he obtained the factorization of the branch curve of these surfaces—without, however, any explicit computation of the branch curve itself.

### 5.3.3.3 Conclusion: Moishezon and Chisini

According to Libgober, "no later than 1986–1987 [...] when I mentioned to him [Moishezon] Chisini's works on wiring diagrams (I ran into one of Chisini's paper accidentally, while browsing at UIC [University of Illinois] library) he did not show any

---

[224] Moishezon uses this expression in: Ibid., p. 341.

signs that he heard of them."[225] According to Moishezon, he discussed Chisini's work with Fabrizio Catanese starting in 1987.[226] Moreover, the first time the works of Moishezon, Chisini, Tibiletti and B. Segre are mentioned together was in a survey paper from 1987 of William Fulton on the topology of algebraic varieties, in the chapter he deals with branched coverings.[227] However, Fulton did not note the similarity between Chisini's method and Moishezon's. Hence, one may observe here an ignorance from Moishezon's side regarding Chisini's works, at least starting 1987. I would like to now discuss the similarities and the differences between Moishezon and Chisini's configurations briefly.

One may claim that both Moishezon and Chisini used similar tools: the braid monodromy (or the "characteristic braid") and degeneration techniques. Both Moishezon and Chisini explicated the regeneration process and how the braid monodromy factorization would change. It is remarkable that both sets of regeneration rules were almost identical. Both also wanted to find "normal" (or "canonical") forms for this factorization. However, it is also clear that Moishezon's reasoning and arguments were very different from Chisini's. Moishezon was aware of the re-writing processes that algebraic geometry went through (and this may be the reason why he used similar expressions regarding this re-writing), while Chisini refused to take part in them. Moishezon relied on symbolical reasoning and on working with algebraic expressions. His knowledge of group theory was evident, while Chisini avoided using its tools. And while Chisini emphasized the usage of material models of strings to verify the results, when Moishezon referred to "empirical information" in mathematics, he related it to various mathematical theories, information to be placed in "an intrinsically consistent scheme".[228] Moreover, while for Moishezon "empirical knowledge" or "empirical evidence"[229] was stronger than "theoretical", both types of knowledge are never obtained from working with material models, but rather by proving theorems, either for restricted classes (as in the case of "empirical knowledge") or for the general case. Intuition, as we saw, was for Moishezon more symbolical, while for Chisini it was much more tactile, as the other modes or reasoning were not to be trusted.

## 5.4   Moishezon and Teicher Cross the Watershed

At the Institute for Advanced Study, at Princeton University, "the academic year 1981–82 was extremely active for the School of Mathematics, in particular because of the seminars associated with the Special Year in Algebraic Geometry. Among the activities for the Special Year were a series of seminars on Moduli Problems organized by David Mumford [. . .]"[230]

---

[225] Private communication with Anatoly Libgober (email from 11.7.2018). Libgober refers to (Chisini 1952b).

[226] Moishezon (1994, p. 174).

[227] Fulton (1987, p. 36–39).

[228] Moishezon (2001, p. 2).

[229] Moishezon (1977, p. 4) and (1983, p. 202).

[230] Annual Report of the Institute for Advanced Study (1982, p. 31–32).

The year 1981 proved to be a year of utter importance for the development of research on the branch curve and their associated braid monodromy factorization and the "special year" mentioned above played an essential role in it. As Donu Arapura notes, "by this point [the year 1981], he [Moishezon] was already well into braid monodromies; he had given a number of seminars on this topic at Columbia."[231] Moishezon was present during the "Special Year in Algebraic Geometry" in Princeton, giving two lectures there on "Braid Monodromies", the first on November 20, 1981 and the second on December 4, 1981.[232] Mina Teicher, who was at the same time a visiting member at the institute,[233] was present during these lectures. Teicher knew Moishezon very well, as he was her advisor for her Master thesis.[234] Teicher recalls that at the end of one of these lectures David Mumford noted that Moishezon's program was extremely technical, and that one could not have expected to obtain the desired results regarding, for example, the structure of the fundamental group of the complement of the branch curve.[235] Katsenelinboigen noted Moishezon's hard feelings while hearing this kind of opinion, adding that "[t]he last several years of his life he experienced a lack of understanding among his colleagues who underestimated his new developments in mathematics, which concerned braid group techniques in complex geometry. [...] Only a couple of years before his death, his innovative ideas gradually found appreciation and acceptance [...] [and] the attitude of his colleagues began to change."[236]

The series of papers Teicher and Moishezon published together may point to the fact that both considered braid monodromy techniques as epistemic (and not just technical) as they worked intensively on invariants, which can be derived from this monodromy. This was done while Moishezon took his sabbatical during the 1983–84 academic year. Their starting point was Bogomolov's watershed conjecture (that every surface $X$ of general type with positive index has an infinite fundamental group $\pi_1(X)$; see Sect. 5.2.2). Indeed, as Donu Arapura recalls, Moishezon "mentioned a problem he was thinking about in connection with a conjecture of Bogomolov [...]. He said he hoped to compute the fundamental group of a certain example by degeneration and braid monodromy techniques."[237] In order to understand Teicher and Moishezon's construction, one has to make a short detour.

---

[231] Private communication with Donu Arapura (email from 25.7.2018).

[232] Annual Report of the Institute for Advanced Study (1982, p. 73, 74).

[233] Ibid., p. 37.

[234] The MA thesis of Teicher "Riemann-Roch theorem for algebraic varieties of dimension 3", under the guidance of Moishezon, was submitted in 1975 at the Tel Aviv University. See: (Teicher 1975). As Teicher recalls, she also wanted Moishezon to be her advisor for the PhD in mathematics, but unfortunately Moishezon left to the US in 1977.

[235] Private communication with Mina Teicher (conversation on 16.10.2018).

[236] Moishezon (2001, p. 217).

[237] Private communication with Donu Arapura (email from 25.7.2018).

In 1982 Yoichi Miyaoka constructed an infinite series of surfaces with positive index,[238] i.e. surfaces for which the inequality of Chern classes holds: $c_1^2 > 2c_2$. These surfaces were constructed using topological tools from the theory of coverings as well as properties of the branch curve. Miyaoka noted that given a projective complex surface $S$ of degree $d$, a projection to $\mathbb{CP}^2$, and the corresponding branch curve $B$, the "natural representation $\pi_1(\mathbb{CP}^2 - B) \to Sym_d$ [exists] where $Sym_d$ denotes the symmetric group of degree $d$."[239] This representation, according to Miyaoka, allows for the construction of a covering of $\mathbb{CP}^2 - B$, denoted as $X^o$, such that $\pi_1(X^o)$ would be the kernel of the mapping: $\ker(\pi_1(\mathbb{CP}^2 - B) \to Sym_d)$. Since this kernel is a normal subgroup of $\pi_1(\mathbb{CP}^2 - B)$, Miyaoka used a well-known theorem from algebraic topology: given a path-connected, locally path-connected space $Y$, then for every normal subgroup $G'$ of $G = \pi_1(Y)$ there corresponds a covering $Y'$ of $Y$ such that $\pi_1(Y')$ is isomorphic to $G'$ (here $G' = \ker(\pi_1(\mathbb{CP}^2 - B) \to Sym_d)$. Using the properties of the singularities of the branch curve,[240] Miyaoka noted that "we have a natural compactification $X$ of $X^o$",[241] called also the *Galois cover* of $S$. Miyaoka then proved, by computing the Chern classes, that "the minimal resolution $\overline{X}$ of $X$" has a positive index,[242] hence obtaining the desired surface.

Using these results, in 1987 Moishezon and Teicher disporved Bogomolov's conjecture. Denoting the homomorphism $\pi_1(\mathbb{C}^2 - B) \to Sym_d$ by $\psi$, they noted that if $X$ is one of Miyaoka's surfaces, when taking the affine part of $X$, denoted as $X^{Aff}$, and a system of generators for $\pi_1(\mathbb{C}^2 - B)$, denoted by $\Gamma_j$, then the following exact sequence holds:

$$1 \to \frac{\ker \psi}{\left\langle \Gamma_j^2 \right\rangle} \to \frac{\pi_1\left(\mathbb{C}^2 - B\right)}{\left\langle \Gamma_j^2 \right\rangle} \to Sym_d$$

such that $\pi_1(X^{Aff})$ is isomorphic to $\frac{\ker \psi}{\left\langle \Gamma_j^2 \right\rangle}$.[243] They hence underlined that "[i]t is evident that $\pi_1(X^{Aff}) \to \pi_1(X)$ is surjective. Therefore, if we prove that $\pi_1(X^{Aff})$ is finite or trivial, so would be $\pi_1(X)$";[244] thus disproving the Bogomolov's watershed conjecture.

Teicher recalls[245] that their aim was in fact to strengthen Bogomolov's watershed conjecture, as Moishezon and she believed the conjecture true. As a result, they thought that the fundamental group $\pi_1(X)$ is infinite, hence—once one would calculate it—one

---

[238] Miyaoka (1983).

[239] Ibid., p. 296–297.

[240] Miyaoka assumes here, as many other mathematicians before him, that the singularities of the branch curve are either nodes or cusps. See: ibid., p. 297, Lemma 3.1.

[241] Ibid., 297–298.

[242] Ibid., 299.

[243] Moishezon/Teicher (1987, p. 604–606).

[244] Ibid., p. 604.

[245] Private communication with Mina Teicher (conversation on 16.10.2018).

should have looked for an invariant of this group. Indeed, in retrospect, Teicher and Moishezon recall in a paper from 1996:

> The "results [of Zariski and Moishezon] were almost the only ones known about $\pi_1(\mathbb{C}^2 - B)$ [...] till the middle of the eighties [of the 20th century]. They gave the impression that these groups are very big (in particular, contain large free groups). Such an impression was partly responsible for the general belief in the following conjecture: The Galois coverings corresponding to generic projections of algebraic surfaces to $\mathbb{CP}^2$ have (as a rule) infinite fundamental groups. This was a partial case of Bogomolov's conjecture which stated that algebraic surfaces of general type with positive index have infinite fundamental groups."[246]

Teicher and Moishezon hence started with one of the examples that Moishezon presented in 1985: the projective embedding of $\mathbb{CP}^1 \times \mathbb{CP}^1$ by the linear system $|a\ell_1 + b\ell_2|$, and its degeneration into a union of planes. When taking $a \geq 6$, $b \geq 5$ and $a$ and $b$ to be relatively prime, the results were astonishing: in 1987 it was proved that not only the associated Galois cover $X$ has a positive index, as was already known from Miyaoka's construction, but that it is simply connected, that is, its fundamental group is trivial.

The proof of this theorem is laborious, involving mainly algebraic procedures and techniques of group theory. Notwithstanding, in contrast to the 1981 paper of Moishezon, the presentation of the braids is remarkably visual: a table depicting the various braids in the different factorizations is given, which visualizes the induced homormorphism of a disc which fixes several points (see Fig. 5.7).[247] However, while in the first chapter of the 1987 paper a presentation of $\pi_1(\mathbb{C}^2 - B)$ is computed using the braid monodromy obtained in 1985 and the techniques of Zariski and van Kampen, the rest of the proof relies completely and solely on techniques from group theory. The main object of study is the exact sequence of group $1 \rightarrow \frac{\ker \psi}{\langle \Gamma_j^2 \rangle} \rightarrow \frac{\pi_1(\mathbb{C}^2 - B)}{\langle \Gamma_j^2 \rangle} \rightarrow Sym_d$; Teicher and Moishezon first obtained a splitting of this sequence, that is, an epimorphism $Sym_d \rightarrow \frac{\pi_1(\mathbb{C}^2 - B)}{\langle \Gamma_j^2 \rangle}$. Using this splitting and the Reidemeister-Schreier method, being a method for finding generators and a presentation for subgroups of a given group, they managed to simplify the presentation of $\pi_1(X^{Aff}) \cong \frac{\ker \psi}{\langle \Gamma_j^2 \rangle}$, being a subgroup of $\frac{\pi_1(\mathbb{C}^2 - B)}{\langle \Gamma_j^2 \rangle}$. Then they proved the commutativity of $\pi_1(X^{Aff})$ and the fact that each generator is of finite order, and eventually, to prove the main theorem that "[f]or $a \geq 3$, $b \geq 2$, $\pi_1(X)$ is finite and commutative. If $a$ and $b$ are relatively prime $\pi_1(X)$ equals 0."[248]

* * *

---

[246] Moishezon/Teicher (1996, p. 330–331).

[247] Moishezon/Teicher (1987, p. 612–613).

[248] Ibid., p. 642.

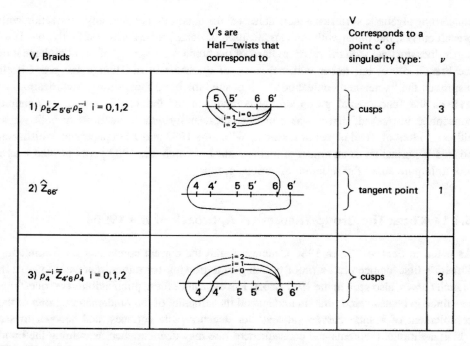

| V, Braids | V's are Half–twists that correspond to | V Corresponds to a point c′ of singularity type: | ν |
|---|---|---|---|
| 1) $\rho_5^i Z_{5'6'} \rho_5^{-i}$  i = 0,1,2 |  | cusps | 3 |
| 2) $\hat{Z}_{66'}$ |  | tangent point | 1 |
| 3) $\rho_4^{-i} Z_{4'6} \rho_4^i$  i = 0,1,2 |  | cusps | 3 |

**Fig. 5.7** A table depicting parts of the braids in the braid monodromy factorization of the regeneration of a singular point called "6-point". The braids in the middle column are defined according to the corresponding diffeomorphism of a disc, exchanging along the drawn path the connected points, while keeping the others fixed (Moishezon/Teicher 1987, p. 612)

The story behind this discovery, as well as the proof itself, is telling, especially since it points towards another shift in how branch curves were treated. Indeed, as Teicher recalls, while she and Moishezon believed that Bogomolov's conjecture is true, the opposite result was proved. This refutation caused them to note that their "work is one of the few initial attempts to penetrate into the mysterious world of branch curves of stable complex morphisms and the corresponding fundamental groups."[249]

However, while relying on the results of Moishezon and on his program as formulated in the beginning of the 1980s, the investigation of the braid monodromy factorization as an object of mathematical inquiry was abandoned in favor of an investigation of the structure of the fundamental group of the branch curve. That is, Moishezon's program of "arithmetic of braids" was no longer at the center of attention, but rather the associated braid monodromy factorization was used as a tool to compute the fundamental group. Moreover, the investigation of this group is presented in order to prove (or disprove) theorems

---

[249] Ibid., p. 602.

concerning algebraic surfaces, which, at least at the outset, do not have any connection with branch curves, but rather with the classification of complex surfaces. Explicitly, the 1987 paper does not deal with the structure itself of the fundamental group of the complement of the branch curve, but rather with subgroup of quotients in it. Though the proof itself disproved the 'watershed conjecture',[250] it posited the braid monodromy factorization, as well as the fundamental group of the complement of the branch curve, as being a conceptual watershed, that is—as a mathematical entity which functions in both ways: either as a tool or as an object of research. While the 1983 and 1985 papers of Moishezon posited factorization as an object of mathematical research, the 1987 paper posited it as a tool to inquire about the structure of various groups.

### 5.4.1   Coda: The Group-Theoretical Approach of the 1990s

As noted in Sect. 4.1.3., in 1986 Catanese found the correct conditions for formulating Chisini's first 'conjecture', a proof pointing to a growing (or rather, renewed) interest in branch curves, also seen in the 1987 paper. But even after these publications, two questions remained to be answered: first, to understand the structure of the fundamental group of the complement of branch curves—at least for degeneratable surfaces; and second, to see which additional theorems and consequences one may deduce when examining the braid monodromy factorization as an object of mathematical research, and not just as a tool for deriving other mathematical invariants.

Concerning the first question: In a series of papers called "Braid Group Techniques in Complex Geometry" numerated I, II, III, IV and V, published between 1988 and 1996, Teicher and Moishezon presented detailed explanations regarding degeneration and regeneration of surfaces, the corresponding braid monodromy and the computation of fundamental groups. The last three papers of the series deal with a surface of degree 9: the Veronese surface $V_3$.[251] Though the degeneration and the associated braid monodromy factorization (of its branch curve) for the general case (that is, for $V_n$, $n > 1$) were already

---

[250] Around the same years, Chen (1987) also proved the existence of simply connected minimal surfaces with positive index, using other techniques, by constructing double covers of $\mathbb{CP}^1 \times \mathbb{CP}^1$ branched over a singular curve, and then taking $X$ as the minimal resolution of the double cover (ibid., p. 147–154). In contrast to the Moishezon-Teicher method, which employed group-theoretical methods to prove directly that $\pi_1(X)$ is trivial (or finite and commutative), Chen's proof of the triviality of $\pi_1(X)$ relied on topological arguments concerning fibrations, and is five lines long. Chen's proof relies on a lemma from Persson (1981, p. 34–35, lemma 3.20) dealing with fibrations; Persson himself notes that "I am indebted to Prof. Moishezon for supplying the arguments" for this lemma (ibid., p. 50). This shows the central role Moishezon played also in this discovery.

[251] The Veronese embedding of degree $n$ of the projective plane is an embedding of the projective complex plane $\mathbb{CP}^2$, when one obtains a projective complex surface of degree $n^2$, defined by $v_n(x_0; x_1; x_2) = \left(\ldots; x_0^i x_1^j x_2^k; \ldots\right)$, when $i + j + k = n$. For the case of the Veronese embedding of degree 3 of $\mathbb{CP}^2$, one obtains a surface of degree 9 embedded in $\mathbb{CP}^9$.

computed by Moishezon in 1985, the explicit computation of the associated fundamental group was not done yet. This is handled in the last paper of the above-mentioned series, together in another paper published in 1995.

The main result of this series of papers concerns the structure of the fundamental group of the complement of branch curve $B_3$ of $V_3$. Denoting $G = \pi_1(\mathbb{C}^2 - B_3)$, it is proven that $G$ is an extension of a solvable group by a symmetric group.[252] A second theorem presents the structure of $G$ explicitly. For two elements $x$, $y$ in a group, denote $[x, y] = xyx^{-1}y^{-1}$. For the braid group $B_n$, $n > 2$, generated by $x_1, \ldots, x_{n-1}$ one denotes by $\widetilde{B}_n$ the following quotient of $B_n$:

$$\widetilde{B}_n = B_n \Big/ \big\langle \big[x_2, x_3^{-1}x_1^{-1}x_2x_1x_3\big] \big\rangle$$

The group $\widetilde{P}_n$ is defined to be the kernel of the map $\widetilde{B}_n \to Sym_n$ (i.e. the subgroup of all the braids in $\widetilde{B}_n$ which are sent to the identity permutation).[253] Denoting now $\widetilde{P}_{n,0}$ the subgroup of all the braids in $\widetilde{P}_n$ whose degree is 0 (i.e. the sum of all exponents is 0), it is proven that $G$ is isomorphic to a quotient of a semi direct product of $\widetilde{B}_9$ and $\widetilde{P}_{9,0}$.[254]

While finding a counter-example to Bogomolov's watershed conjecture involved dealing with the calculation of the quotient $\dfrac{\ker \left( \pi_1(\mathbb{C}^2 - B) \to Sym_d \right)}{\langle \Gamma_j^2 \rangle}$, the goal of the series of papers of "Braid Group Techniques in Complex Geometry" was far more ambitious, in the sense that Teicher and Moishezon redirected their efforts finding not a single group, but rather the general structure of the fundamental group of the complement of the branch curve, a structure which was searched for since Zariski. Indeed, while Zariski and van Kampen did manage to find a *representation* of it in terms of generators and relations, it was Teicher and Moishezon who succeeded (for numerous cases) to find the *structure* of the group in terms of a normal series.

Nevertheless, one cannot say that what was presented was situated solely in a structural conception of algebraic geometry. While the approach presented in these papers was

---

[252] Moishezon/Teicher (1996). In particular, it is proven that there exists a series of normal subgroups in the form of $1 \lhd H_0' \lhd H_0 \lhd H \lhd G$.

[253] The definition given by Moishezon and Teicher in (1994, p. 356) is slightly different but equivalent to the above definition.

[254] These results were later generalized to families of degeneratable surfaces, including all embedding of the Veronese surface, embedding of $\mathbb{CP}^1 \times \mathbb{CP}^1$, and surfaces being complete intersection. See: (Moishezon 1993). The two theorems were afterwards formulated in a simpler way in (Auroux et al. 2004), considering also for symplectic 4-dimensonal manifolds, presenting $G$ for numerous (degeneratable) surfaces as a quotient of the following group:

$$\widetilde{B}_n^{(2)} = \big\{ (x, y) \in \widetilde{B}_n \times \widetilde{B}_n, \deg(x) = \deg(y), \sigma(x) = \sigma(y) \big\}$$

where $\sigma$ is the map from $\widetilde{B}_n \to Sym_n$.

indeed highly group-theoretical, it did not reject the usage of illustrative and diagrammatical elements and presented an image of algebraic geometry as a discipline employing various techniques and methods, and not just the tools of algebra or group theory.

* * *

While the above series of papers concentrated on an already well researched object (the fundamental group), when one considers the second question—concerning the research of braid monodromy factorizations—then it is noticeable that one obtained during the 1990s results, which reshaped the mathematical configuration itself, in which these factorizations were researched. This is to be seen in a paper of Moishezon published posthumously in 1994, where the braid monodromy factorization was reconsidered as an object to investigate, and not just as a tool. Moishezon's starting point was a statement, which Chisini claimed to prove in 1955, indicating that every factorization of positive braids, equal to $\Delta^2$, is in fact equivalent to a braid monodromy factorization of an algebraic curve. While Chisini claimed to have proven this statement (see Sect. 4.2.2), Moishezon, after discussions with Fabrizio Catanese,[255] which took place in 1987, proved that the statement is false. The way to disprove this statement was to look explicitly at the factorizations themselves, and to examine changes done on them.

To recall briefly the statement made by Chisini, as Moishezon stated it, Chisini claimed that given a factorization of $\Delta^2$,[256] which satisfies a certain combinatorial condition[257] and whose only factors are (possibly a conjugation of) braids of degree 3, 2 or 1, then this factorization corresponds to a factorization of an algebraic curve, when the braids of degree 3, 2 or 1 correspond to the cusps, nodes and branch points (with respect to a projection) of the curve respectively. How did Moishezon disprove this claim?

In his paper from 1994, Moishezon recalled that one may have two types of equivalence between factorizations. Given a factorization $h = g_1 \cdot g_2 \cdot \ldots \cdot g_r$ in the braid group $B_m$, a *Hurwitz move* "changes two neighboring factors [...] as follows:"[258]

$$g_i \cdot g_{i+1} \longmapsto \left(g_i g_{i+1} g_i^{-1}\right) \cdot g_i$$

[255] Moishezon (1994, p. 174)

[256] It is essential to emphasize that Chisini called this factorization the "characteristic bundle" (see Sect. 4.1.1). The notation $\Delta^2$ was not even used in the context of braid theory during the 1930s, when Chisini developed his theory, and he neither used this notation (nor employed any algebraic tools of braid theory).

[257] The combinatorial condition is $d + 2c + 3 < \frac{1}{2} n(n + 3)$ (when $n =$ degree of the curve, $c =$ number of cusps, $d =$ number of nodes).

[258] Ibid., p. 152. Here Moishezon obviously refers to Hurwitz, who was the first to introduce these operations for elements in the symmetric group. See Sect. 5.2.

$$g_i \cdot g_{i+1} \longmapsto g_{i+1} \cdot \left(g_{i+1}^{-1} g_i g_{i+1}\right).$$

Moreover, two factorizations are called *Hurwitz equivalent* "if one can be obtained from the other by a finite sequence of Hurwitz moves." For the second equivalence type, Moishezon takes an element $z$ in $B_m$, such that $z$ commutes with $h$, and denotes $h_z = z^{-1}g_1z \cdot z^{-1}g_2z \cdot \ldots \cdot z^{-1}g_rz$; $h_z$ is called *conjugation equivalent* to $h$.

Having defined these two equivalence types, Moishezon calls a factorization $\Delta^2 = g_1 \cdot g_2 \cdot \ldots \cdot g_r$ in $B_m$ *geometric* if it is "Hurwitz and conjugation equivalent to the basic factorization $\Delta^2 = \prod_{m \text{ times}}(X_1 \cdot \ldots \cdot X_{m-1})$", and *analytic*, "if it corresponds to the braid monodromy of an algebraic curve." Immediately after this differentiation, he notes that it is clear "that any analytic factorization is geometric", since the braid monodromy factorization associated to a complex plane curve is equivalent to $\Delta^2$ by definition. But "[t]he inverse is not true even for cuspidal factorizations, and one of the goals of our work is to show it";[259] hence disproving Chisini's statement. This is since Chisini's 'theorem' consisted of stating the equivalence of the two classes of the factorizations.

Moishezon's insight was to notice the following: "The parameter space of degree $m$ algebraic curves in $\mathbb{CP}^2$ has a finite stratification such that for any two curves $S_1$, $S_2$ of the same stratum the corresponding embeddings $S_1$, $S_2 \subset \mathbb{CP}^2$ are topologically equivalent [...]. [Hence] for any $m$ there exists only a finite number of Hurwitz and conjugation equivalence classes of analytic cuspidal factorizations of $\Delta^2$."[260] Hence, finding infinite sequence of non-equivalent factorizations would be a contradiction to Chisini's theorem, since only a finite number of them would correspond to algebraic curves. Moishezon began with a specific braid monodromy factorization: the braid monodromy factorization in the braid group $B_{m_0}$ ($m_0$ being an integer) of the branch curve $B$ of the Veronese surface $V_3$. From this factorization he constructed an infinite series of other factorizations (in the braid group $B_{3m_0}$), denoted by $\widetilde{\mathcal{E}}_{m,k}$, when $k$ is any integer. The main point of the proof is to compute the associated 'fundamental group' to each factorization. However, since it is not clear whether the $\widetilde{\mathcal{E}}_{m,k}$'s are indeed factorizations corresponding to algebraic curves, though one obtains an infinite series of groups $G_k$ (applying the Zariski-van Kampen method), it cannot be said that these groups are fundamental groups of the complement of existing algebraic curves. The key point in the argument is that equivalent factorizations induce isomorphic 'fundamental groups'. Using explicitly properties of $\pi_1(\mathbb{C}^2 - B)$, which were only stated but published two years later,[261] Moishezon proved that for every two different integers $k_1$, $k_2$, the groups $G_{k_1}$, $G_{k_2}$ are *not* isomorphic.[262] Hence, for every $k_1 \neq k_2$, the

---

[259] Ibid., p. 153. Note that there is a typo in the definition of *geometric* factorization: there it is wrongly written that $\Delta^2 = \prod_{m-1 \text{ times}}(X_1 \cdot \ldots \cdot X_{m-1})$. This is however not correct, since the degree of $\Delta^2$ is $m(m-1)$.

[260] Ibid., p. 155.

[261] See: Moishezon / Teicher (1996).

[262] Moishezon (1994, p. 167, 173–174).

factorizations $\widetilde{\mathcal{E}}_{m,k_1}$ and $\widetilde{\mathcal{E}}_{m,k_2}$ are not equivalent. However, the factorizations $\widetilde{\mathcal{E}}_{m,k}$ do satisfy the combinatorial condition of Chisini. Thus, according to Chisini's statement, they all should have corresponded to algebraic curves; but as Moishezon's noted, this ends in contradiction.

While Moishezon did compute explicitly the structure of the groups $G_k$, he did not give a single example for which values of $k$ the factorizations $\widetilde{\mathcal{E}}_{m,k}$'s do *not* correspond to algebraic curves. Moreover, the obtained factorizations are factorizations of $\Delta^2$ in the braid group $B_{54}$ with 378 factors of degree 3 and 756 factors of degree 2, but nowhere in the paper did Moishezon even prove that there *exists* a curve of degree 54 with 378 cusps and 756 nodes. Indeed, proving that there is at least one such curve would have indicated that there might be one factorization among the $\widetilde{\mathcal{E}}_{m,k}$'s which correspond to an algebraic curve, while an infinite many others do not correspond to such curves, which would have made the result even stronger.

However, the crux of Moishezon's argument is not only to look at the structure of the groups $G_k$, but also to perform a differentiation of types of factorizations. That is, Moishezon took the factorization as an object of research, and notes that once the configuration of this object changes—i.e. when it either corresponds to a nodal-cuspidal curve or when it is symbolically equivalent to $\Delta^2$—one cannot treat the two configurations as equivalent. While Chisini stated that the two ways of considering the curve (the analytical way and as a factorization—are the same, Moishezon separated them, showing that while any analytic factorization is geometric (i.e. equivalent to $\Delta^2$), the inverse is not correct. One might say that Moishezon made the 'limits' of these two configurations clearer, or, to use a formulation of Moishezon, the boundaries of the "translation" between them were explicated. In a way, what is created are two distinct configurations, which are not necessarily equivalent to each other. The question is: if the class of geometric factorizations is bigger than analytic factorizations, do the former factorizations even correspond to more general topological objects (or curves), which are not necessarily algebraic? The answer— on which I will elaborate below—is positive, but it takes us outside the realm of algebraic curves. For example, Auroux and Katzarkov interpreted the disproval of Chisini's claim by considering existence of quasiholomorphic curves, not isotopic to algebraic ones, curves for which Chisini's 'theorem' does hold.[263] Chisini began with his 'theorem' that every algebraic factorization does correspond to an algebraic curve, but the refutation of this 'theorem' prompts the opening of a different configuration, when the object which corresponds to algebraic factorizations is not necessarily an algebraic curve, but rather a quasiholomorphic one.

The paper of Moishezon from 1994 stressed the need to researching braid monodromy factorizations, and more generally, factorizations of algebraic elements, on their own. To elaborate on the research of Auroux and Katzarkov: If Moishezon proved the splitting of the set of factorization into two, Auroux and Katzarkov aimed to bridge this splitting. They

---

[263] Auroux / Katzarkov (2000, see esp. theorem 3).

showed that if one replaces algebraic curves by quasiholomorphic curves, allowing also negative nodes as singularities, that is, nodes defined locally not as $(y - x)(y + x) = 0$ but rather by the equation $(y - \bar{x})(y + \bar{x}) = 0$,[264] then one does have an equivalence between factorizations of $\Delta^2$ and (factorizations associated to) quasiholomorphic curves. Catanese, reviewing this work, calls the introduction of the negative nodes a "price they [Auroux and Katzarkov] have to pay". Indeed, due to considering also negative nodes, the associated factorization "is not unique, because it may happen that two consecutive nodes, one positive and one negative, may disappear, and the corresponding two factors disappear from the factorization. In particular $\pi_1(\mathbb{CP}^2 - B)$ is no longer an invariant [. . .]."[265] This means, that Moishezon's split can be resolved, but for that a redefinition of the objects of the configuration is needed (with Auroux's and Katzarkov's introduction of negative braids and quasiholomorphic curves).

* * *

While this split was somewhat resolved, other results redelineated the limits of this configuration: it was proven in 2005 that there is no algorithm which decides whether two given braid monodromy factorizations are equivalent, hence one is unable to actually compare two factorizations;[266] to paraphrase, the object itself is almost impossible to work with.[267] The solution was, as we saw, to extract the information contained in these factorizations by indirect means, via the introduction of more manageable (but less powerful) invariants.

Nevertheless, Moishezon introduced—as Chisini and Dedò before him—new epistemic configurations to research the branch curve. This is clear due to the emphasis put on the braid monodromy factorization as a *symbolical* factorization, and on the equivalence of different factorizations. The role that visual or material practices, which might have constituted an essential part of the reasoning of the proof, was reduced in contrast to Chisini. While examples for the usage of visual means could be found (Livne's drawing of the braid, as in Fig. 5.5, or Moishezon's and Teicher's table of braids, as in Fig. 5.7), these were not used to prove propositions, but rather to illustrate objects and relations—though the role of this illustration was essential. Moreover, as we saw, during the first stages of the research of Livne and Moishezon on braid monodromy factorization, the earlier research of Chisini and his students was not known and hence was not taken into account. As seen above, in 1987 the results of both traditions were mentioned side by side; hence one may question why Moishezon was referring to Chisini's results only in 1994—and also in a very

---

[264] To recall: if $x = a + bi \in \mathbb{C}$ then $\bar{x} = a - bi \in \mathbb{C}$.

[265] Catanese (2008, p. 157). Moreover, as Catanese notes immediately afterwards, even if one looks at the appropriate quotient of the fundamental group to induce invariants of symplectic structures, this (and the fundamental group itself, as Catanese claims) allows us only to detect homology invariants of the projected four-manifold, that is, invariants which were already known before.

[266] Explicitly it was proven that the equivalence problem is undecidable (Liberman / Teicher 2005).

[267] Regarding modern approaches to these factorizations and how to associate them to curves, see: (Kharlamov / Kulikov 2003).

partial way, not taking the results and his approaches into account. This might be due to a language barrier, though Moishezon had direct access to Chisini's writings via his discussions with Catanese. Nevertheless, the refutation of the watershed conjecture, the pinpointing of the explicit structure of the fundamental group of the branch curve for several families of degeneratable surfaces as well as clarifying the distinction between analytic and geometric factorizations—all of this show that Moishezon and Teicher went beyond the configuration opened by Chisini.

# Epilogue: On Ramified and Ignored Spaces

**6**

Reiterating Enriques's statement that "algebraic curves [...] are created by God [while] surfaces, instead, are the work of the devil,"[1] the cost of bringing the heaven-made curves and the allegedly demonical surfaces together involved, one might say, a plethora of mathematical configurations. Those configurations consisted of various techniques and notations, each investigating the branch curve in its own way, and hence each configuration constituted the branch curve differently. Such an emergent assemblage provides a more faithful image of mathematical research than the grand narrative of the twentieth century mathematical formalism. This plethora of configurations is also to be seen with the (some more and some less) successful rewriting and translation projects, as if by every such project, there is either a complete erasure of older traditions or a total translation of previous results and practices that are thus absorbed. But as we have seen, neither a total and faithful translation nor a complete rewrite ever took place between these configurations. Every such rewriting project may have prompted a marginalization of more 'minor' traditions, either by forgetting or ignoring them, or by integrating them allegedly seamlessly into newer configurations. The history of the various research configurations dealing with branch curves therefore shows how mathematicians create and employ practices of identity, in the sense of identifying practices and objects of former research configurations as 'naturally' belonging or leading to later ones. But the integration of the past into the 'present' is nothing more than an integration into another epistemological configuration, which is not the same as that which preceded it;[2] this, in turn, makes the former epistemological configuration something ephemeral. In this sense, I follow Lucia Turri, who claims that the "history of mathematics is [...] a process of re-emersion and re-actualization of its

---

[1] Enriques (1949, p. 464). See the motto of Sect. 1.2.

[2] On presupposing the category of the 'same' when researching the history of mathematical objects and notions, see Sect. 1.2.1.

different moments. [According to this it is] possible to understand the apparently enigmatic statement that [Jean] Cavaillès wrote in [. . .] *Méthode axiomatique et formalisme*: 'There is nothing less historical [. . .] than the history of mathematics. But nothing less reducible, in its radical singularity'."[3] Here Cavaillès means that the becoming of mathematics is so singular, that it cannot be accounted for by only taking into account the 'usual' historical factors – social, political, material – but one must consider also and in particular the internal developments of the mathematical domain in question.[4] Hence one may claim that mathematics is autonomous. Moreover, how mathematics develops cannot be predicted in advance, and its necessity, so Cavaillès, is only revealed retroactively. Cavaillès underlines that the history of mathematics should not be considered a "chronology of results."[5] This can be seen also by noting that many of the configurations presented herein attempted—in vain—to present themselves as a reactualization, accumulation and inclusion of all of the research done to that point, thereby eliminating the need to consider the contingent histories of those past configurations. But exactly this contingent but at the same time necessary character of mathematics is what Cavaillès calls to take into consideration.[6]

This book, therefore, tries to show that such an inclusion and the resulting image of continuity is distorted and misleading. What the previous chapters show is how during the twentieth century, the investigations of the branch curve not only led to new mathematical insights, but also went through shifts in material and notational practices, as well as in the conception and terminology of the curve itself. Such shifts produce a more differentiated history for algebraic geometry, which cannot be forced into the historical narrative of twentieth century formalism or the rewriting projects of algebraic geometry.

With this in mind, I aimed in this book to present a richer picture of mathematics, nuancing the narrative of modernism in mathematics in the twentieth century—a narrative presented by Herbert Mehrtens as one of "dominance of mathematical modernism with its preference for general theory, symbolic formalism, and the treatment of mathematical theories as worlds of their own without any immediate relation to the physical world around us"—as being closely related to "the neglect of 3-D models (and 2-D diagrams) during most of the twentieth century."[7] While I do not aim to summarize the arguments

---

[3]Turri (2011, p. 74). The full citation from Cavaillès's *Méthode axiomatique et formalisme* is as follows: "Il n'y a rien de si peu historique—au sens de devenir opaque, saisissable seulement dans une intuition artistique—que l'histoire mathématique. Mais rien d'aussi peu réductible, dans sa singularité radicale." (Cavaillès 1938, p. 176). On Cavaillès's philosophy of mathematics, see for example: (Benis-Sinaceur 1987; Granger 1996; Cassou-Noguès 2001).

[4]This however does not mean that one should ignore those social, political and material factors, as these certainly influence the mathematical configuration.

[5] . . . But rather taken in "its radical singularity". Here I follow Krömer (2007, p. 3–4): "mathematics has only a history exceeding a pure chronology of results when the acts constituting it, for instance the modifications of the conceptual framework, are taken into account."

[6]See the conversation between Cavaillès and Albert Lautman "Mathematical Thought" in 1939, in: Cavaillès / Lautman (1946)

[7]Mehrtens (2004, p. 278).

from Mehrtens' 1990 book *Moderne Sprache Mathematik*,[8] one can present the history of algebraic geometry as reflecting the narrative itself of mathematics becoming modern: the rewritings of Weil, Zariski and later Grothendieck represent generalizations and abstractions of the objects of algebraic geometry, where the researched objects became less and less visual or 'intuitive.'[9] But Mehrtens's book was also criticized for presenting only this narrative, ignoring not only more local research directions, but also French and Italian traditions and schools.[10] That being said, taking the research on the branch curve as a cross-section of algebraic geometry during the twentieth century, when several epistemic configurations may also overlap, provides yet another way of considering the history of this field, perhaps even a more subtle one. Indeed, as Moritz Epple notes, there are several paths to modernity: different mathematical milieus and configurations may "represent different shades of modernism in mathematics."[11] Moreover, within the works of several mathematicians, one may find modern as well as countermodern elements.[12] Some of these various paths to modernity did not have as a goal the banning of diagrams or three-dimensional models. Moreover, taking into consideration the more 'minor' traditions, local 'schools' or ephemeral configurations (such as Chisini's or Moishezon's research on braid monodromy, or even d'Orgeval's work in Oflag X B), which existed alongside the 'major' currents, presents a more fine-tuned image of how mathematical knowledge emerged and was developed in algebraic geometry. That diagrams (and even three-dimensional models), as well as diagrammatic reasoning, did play an essential role is to be seen with the research of Chisini and his colleagues,[13] but also with that of Enriques in 1899, and of Moishezon in the 1980s. Indeed, Enriques emphasized the infinity of ways to develop intuition, also stressing visual intuition and imagination. Moishezon—whose work with Teicher on the structure of the fundamental group can certainly be characterized as modern—also used diagrams of braids to explicate several of the involved procedures.

The different ways in which formalization and diagrams were interlaced reflect the different ways in which each epistemic configuration considered and constructed the branch curve. To see this, it might be worth reflecting on the question of finding the

---

[8] Mehrtens (1990).

[9] Here 'intuitive' is a term similar to 'rigor', both being epistemic values which are dependent on the time, place and mathematical configuration and community in which they are employed; at the same time, these values are also influenced by how former configurations used these and similar terms.

[10] In this respect, see Gray's critique on Mehrtens's book (2008, p. 9–12). Gray notes that "the book's tight focus on Germany is unfortunate" (ibid., p. 12) and that the "explanations of major social movements [modernism and countermodernism as presented in *Moderne Sprache Mathematik*] do not sit comfortably with individuals and their actions, and while Mehrtens does well to give his central figures the autonomy they have, ultimately they are small parts of a machine that seems to have a logic of its own [...]." (ibid., p. 11).

[11] Epple (2004, p. 160).

[12] As Epple notes, the same mathematical configuration may contain both modern and countermodern elements and a "grey scale will appear between the white modems and the black counter-modems" (1997, p. 192).

[13] Though one can indeed question whether they considered themselves as modern.

necessary and sufficient conditions for a curve to be a branch curve—conditions posited as representing the object to be found in the most exact way. This question arose in several of the research configurations addressed in this book; but the answers (for example, Enriques's invariance conditions or Beniamino Segre's positional approach) were not only formulated in different mathematical languages but also with the help of various techniques and objects, changing from one configuration to the next (mappings to the symmetric group or adjoint curves).

As I noted in the introduction, with each epistemic configuration one obtains a different 'branch curve'—these branch curves are not always necessarily related to each other, nor can they always be transformed or translated to other 'branch curves' in other configurations; explicitly, not every epistemic configuration takes into account (all of) the results of the former configurations while researching branch curves. Hence, one can talk about the 'branch curve' only within a specific, local epistemic setting. One of these specific settings may be defined by searching for necessary and sufficient conditions; other examples can be noted throughout this book. However, these configurations, while each approaching the branch curve differently, may also be considered as 'circumventing' or even 'evading' dealing with the branch curve itself. Even Monge, while emphasizing that the "apparent contour" is the only curve one need to draw exactly, hardly ever drew it, and did not consider more complicated cases, which might have led to an investigation of branch curves with nodes and cusps. This is also to be seen in Zariski's smoothing procedure of cusps of a singular curve: Zariski found the fundamental group of the complement of a sextic with 6 cusps, which do not lie on a conic, not by finding the equation of the curve itself, but rather by starting with another curve: the sextic with 9 cusps, and smoothing 3 of them—a smoothing procedure that had only been proven as feasible, but not shown in practice. Another example is the calculation of the fundamental group of the complement of the branch curve as done by Zappa, using degenerations of this curve, although an equation of the curve was not found (or even sought). The computation of the regeneration to write down the braid monodromy factorization by Chisini and his students, or by Moishezon and Teicher, was only done locally; a regeneration of the entire curve was never presented.

Each such local configuration had its own mode of representation of the branch curve; in other words, representation creates a mathematical object that is individuated only in a local epistemic configuration. As Hans-Jörg Rheinberger claims, "[t]here is no such thing as a simple representation of a scientific object in the sense of an adequation or approximatione of something out there, either conceptually or materially."[14] But while representation enables choosing certain aspects or properties of the object (to represent them or with them), it also allows a certain freedom, by opening new spaces of inscription or reasoning. Since representation as a symbolic or diagrammatic framework is not fixed from the beginning—and this is certainly to be seen in Enriques's plurality of epistemic configurations—representation is also dynamic, in the sense that new notational practices,

---

[14]Rheinberger (1997, p. 104).

symbolical means and even new concepts may be introduced.[15] This understanding of representation as local or partial can have two consequences.

First, since several configurations exist, one may consider a transformation or translation between two configurations; but due to the local and partial nature of the chosen representation, this translation will not necessarily be a complete, infallible correspondence. I use here the term 'translation' not by chance, but rather as the term used by the players. Indeed, it was used explicitly by Monge and Moishezon, and implicitly by Klein and Weil when both noted the 'Tower of Babel' which might emerge from algebraic geometry—to recall a few of the players mentioned in this book (see Sects. 2.1, 5.3.3.2, 3. 1.3 and 3.3 resp.).[16] The proposed 'translations' were not complete ones, as if there were a bijective correspondence between the two systems of representation.[17] It was more that the two mathematical languages to be translated were considered as interacting in a specific "trading zone", to use Peter Galison's term: "an intermediate domain in which procedures could be coordinated locally even where broader meanings clashed."[18] Trading zones express "different subcultures of theorizing" and how the interaction between the different subcultures develops; seeing "translation" between *local but different* mathematical languages as a "trading zone" concentrates on the local coordination between those languages, and certainly not on their homogenization by establishing a "universal language" (the aim of a "global translation" project).[19] The "translation," or rather the transformation, occurs between local and ephemeral epistemic configurations. I stress here the ephemerality of those configurations, since once such a translation occurs, one of the configurations might be revealed as flawed or as having gaps. This also highlights the potentially fallible nature of a local epistemic configuration: the fallibility of certain practices is often revealed retroactively, when the configurations in which those practices are couched are reconsidered by and translated into newer configurations. More to the point, the retroactive exposure of a possible fallibility of practices implies that establishing a local trading zone will not always lead to harmonious communication: this is seen in the case of Chisini's 'theorem' from 1954–55, which was refuted by Moishezon when he reformulated Chisini's results with "the language of (positive) factorizations."[20]

Moreover, the fact that there could be more than one system of representation for several epistemic configurations may also imply that when having two different configurations, the epistemic status of the represented objects can change when moving from one system to another. For example, for the classification project of Castelnuovo and Enriques, one of the

---

[15] I follow here (Ferreirós 2016, p. 49–50).

[16] On "Translation" as an actors' term and the metaphor of the Tower of Babel in mathematical discourse, see (Friedman 2020).

[17] On the theme of rewriting of mathematics and in the history of mathematics, see (Schappacher 2011) and below.

[18] Galison (1997, p. 46).

[19] Galison (1999, p. 145, 157)

[20] Moishezon (1983, p. 202).

ways to classify and investigate surfaces was by looking at them as 'multiple planes,' that is, branched covers of the complex projective plane. With the Kodaira classification of complex surfaces this changed, and accordingly, so did the status of the branch curve.

A second consequence of this view of representation is that one may note a strange loop: trying to represent a mathematical object within an epistemic configuration may open a new space for inscription and notation. These spaces consolidate the practices and the mathematical communities involved (as in the case of Chisini, Dedò and Tibiletti) but at the same time, they can turn out to be productive, in the sense that they prompt the emergence of *other* mathematical objects and techniques, and *other* methods of representation. As Karin Krauthausen puts it, notations as "processes on paper sometimes develop productivity precisely because they 'overtake' the one who writes or the draftsman to a certain extent, confronting him or her in this way with something 'unforeseen'."[21] This aspect of the 'unforeseen' indicates that these processes operate as an open system, in the sense that different traditions and configurations may interact with and influence each other, prompt representations with other means, and potentially (even if retroactively) point out, as indicated before, the fallibility of methods and techniques of former configurations.

Two examples of this process from the research on the branch curve can be provided. The first concerns Enriques's calculation from 1912 of the moduli of algebraic surfaces. This calculation was (also) done by counting invariants of the branch curve. However, Enriques had to presuppose the completeness of systems of surfaces in $\mathbb{CP}^3$, and this was criticized over the years (by Zariski and Mumford, for example). However, it was only in 1974 that this assumption was proven by Wahl to be incorrect, but the space of representation, where the counter-example was to be found, was completely different from its initial configuration, now using the language of schemes and infinitesimal deformations. Surprisingly, the counter-example also involved branch curves, but Enriques's way of considering branch curves differed from that of Wahl, i.e., their epistemic configurations were different.

Taking as a second example Moishezon's above-mentioned refutation of Chisini's 'theorem' regarding the equivalence between algebraic curves and factorizations of braids, one can note that Chisini's epistemic configuration (further developed by Dedò's notation) may also lead to erroneous results or dead ends. Moishezon's counter-example, relying on calculations related to a specific factorization, showed a split between two different spaces of representation that had previously been considered equivalent. As we saw in Sect. 5.4.1, this split can be resolved, but a redefinition of the objects of the configuration is needed, consisting of Auroux's and Katzarkov's introduction of negative nodes.

* * *

If we reflect for a moment on Auroux's and Katzarkov's negative braids, then, to recall, Catanese called this proposal of changing (or expanding) the epistemic configuration the "price they have to pay."[22] Catanese was referring to the fact that the fundamental group

---

[21] Epple / Krauthausen (2010, p. 131).

[22] Catanese (2008, p. 157).

ceases to be an invariant. But if one can generalize this comment, then there is a 'price to pay' with every choice of configuration and representation. Aleida Assmann, in her book *Formen des Vergessens*, addresses the question of the economics of memory and its criteria of selection.[23] Assmann classifies techniques of forgetting (such as erasing, silencing, ignoring, denying. . .) as well as forms of forgetting (such as automatic, selective, repressive. . .).[24] A choice of notation or of a certain technique to be used and the resulting price may also be seen as a form of forgetting. This price may not only be mathematical— as in the loss of certain invariants or the introduction of additional new objects (e.g. negative nodes)—but can also be manifested as being forgotten or ignored. The landscape of the research on branch curves is, to a certain extent, sown with instances of ignoration.

For example, in as early as the nineteenth century, Salmon declared himself the first to deal with the question of finding the properties of ramification curves, not being aware of Bobillier's research. Bobillier himself was not aware of the results achieved before him, by Poncelet and Monge, although he (and Gergonne) corrected this after being reminded of them by Poncelet. Moreover, during the twentieth century one can also detect a variety of forms of ignorance, some caused by operating outside the mainstream research of algebraic geometry, some by political circumstances, some by structured and deliberate ignoring, and others by marginalization or by only a partial acceptance, such as when the rewriting projects did not consider the entire body of results achieved by the former traditions of research.[25]

The most explicit example of ignoration during the twentieth century to be seen in this book is the research of Chisini, Dedò and Tibiletti. As Norbert Schappacher notes with respect to the rewriting project of algebraic geometry by Zariski and Weil, "[t]here are different ways to tell the story of a rewriting."[26] Chisini, Dedò and Tibiletti deliberately chose not to participate in the rewriting project, by ignoring the algebraic tools which were already available for use—and here one can note a double-sided ignorance as a mutual non-belonging to the rewriting project, which can be seen with how Chisini in 1933 either ignored Emil Artin's research—as part of Chisini's mistrust of algebraic reasoning—or did not know about it. Dedò, however, knew of Artin's research (as Chisini did in 1952), but chose explicitly not to consider either Artin's notation or his algebraic techniques, which were far more advanced at that time than what was being used in the research of braids by Chisini and his students. This ignoring of modern approaches to braid theory (and, in a certain sense, this ignoring of or shift outside modernity) was also to be seen with the emphasis given to proofs with material models, a practice which was unique (at that time and in this context) to Chisini and his students. The decision to ignore other, more modern

---

[23] Assmann (2018, p. 44).

[24] Ibid., p. 11–68.

[25] Concerning rewriting and overwriting as a technique of forgetting, see: ibid., p. 23.

[26] Schappacher (2011, p. 3285).

research traditions was also partially political, as we saw in Chap. 4. On the other hand, the mathematical community during the 1960s outside of Italy also did not show (or hardly showed) any interest in Chisini's results. Even when Chisini's conclusions and those of Dedò and Tibiletti were mentioned in 1987 alongside more modern research on branch curves, there was barely any impact in the following years or any attempt to reconceptualize their results. While Lanteri and Catanese did point out the mistakes in Chisini's 1944 theorem, other aspects of his research remained somewhat hidden from the mathematical community outside Italy. This was also seen with how Moishezon formulated his regeneration rules, which were similar to those of Chisini.[27] Even in 1987, after Moishezon learnt of Chisini's research, from Libgober as well as from Catanese, and even after Fulton's call to consider Chisini and Tibiletti's research, that "[i]t should be rewarding to take another look at their methods and results,"[28] which was mentioned immediately after the publication of Moishezon's and Teicher's results, the research of Chisini, Dedò and Tibiletti was hardly considered by mathematicians (or by historians of mathematics). Moishezon only disproved Chisini's theorem from 1954–55 in 1994, explicitly referring to Chisini's papers from 1954 and 1955. However, this is not to say that Moishezon adopted the methods and practices of proof of Chisini's configuration—indeed, the use of material models was not even known to Moishezon.

There are obviously other, perhaps more subtle instances of ignoration: Salmon using Bobillier's nomenclature without citing or actually knowing Bobillier's work is one example. It is not uncommon for mathematicians to be forgotten while their results continue to be known. But can one claim that not citing a certain result or not giving a reference indicates intentional marginalization, or that it causes marginalization of branches of knowledge? The complexity of the question concerning intentionality can be seen in how the result of Beniamino Segre from 1930 (on the necessary and sufficient conditions as a characterization of the branch curve) was forgotten. This result was overshadowed not only by Zariski's work during the same period,[29] but also by the fact that Zariski himself mentioned Segre's result (on these conditions) only once in his book *Algebraic Surfaces*. Indeed, how ignored, and consequently 'minor' and marginalized configurations are produced as such also concerns decisions about which kinds of mathematical knowledge should be supported: these decisions "are also decisions about what kinds of ignorance should remain in place."[30]

Either ignoring certain mathematical traditions or 'just' not being aware of them shows that, according to Londa Schiebinger, "[i]gnorance is often not merely the absence of

---

[27] I do not claim that the regeneration rules of Chisini and Moishezon were identical: they were both embedded in different epistemic configurations.

[28] Fulton (1987, p. 39).

[29] To recall, Zariski published much more than Segre on this subject between 1927 and 1937.

[30] Proctor (2008, p. 26).

knowledge but an outcome of cultural struggles."[31] With this remark, Schiebinger follows
Marilyn Frye, who notes that "[i]gnorance is not something simple: it is not a simple lack,
absence or emptiness, and it is not a passive state. [. . .] [It] is a complex result of many acts
and many negligences."[32] Can one take as another example of those "cultural struggles"
and marginalized spaces the emigration of Moishezon or the captivity of d'Orgeval? Which
unique epistemic configurations resulted from those forced events or from those acts of
ignoration and marginalization? In this book, we saw that the geography of ignorance can
be created via struggles between mathematical cultures and approaches, either happening at
the same time (through, for example, a lack of citations of the results of colleagues) or
retroactively (when a certain image of mathematics is declared, positing the other, older
traditions as obsolete and unreliable). For precisely this reason, it can be claimed that
"ignorance is frequently constructed and actively preserved,"[33] and that when an epistemic
configuration becomes marginalized or minor, this can be viewed as an active production.[34]
The examples of the knowledge obtained by Beniamino Segre and Bobillier, which were
later forgotten, and more clearly, the results of 'Chisini's school,' show that also ignorance
has a history,[35] which should be included within the mathematical history of ephemeral
epistemic configurations. The above accounts thus show that there are no simple projects of
'rewriting' or simple translations and transformations, as if changing the language entirely
would also smooth out the errors and rebuild the shaky foundations, replacing them with
more stable ones; it is already clear from the rewriting enterprises that mathematics is not a
cumulative science—it is also a landscape that holds not only abundant areas of knowledge, but also craters of ignorance, which are also created retroactively and actively. In this
sense, considering the plurality of branch curves in an examination of the history of
algebraic geometry, aims—so is my hope—to reverse these craters of marginalization
and ignorance.

---

[31] Schiebinger (2008, p. 152).

[32] Frye (1983, p. 118).

[33] Tuana (2008, p. 109).

[34] Proctor (2008, p. 9).

[35] Cf. Tuana (2008, p. 110): "Epistemologies that view ignorance as an arena of not-yet-knowing will
also overlook those instances where knowledge once had, has been lost. What was once common
knowledge or even common scientific knowledge can be transferred to the realm of ignorance not
because it is refuted and seen as false, but because such knowledge is no longer seen as valuable,
important, or functional."

# Bibliography

Abhyankar, Shreeram, 1957, "Coverings of Algebraic Curves", *American Journal of Mathematics* 79 (4), p. 825-856.

Abhyankar, Shreeram, 1959, "Tame coverings and fundamental groups of algebraic varieties, Part I: Branch loci with normal crossings, Applications: theorems of Zariski and Picard", *American Journal of Mathematics* 81, p. 46-94.

Abhyankar, Shreeram, 1960, "Cubic surfaces with a double line", *Memoirs of the College of Science, University of Kyoto, Series A Mathematics* 32, p. 455–511.

Alexander, James W., 1920, "Note on Riemann spaces", *Bulletin of the American Mathematical Society* 26, p. 370–372.

Alexander, James W., 1928, "Topological invariants of knots and links", *Transactions of the A.M.S.* 30, p. 275-306.

Alexander, James W. / Briggs, G. B., 1927, "On Types of Knotted Curves", *The Annals of Mathematics, Second Series* 28 (1/4), p. 562-586.

Allmendinger, Henrike, 2014, *Felix Kleins Elementarmathematik vom höheren Standpunkte aus: Eine Analyse aus historischer und mathematikdidaktischer Sicht*, Siegener Beiträge zur Geschichte und Philosophie der Mathematik 3, Siegen: Universitätsverlag Siegen.

Annual Report of the Institute for Advanced Study, 1982, *Annual Report for the Fiscal Year July 1, 1981-June 30, 1982*, The Institute for Advanced Study, Princeton: Princeton University Press.

Assmann, Aleida, 2018, *Formen des Vergessens*, Göttingen: Wallstein.

Artin, Emil, 1926, "Theorie der Zöpfe", *Abhandlungen aus dem Mathematischen Seminar der Universität Hamburg* 4, p. 47–72.

Artin, Emil. 1947, "Theory of Braids", *Annals of Mathematics*, Second Series 48(1), p. 101–126.

Artin, Emil. 1950, "The Theory of Braids", *American Scientist* 38(1), p. 112–119.

Artin, Michael / Mazur, Barry, 2009, "A Discussion of Zariski's Work Zariski's Topological and Other Early Papers", in: Parikh, Carol, *The Unreal Life of Oscar Zariski*, Boston: Springer, p. 137–150.

Auroux, Denis / Katzarkov, Ludmil, 2000, "Branched coverings of $CP^2$ and invariants of symplectic 4-manifolds", *Inventiones mathematicae* 142, p. 631–673.

Auroux, Denis / Donaldson, Simon K. / Katzarkov, Ludmil / Yotov, Miroslav, 2004, "Fundamental groups of complements of plane curves and symplectic invariants", *Topology* 43, p. 1285-1318.

Badescu, Lucian, 2001, *Algebraic Surfaces*, New York: Springer.

Baker, Roger / Christenson, Charles / Orde, Henry (eds.), 2004, *Collected Papers of Bernhard Riemann (1826–1866)*, Heber City: Kendrick Press.

Barbin, Évelyne, 2019a, "Monge's Descriptive Geometry: His Lessons and the Teachings Given by Lacroix and Hachette", in: Barbin, Évelyne / Menghini, Marta / Volkert, Klaus (eds.), *Descriptive Geometry, The Spread of a Polytechnic Art. International Studies in the History of Mathematics and its Teaching*. Cham: Springer, p. 3–18.

Barbin, Évelyne, 2019b, "Descriptive Geometry in France: Circulation, Transformation, Recognition (1795–1905)", in: Barbin, Évelyne / Menghini, Marta / Volkert, Klaus (eds.), *Descriptive Geometry, The Spread of a Polytechnic Art. International Studies in the History of Mathematics and its Teaching*. Cham: Springer, p. 19–38.

Bari Kolata, Gina, 1978, "Anti-Semitism alleged in Soviet mathematics", *Science* 202, p. 1167–1170.

Barth, Wolf / Peters, Chris A. M. / Van de Ven, Antonius, 1984, *Compact complex surfaces*. Berlin and Heidelberg: Springer.

Barthel, Gottfried / Hirzebruch, Friedrich / Höfer, Thomas, 1987, *Geradenkonfigurationen und algebraische Flächen*, Braunschweig / Wiesbaden: Vieweg.

Belhoste, Bruno, 2003, *La formation d'une technocratie. L'École polytechnique et ses élèves de la Révolution au Second Empire*, Paris: Belin.

Bélanger, Mathieu, 2010. *Grothendieck et les topos: rupture et continuité dans les modes d'analyse du concept d'espace topologique*. PhD thesis, Université de Montréal.

Benis-Sinaceur, Hourya. 1987, "Structure et Concept Dans l'épistémologie Mathématique de Jean Cavaillès." *Revue d'histoire Des Sciences*, 40(1), p. 5–30.

Berglund, Krista, 2012, *The Vexing Case of Igor Shafarevich, a Russian Political Thinker*, Basel: Birkhäuser.

Bobillier, Étienne, 1827-1828a, "Démonstration de quelques théorèmes sur les lignes et surfaces algébriques de tous les orders", *Annales de mathématiques pures et appliquées* 18, p. 89–98.

Bobillier, Étienne, 1827-1828b, "Recherche sur les lois générales qui régissent les lignes et surfaces algébriques", *Annales de mathématiques pures et appliquées* 18, p. 253–269.

Bobillier, Étienne, 1828-1829a, "Recherches sur les lois générales qui régissent les surfaces algébriques", *Annales de mathématiques pures et appliquées* 19, p. 138–150.

Bobillier, Étienne, 1828-1829b, "Géométrie de situation. Théorèmes sur les polaires successives", *Annales de Mathématiques pures et appliquées* 19, p. 302–307.

Boedecker, C., 1882, "Die Krümmung der Schienen beim Langschwellen-Oberbau in Hauptbahn-Strecken", *Centralblatt der Bauverwaltung* 29, p. 264-266.

Borenstein, Eliot, 2019, *Plots against Russia. Conspiracy and Fantasy after Socialism,* Ithaca / London: Cornell University Press.

Bottazzini, Umberto / Conte, Alberto / Gario, Paola, 1996, *Riposte Armonie: Lettere di Federigo Enriques a Guido Castelnuovo*, Torino: Bollati Boringhieri.

Bottazzini, Umberto / Gray, Jeremy, 2013, *Hidden Harmony – Geometric Fantasies. The Rise of Complex Function Theory*, New York: Springer.

Bourbaki, Nicholas, 1950, "The Architecture of Mathematics", *The American Mathematical Monthly* 57(4), p. 221-232.

Bouriau, Christophe / Bravermann, Charles / Mertens, Aude (eds.), 2016, *Kant et ses grands lecteurs. L'intuition en question*, Nancy: Presses Universitaires de Nancy.

Brasselet, Jean-Paul, 1998, "From Chern classes to Milnor classes. A history of characteristic classes for singular varieties", in: Brasselet, Jean-Paul / Suwa, Tatsuo (eds.), *Singularities – Sapporo 1998*, Tokyo: Mathematical Society of Japan, p. 31-52.

Brechenmacher, Frédéric, 2006, "Les matrices: formes de representation et pratiques operatoires (1850-1930)", p. 1-65, available at: https://hal.archives-ouvertes.fr/hal-00637378/fr/ (accessed on 24 November 2021).

Brechenmacher, Frédéric, 2007, "La controverse de 1874 entre Camille Jordan et Leopold Kronecker", *Revue d'histoire des mathématiques* 13, p. 187–257.

Brechenmacher, Frédéric, 2022, "Knowing by Drawing: Geometric Material Models in nineteenth Century France", in: Friedman, Michael / Krauthausen, Karin (eds.), *Model and Mathematics*, Cham: Birkhäuser p. 53–143.

Brechenmacher, Frédéric / Jouve, Guillaume / Mazliak, Laurent / Tazzioli, Rossana (eds.), 2016, *Images of Italian mathematics in France. The Latin Sisters, from Risorgimento to Fascism.* Birkhäuser, Basel.

Brigaglia, Aldo / Ciliberto, Ciro, 1995, *Italian Algebraic Geometry between the Two World Wars*, Ontario: Queen's Papers.

Brigaglia, Aldo, 2001, "The creation and the persistence of national schools: The case of Italian algebraic geometry", in: Bottazzini, Umberto/ Delmedico, Amy (eds.), *Changing Images in Mathematics*, Lonodn: Routledge, p. 187–206.

Brill, Alexander, 1887, "Über die Modellsammlung des mathematischen Seminars der Universität Tübingen" (Vortrag vom 7. November 1886), *Mathematisch-Naturwissenschaftliche Mitteilungen* 2, p. 69–80.

Brill, Ludwig, 1888, *Katalog mathematischer Modelle für den höheren mathematischen Unterricht*, 4th ed., Darmstadt: Brill.

Bulletin of the Atomic Scientists, "Evgeny Levich - An Appeal", December 1973.

Bussotti, Paolo / Pisano, Raffaele, 2015, "The Geometrical Foundation of Federigo Enriques' Gnoseology and Epistemology", *Advances in Historical Studies* 4, p. 118–145.

Bukhshtaber, Viktor M. / Novikov, Sergeï P., 1995, "Boris Gershevich Moishezon (obituary)", *Russian Mathematical Surveys* 50(3), p. 613-614.

Capelo, Antonio Candido / Ferrari, Mario, 1982, "La 'cuffia' di Beltrami: storia e descrizione", *Bollettino di Storia delle Scienze Matematiche* 2(2), p. 233–247.

Capristo, Annalisa, 2005, "The Exclusion of Jews from Italian Academies", in: Zimmerman, Joshua D. (ed.), *Jews in Italy under Fascist and Nazi Rule, 1922–1945*, Cambridge: Cambridge University Press, p. 81–95.

Casnati, Gianfranco et al. (eds.), 2016, *From Classical to Modern Algebraic Geometry. Corrado Segre's Mastership and Legacy*, Cham: Birkhäuser.

Cassou-Noguès, Pierre, 2001, *De l'expérience mathématique: essai sur la philosophie des sciences de Jean Cavaillès*, Paris: Vrin.

Castelnuovo, Guido, 1896, "Sulle superficie di genere zero", *Memorie di Matematica e Fisica della Società italiana delle Scienze, detta dei XL*, (3) 10, p. 103–123.

Castelnuovo, Guido / Enriques, Federigo, 1900, "Sulle condizioni di razionalità dei piani doppi", *Rendiconti del Circolo Matematico di Palermo* XIV, p. 290-302.

Castelnuovo, Guido / Enriques, Federigo, 1896, "Sur quelques récents résultats dans la théorie des surfaces Algébriques", *Mathematische Annalen* 48, p. 241–316.

Castelnuovo, Guido / Enriques, Federigo, 1908, "Grundeigenschaften der algebraischen Flächen", *Encyklopädie der mathematischen Wissenschaften*, vol. III, Leipzig: Teubner, p. 635–673.

Castelnuovo, Guido / Enriques, Federigo, 1914, "Die algebraischen Flächen vom Gesichtspunkte der birationalen Transformationen aus", in: *Encyklopädie der mathematischen Wissenschaften*, vol. III, Leipzig: Teubner, 674–768.

Castelnuovo, Guido, 1928, "La geometria algebrica e la scuola italiana", *Atti del Congresso Internazionale dei Matematici*, vol. I, Bologna: Zanichelli, p. 191–201.

Catanese, Fabrizio, 1986, "On a problem of Chisini", *Duke Mathematical Journal* 53(1), p. 33–42.

Catanese, Fabrizio, 2008, "Differentiable and Deformation Type of Algebraic Surfaces, Real and Symplectic Structures", in: Catanese, Fabrizio / Tian, Gang (eds.), *Symplectic 4-Manifolds and Algebraic Surfaces*, Berlin/Heidelberg: Springer, p. 55–167.

Cavaillès, Jean, 1938, *Méthode axiomatique et formalisme*, Paris: Hermann.

Cavaillès, Jean / Lautman, Albert, 1946, "La pensée mathématique. Séance du 4 février 1939", *Bulletin de la Société Française de Philosophie* 40(1), p. 1–39. English translation at: https://www.urbanomic.com/document/mathematical-thought/ (accessed on: 4.7.2022)

Chen, Zhijie, 1987, "On the geography of surfaces. Simply Connected Minimal Surfaces with Positive Index", *Mathematische Annalen* 277, p. 141–164.

Chern, Shiing-Shen, 1946, "Characteristic classes of Hermitian Manifolds", *Annals of Mathematics, Second Series* 47 (1), p. 85–121.

Ciliberto, Ciro / Flamini, Flaminio, 2011, "On the branch curve of a general projection of a surface to a plane", *Transactions of the American Mathematical Society* 363–7, p. 3457–3471.

Ciliberto, Ciro / Gario, Paola, 2012, "Federigo Enriques. The First Years in Bologna", in: Coen, Salvatore (ed.), *Mathematicians in Bologna 1861-1960*, Basel: Springer, p. 105-142.

Cimorelli, Daria (ed.), 2014, *Objets Mathématiques*, Milano: Silvana Editoriale.

Chisini, Oscar, 1917, "Sulla riducibilità della equazione tangenziale di una superficie dotata di curva doppia", *Rendiconti della R. Accaemia Nazionale dei Lincei* (5) 26, p. 543-548.

Chisini, Oscar, 1933, "Una suggestiva rappresentazione reale per le curve algebriche piane", *Rendiconti del R. Istituto Lombardo Accademia di Scienze e Lettere* (2) 66, p. 1141–1155.

Chisini, Oscar, 1934, "Un teorema di esistenza dei piani multipli", *Rendiconti Accad. Lincei* 19, p. 688-693, 766-772.

Chisini, Oscar, 1937, "Forme canoniche per il fascio caratteristico rappresentativo di una curva algebrica piana", *Rendiconti del R. Istituto Lombardo Accademia di Scienze e Lettere* 2(70), p. 49–61.

Chisini, Oscar, 1938, "Un più generale teorema di esistenza dei piani multipli", *Rendiconti Accad. Lincei* 27, p. 535–537.

Chisini, Oscar, 1944, "Sulla identita birazionale delle funzioni algebriche di due variabili dotate di una medesima curva di diramazione", *Rendiconti del R. Istituto Lombardo Accademia di Scienze e Lettere* 77, p. 339–356.

Chisini, Oscar, 1947a, "Accanto a Federigo Enriques", *Periodico di matematiche* (4) 25, p. 117–123.

Chisini, Oscar, 1947b, "Sulla identità birazionale delle funzioni algebriche di più variabili dotate di una medesima varietà di diramazione", *Istituto Lombardo (Rend. Sc.)* 80, p. 3–6.

Chisini, Oscar, 1952a, "Sulla costruzione a priori delle trecce caratteristiche", *Annali di Matematica Pura ed Applicata* (33) 1, p. 353–366.

Chisini, Oscar, 1952b, "Courbes de diramation des plans multiples et tresses algébriques", *Colloque de géométrie algébrique*, Paris: Masson, p. 11–27.

Chisini, Oscar, 1954–1955, "Il teorema di esistenza delle trecce algebriche", *Rendiconti della R. Accaemia Nazionale dei Lincei*, Series 8, vol. 17, p. 143–149, 307–311; series 8, vol. 18, p. 8–13.

Columbia University Record Archives, 1993, "Boris Moishezon, Mathematician, 55", in: Columbia University Record 19(1), September 3, 1993, online at: http://www.columbia.edu/cu/record/archives/vol19/vol19_iss1/record191.38 (accessed on: 4.1.2020)

Conte, Alberto / Ciliberto, Ciro, 2004, "La seconda rivoluzione scientifica: matematica e logica. La scuola di geometria algebrica italiana", online at: http://www.treccani.it/enciclopedia/la-seconda-rivoluzione-scientifica-matematica-e-logica-la-scuola-di-geometria-algebrica-italiana_%28Storia-della-Scienza%29/, (accessed on: 31.12.2019).

Corry, Leo, 1989, "Linearity and Reflexivity in the Growth of Mathematical Knowledge", *Science in Context, 3*(2), 409-440.

Corry, Leo, 2004, *Modern Algebra and the Rise of Mathematical Structures*, Basel: Birkhäuser.

Clebsch, Alfred, 1870, "Ueber den Zusammenhang einer Klasse von Flächenabbildungen mit der Zweiteilung der Abelschen Funktionen", *Mathematische Annalen* 3, p. 45–75.

Clebsch, Alfred, 1872, "Zur Theorie der Riemannschen Flächen", *Mathematische Annalen* 6, p. 216–230.

de la Gournerie, Jules, 1855, *Discours sur l'art du trait et la géométrie descriptive*, Paris: Malet-Bachelier.

De Cruz, Helen / De Smedt, Johan, 2013, "Mathematical symbols as epistemic actions", *Synthese* 190, p. 3–19.

De Toffoli, Silvia, 2021, "Groundwork for a Fallibilist Account of Mathematics", *Philosophical Quarterly* 7 (4), p. 823-844.

Dedekind, Richard / Weber, Heinrich, 1882, "Theorie der algebraischen Funktion einer Veränderlichen", *Journal für die reine und angewandte Mathematik* 92, p. 181-290.

Dedò, Modesto, 1950, "*Algebra delle trecce caratteristiche:* Relazioni fondamentali e loro applicazioni", *Rendiconti del R. Istituto Lombardo Accademia di Scienze e Lettere* 83, p. 227–258.

Degtyarev, Alex, 2012, *Topology of Algebraic Curves. An Approach via Dessins d'Enfants*, Berlin/Boston: De Gruyter.

Delcourt, Jean, 2011, "Analyse et géométrie, histoire des courbes gauches de Clairaut à Darboux", *Archive for the History of the Exact Sciences* 65, p. 229–293.

Deligne, Pierre, 1981, "Le groupe fondamental du complément d'une courbe plane n'ayant que des points doubles ordinaires est abélien", *Séminaire Bourbaki*, vol. 1979/80, exposés 543-560, Séminaire Bourbaki, no. 22, talk no. 543, p. 1-10.

Dhombres, Jean (ed.), 1992, *L'École normale de l'an III. Vol. 1, Leçons de mathématiques. Laplace - Lagrange – Monge*, Paris: Dunod.

Dhombres, Jean / Dhombres, Nicoles, 1989, *Naissance d'un nouveau pouvoir: sciences et savants en France, 1793–1824,* Paris: Payot.

Dieudonné, Jean, 1985, *History of Algebraic Geometry*, Monterey: Wadsworth.

Dolgachev, Igor, 2018, "Rostislavovich Shafarevich: in Memoriam", online at: https://arxiv.org/pdf/1801.00311 (accessed on: 4.12.2020).

d'Orgeval, Bernard, 1938, "Une construction des plans multiples représentatifs des surfaces algébriques de genres 1", *Comptes rendus de l'Académie des Sciences.* 206, p. 1866-1867.

d'Orgeval, Bernard, 1942, "Remarques sur la détermination des plans multiples représentant une surface algébrique", *Comptes rendus de l'Académie des Sciences.* 215, p. 341-342.

d'Orgeval, Bernard, 1945 [1943], *Sur les surfaces algébriques dont tous les genres sont 1*, Thèses de l'entre-deux-guerres no. 254, Paris: Gauthier-Villars.

d'Orgeval, Bernard, 1950, "Sur la dégénérescence des surfaces algébriques en systèmes de plans et la dégénérescence des courbes de diramation des plans multiples", *Bulletin de la Société Royale des Sciences de Liège* 19, p. 351-355.

d'Orgeval, Bernard, 1953, "Courbe de diramation de certains plans multiples", *Bulletin de la Société Royale des Sciences de Liège* 22, p. 188-194.

du Val, Patrick, 1979, "Beniamino Segre", *Bulletin of the London Mathematical Society* 11 (2), p. 215–235.

Durand, Yves, 1994, "Universitaires et universités dans les camps de prisonniers de guerre", *Les Facs Sous Vichy*, p. 169-188.

Dyck, Walter von, 1892, *Katalog mathematischer und mathematisch-physikalischer Modelle, Apparate und Instrumente*, Munich: Wolf & Sohn.

Eckes, Christophe, 2020, "Captivité et consécration scientifique. Reconsidérer la trajectoire académique du mathématicien prisonnier de guerre Jean Leray (1940-1947)", *Genèses* 4 (n° 121), p. 31-51.

Eckes, Christophe, 2021, "Recenser des articles mathématiques pour l'occupant: une étude sur les comportements de mathématiciens français sollicités par les autorités d'occupation allemandes", *Revue d'histoire des mathématiques* 27, p. 1-94.

Elkana, Yehuda, 1981, "A Programmatic Attempt at an Anthropology of Knowledge", in: Mendelsohn, Everett / Elkana, Yehuda (eds.), *Sciences and Cultures. Sociology of Sciences*, vol. 5, Dordrecht: Reider, p. 1-76.

Enriques, Federigo, 1893, *Ricerche di geometria sulle superficie algebriche*, Torino: Clausen.

Enriques, Federigo, 1894a, *Conferenze di Geometria*, Bologna: Litografia.

Enriques, Federigo, 1894b, "Sui fondamenti della geometria proiettiva", *Rendiconti dell'Istituto Lombardo di scienze, lettere ed arti*, s. 2 XXVII, p. 550–567.

Enriques, Federigo, 1896a, "Sui piani doppi di genere uno", *Memorie della Società Italiana di Scienze* (series III) X, p. 201-222.

Enriques, 1896b, "Introduzione alla geometria sopra le superficie algebriche", *Memorie della Società Italiana delle Scienze* 10, p. 1–81.

Enriques, Federigo, 1897a, "Le Superficie algebriche di genere lineare $p^{(1)} = 2$", *Rendiconti Accademia Nazionale dei Lincei* (s. 5), vol. VI, p. 139–144.

Enriques, Federigo, 1897b, "Sulle superficie algebriche di genere lineare $p^{(1)} = 3$", *Rendiconti Accademia Nazionale dei Lincei* (s. 5), vol. VI, p. 169-174.

Enriques, Federigo, 1900, *Questioni riguardanti la geometria elementare*. Bologna: Zanichelli.

Enriques, Federigo, 1908, "Sui moduli delle superficie algebriche", *Rendiconti della R. Accaemia Nazionale dei Lincei* (s. 5) 17, p. 690-694.

Enriques, Federigo, 1912a, "Sur le théorème d'existence pour les fonctions algébriques de deux variables indépendantes", *Comptes rendus des séances hebdomadaires de l'Académie des sciences* 154, p. 418–421.

Enriques, Federigo, 1912b, "Sui moduli d'una classe di superficie e sul teorema d'esistenza per funzioni algebriche di due variabili", *Atti della Accademia delle scienze di Torino* 47, p. 300–307.

Enriques, Federigo, 1914, "Sulla classificazione delle superficie algebriche e particolarmente sulle superficie di genere lineare $p^{(1)} = 1$ ", *Rendiconti della R. Accaemia Nazionale dei Lincei* (s. 5) 23, p. 206-214, 291-297.

Enriques, Federigo, 1915, *Teoria geometria delle equazioni e delle funzioni algebriche*, vol. 1, ed. By Chisini, Oscar, Bologna: Zanichelli.

Enriques, Federigo, 1922, *Per la storia della logica*, Bologna: Zanichelli.

Enriques, Federigo, 1923, "Sulla costruzione delle funzioni algebriche di due variabili possedenti una data curva di diramazione", *Annali di matematica pura ed applicata*, ser. 4, 1, p. 185–198.

Enriques, Federigo, 1938, *Le matematiche nella storia e nella cultura*. Bologna: Zanichelli.

Enriques, Federigo, 1949, *Le superficie algebriche* (edited by Castelnuovo, Guido), Bologna: Zanichelli, 1949.

Epple, Moritz, 1995, "Branch Points of Algebraic Functions and the Beginnings of Modern Knot Theory", *Historia Mathematica* 22, p. 371–401.

Epple, Moritz. 1997. "Styles of Argumentation in Late 19th Century. Geometry and the Structure of Mathematical Modernity", in: Otte, Michael / Panza, Marco (eds.), *Analysis and Synthesis in Mathematics: History and Philosophy*, Dordrecht: Kluwer, p. 177-198.

Epple, Moritz, 1999, *Die Entstehung der Knotentheorie: Kontexte und Konstruktionen einer modernen mathematischen Theorie*, Braunschweig / Wiesbaden: Vieweg.

Epple, Moritz, 2004, "Knot Invariants in Vienna and Princeton during the 1920s: Epistemic Configurations of Mathematical Research", *Science in Context* 17(1/2), p. 131–164.

Epple, Moritz, 2011, "Between Timelessness and Historiality: On the Dynamics of the Epistemic Objects of Mathematics", *Isis* 102(3), p. 481-493.

Epple, Moritz / Krauthausen, Karin, 2010, "Zur Notation topologischer Objekte: Interview mit Moritz Epple", in: Krauthausen, Karin / Nasim, Omar W. (eds.), *Notieren, Skizzieren: Schreiben und Zeichnen als Verfahren des Entwurfs*. Zürich: Diaphanes, p. 119–138.

Ehrhardt, Caroline, 2012, *Itinéraire d'un texte mathématique: les réélaborations d'un mémoire d'Evariste Galois au dix-neuvième siècle*, Paris: Hermann.

Feest, Uljana / Sturm, Thomas, 2011, "What (Good) is Historical Epistemology? Editors' Introduction", *Erkenntnis* 75, p. 285–302.

Feingold, Henry L., 2007, *Silent No More. Saving the Jews of Russia, the American Jewish Effort, 1967-1989*, Syracuse: Syracuse University Press.

Ferreirós, José, 2016, *Mathematical Knowledge and the Interplay of Practices*, Princeton: Princeton University Press.

Feustel, Jan-Michael / Holzapfel, Rolf-Peter, 1983, "Symmetry Points and Chern Invariants of Picard Modular Surfaces", *Mathematische Nachrichten* 111, p. 7-40.

Finzi, Roberto, 2005, "The Damage to Italian Culture: The Fate of Jewish University Professors in Fascist Italy and After, 1938–1946", in: Zimmerman, Joshua D. (ed.), *Jews in Italy under Fascist and Nazi Rule, 1922–1945*, Cambridge: Cambridge University Press, p. 96–113.

Flood, Raymond, 2006, "Mathematicians in Victorian Ireland", *Journal of the British Society for the History of Mathematics* 21(3), p. 200–211.

Fokkink, Robert, 2004, "A forgotten mathematician", *European Mathematical Society Newsletter* 52, p. 9-14.

Forsyth, Andrew Russell, 1893, *Theory of functions of a complex variable*, Cambridge: Cambridge University Press.

Freiman, Grigori, 1980, *It Seems I Am a Jew. A Samizdat Essay*, trans. Melvyn B. Nathanson, Carbondale/Edwardsville: Southern Illinois University Press.

Frenkel, Edward, 2013, *Love and Math*, New York: Perseus Books Group.

Friedman, Michael (Stanford University), 2000, "Geometry, construction, and intuition in Kant and his successors", In: Sher, Gila / Tieszen, Richard (eds.), *Between Logic and Intuition: Essays in Honor of Charles Parsons*, Cambridge: Cambridge University Press, p. 186-218.

Friedman, Michael (Stanford University), 2012, "Kant on Geometry and Spatial Intuition." *Synthese*, vol. 186 (1), p. 231–255.

Friedman, Michael / Leyenson, Maxim / Shustin, Eugenii, 2011, "On ramified covers of the projective plane I: Interpreting Segre's theory (with an appendix by Eugenii Shustin)", in: *International Journal of Mathematics* 22 (5), p. 619-653.

Friedman, Michael, 2018, *A History of Folding in Mathematics Mathematizing the Margins*, Basel: Birkhäuser.

Friedman, Michael, 2019a, "A plurality of (non)visualizations: Branch points and branch curves at the turn of the 19th century", *Revue d'histoire des mathématiques* 25, p. 109–194.

Friedman, Michael, 2019b, "How to notate a crossing of strings? On Modesto Dedò's notation of braids", *Archive for History of Exact Sciences,* doi:https://doi.org/10.1007/s00407-019-00238-8.

Friedman, Michael, 2019c, "Mathematical formalization and diagrammatic reasoning: the case study of the braid group between 1925 and 1950", *British Journal for the History of Mathematics* 34(1), p. 43–59.

Friedman, Michael, 2020, "On Mathematical Towers of Babel and "Translation" as an Epistemic Category", *The Mathematical Intelligencer*, doi:https://doi.org/10.1007/s00283-020-09969-x

Friedman, Michael, 2022, "On 'contour apparent', 'courbe de contact' and ramification curves: Duality between a principle and a tool", in: Krömer, Ralf / Haffner, Emmylou / Volkert, Klaus, *Duality as Archetype of Mathematical Thinking*. Basel: Birkhäuser (accepted for publication).

Friedman, Michael / Krauthausen, Karin (eds.), 2022, *Model and Mathematics*, Cham: Birkhäuser.

Friedman, Murray / Chernin, Albert D. (eds.), 1999, *A Second Exodus: The American Movement to Free Soviet Jews*, Hanover, NH: Brandeis University Press.

Friedman, Robert / Moishezon, Boris / Morgan, John, 1987, "On the $C^\infty$ invariance of the canonical classes of certain algebraic surfaces", *Bulletin of the American Mathematical Society* (N.S.) 17, p. 283-286.

Frye, Marilyn, 1983, "On Being White: Thinking toward a Feminist Understanding of Race and Race Supremacy," in: ibid., *The Politics of Reality: Essays in Feminist Theory*, Berkeley: Crossing Press, p. 110–127.

Fuchs, D. B., 2007, "On Soviet mathematics of the 1950s and 1960s", in: Zdravkovska, Smilka / Duren, Peter L. (eds.), *Golden Years of Moscow Mathematics*, Rhode Island: American Mathematical Soc., p. 213–222.

Fulton, William, 1980, "On the Fundamental Group of the Complement of a Node Curve", *Annals of Mathematics* 111(2), p. 407–409.

Fulton, William, 1987, "On the topology of Algebraic Varieties", *Proceedings of the AMS Bowdoin conference in Algebraic Geometry, 1985, Proceedings Symp. in Pure Mathematics* 46, Providence: American Mathematical Society, p. 15–46.

Galison, Peter, 1997, *Image and Logic—A Material Culture of Microphysics*. Chicago: University of Chicago Press.

Galison, Peter, 1999, "Trading Zone: Coordinating Action and Belief ", in: Biagioli, Mario (ed.), *The Science Studies Reader*, New York: Routledge, p. 137-160.

Gario, Paola, 2014. "The birth of a School", *Lettera Matematica International* 1, p. 173–176.

Gario, Paola, 2016, "Segre, Castelnuovo, Enriques: Missing Links", in: Casnati, Gianfranco et al. (ed.) *From Classical to Modern Algebraic Geometry: Corrado Segre's Mastership and Legacy*, p. 289–324, Cham: Birkhäuser.

Garside, Frank Arnold, 1965, *The theory of knots and associated problems*, PhD. Thesis, Oxford University, 1965.

Garside, Frank Arnold 1969, "The braid group and other groups", *The Quarterly Journal of Mathematics* 20(1), p. 235–254.

Gayme, Évelyne, 2015, "Les OFLAGS, centres intellectuels", *Inflexions* 29, no. 2, p. 125-132.

Giacardi, Livia M., 2015, "Models in mathematics teaching in Italy (1850–1950)". In: Bruter C (ed.), *Proceedings of second ESMA conference, mathematics and art III*. Paris: Cassini, p. 9–33.

Giacardi, Livia, 2012, "Federigo Enriques (1871-1946) and the training of mathematics teachers in Italy". In: Coen, Salvatore (ed.), *Mathematicians in Bologna 1861-1960*, Basel: Springer, p. 209-275.

Giardino, Valeria, 2018, "Tools for Thought: The Case of Mathematics", *Endeavour* 2 (42), p. 172–179.

Gilain, Christian, 1991, "Sur l'histoire du théorème fondamental de l'algèbre: théorie des équations et calcul integral", *Archive for History of Exact Sciences* 42, p. 91–136.

Gohberg, Israel, 1989, "Mathematical Tales", in: H. Dym, S. Goldberg, M. A. Kaashoek, P. Lancaster (eds.), *The Gohberg Anniversary Collection*, vol. I, Basel: Birkhäuser, p. 17–56.

Goldstein, Catherine, 1995, *Un théorème de Fermat et ses lecteurs*, Saint-Denis: Presses universitaires de Vincennes.

Goldstein, Catherine, 2010, "Des passés utiles: mathématiques, mathématiciens et histoires des mathématiques", *Noesis* 17, p. 135-152.

Goldstein, Catherine, 2018, "Long-term history and ephemeral configurations." In: Boyan, Sirakov / Ney De Souza, Paulo / Viana, Marcelo (eds.), *Proceedings Of The International Congress Of Mathematicians 2018*, Rio de Janeiro: World Scientific Publishing, p. 487-522.

González-Meneses, Juan, 2011, "Basic results on braid groups", *Annales Mathématiques Blaise Pascal* 18(1), p. 15-59.

Gow, Rod, 1997, "George Salmon 1819–1904: his mathematical work and influence", *Bulletin of the Irish Mathematical Society* 39, p. 26–76.

Granger, Gilles Gaston, 1996, "Jean Cavaillès et l'histoire", *Revue d'histoire des sciences* 49(4), p. 569-582.

Grauert, Hans / Remmert, Reinhold, 1958, "Komplexe Räume", *Mathematische Annalen* 136, p. 245-318.

Gray, Jeremy, 1994, "German and Italian algebraic geometry", *Supplementi di Rendiconti di circolo matematico di Palermo* 36 (2), p. 151-183.

Gray, Jeremy, 1997, "Algebraic Geometry between Noether and Noether — a forgotten chapter in the history of Algebraic Geometry", *Revue d'histoire des mathématiques* 3 (1), p. 1-48.

Gray, Jeremy, 1999, "The classification of algebraic surfaces by Castelnuovo and Enriques", *The Mathematical Intelligencer* 21 (1), p. 59–66.

Gray, Jeremy, 2008, *Plato's ghost: the modernist transformation of mathematics*, Princeton: Princeton University Press.

Gray, Jeremy, 2015a, "Grothendieck and the transformation of algebraic geometry", *Metascience* 24, p. 135–140.

Gray, Jeremy 2015b, *The Real and the Complex: A History of Analysis in the 19th Century*, Cham: Springer.

Grothendieck, Alexander, 1958, "La théorie des classes de Chern", *Bulletin de la Société Mathématique de France* 86, p. 137–154.

Grothendieck, Alexander, 1967, "Éléments de géométrie algébrique: IV. Étude locale des schémas et des morphismes de schémas, Quatrième partie", *Publications Mathématiques de l'IHÉS* 32, p. 5-361.

Grothendieck, Alexander, 1971, *Revêtements étales et groupe fondamental. Seminaire de Geometrie Algebrique du Bois Marie 1960/61 (SGA1)*, Lecture Notes in Math., vol. 224, Berlin: Springer.

Grothendieck, Alexander, 1985–1987, *Récoltes et Semailles*. Montpelier: Université des Sciences et Techniques du Languedoc.

Guerraggio, Angelo / Nastasi, Pietro, 2006, *Italian mathematics between the two world wars*, Basel: Birkhäuser.

Guerraggio, Angelo / Nastasi, Pietro, 2018, *Matematici da epurare. I matematici italiani tra fascismo e democrazia*, Milan: Centro PRISTEM, Università Bocconi, EGEA.

Hachette, Jean Nicolas Pierre, 1806, "Analyse appliquée à la géométrie. De la courbe de contact d'une surface conique avec une surface dont l'equation est du degre ", *Correspondence de l'École impériale polytechnique* 6, p. 188–191.

Haffner, Emmylou, 2014, *The "Science of Numbers" in action in Richard Dedekind's works: between mathematical explorations and foundational investigations*, Phd Thesis, Paris Diderot.

Halverscheid, Stefan / Labs, Oliver, 2019, "Felix Klein's Mathematical Heritage Seen Through 3D Models", In: Weigand, Hans-Georg / McCallum, William / Menghini, Marta / Neubrand, Michael / Schubring, Gert (eds.), *The Legacy of Felix Klein*. ICME-13 Monographs. Cham: Springer, p. 131–152.

Harbater, David, 1994, "Abhyankar's Conjecture on Galois Groups Over Curves", *Inventiones mathematicae* 117, p. 1-25.

Harbater, David / Obus, Andrew / Pries, Rachel / Stevenson, Katherine, 2018, "Abhyankar's conjectures in Galois theory: Current status and future directions", *Bulletin (new series) of the American mathematical society* 55(2), p. 239–287.

Hartmann, Uta, 2009, *Heinrich Behnke (1898-1979) Zwischen Mathematik und deren Didaktik*, Frankfurt am Main et al.: Peter Lang.

Hashagen, Ulf, 2003, *Walther von Dyck (1856-1934): Mathematik, Technik und Wissenschaftsorganisation an der TH München*, Franz Steiner Verlag.

Haubrichs dos Santos, Cléber, 2015, *Étienne Bobillier (1798-1840) : parcours mathématique, enseignant et professionnel*, Phd thesis, Université de Lorraine and l'Universidade federal do Rio de Janeiro.

Heegaard, Poul, 1898, *Forstudier til en topologisk Teori for de algebraiske Fladers Sammenhang*, Kopenhagen: Det Nordiske Forlag.

Heegaard, Poul, 1916, "Sur l''Analysis situs' ", *Bulletin de la S. M. F.* 44, p. 161–242.

Hirzebruch, Friedrich, 1958, "Automorphe Formen und der Satz von Riemann-Roch", *International Symposium on Algebraic Topology,* Mexico City: Univ. Nacional Autonoma de Mexico and Unesco, p. 129-144.

Hirzebruch, Friedrich, 1983, "Arrangements of Lines and Algebraic Surfaces", *Arithmetic and Geometry. Progress in Mathematics*, II (36). Birkhäuser: Boston, p. 113-140.

Hirzebruch, Friedrich, 1985, "Algebraic surfaces with extreme Chern numbers (Report on the thesis of Th. Höfer, Bonn 1984)", *Russian Mathematical Surveys* 40 (4), p. 135-145.

Hurwitz, Adolf, 1891, "Über Riemannsche Flächen mit gegebenen Verzweigungspunkten", *Mathematische Annalen* 39, p. 1–61.

Holzapfel, Rolf-Peter, 1980, "A class of minimal surfaces in the unknown region of surface geography", *Mathematische Nachrichten* 98, p. 211–232.

Israel, Giorgio, 2004, "Italian Mathematics, Fascism and Racial Policy", in: Emmer, Michele (ed.), *Mathematics and Culture I*, Berlin et al.: Springer, p. 21–48.

Jung, Heinrich W. E., 1951, *Einführung in die algebraische Theorie der Funktionen von zwei Veränderlicher*, Berlin: Akademie Verlag.

Kharlamov, Viatcheslav M. / Kulikov, Viktor, 2003, "On braid monodromy factorizations", *Izv. Math.* 67(3), p. 499–534.

Klein, Felix, 1872, *Vergleichende Betrachtungen über neuere geometrische Forschungen*, Erlangen: A. Duchert.

Klein, Felix, 1873, "Ueber Flächen dritter Ordnung (Dazu gehörig mehrere lithographirte Tafeln)", *Mathematisch Annalen* 6, p. 551–581.

Klein, Felix, 1882, *Über Riemann's Theorie der Algebraischen Functionen und ihrer Integrale.* Leipzig: Teubner.

Klein, Felix, 1908, *Elementarmathematik vom höheren Standpunkte aus. Erster Band.* Leipzig: Teubner.

Klein, Felix, 1911 [1893], *The Evanston Colloquium. Lectures on mathematics*, New York: American Mathematical Society.

Klein, Felix, 1926, *Vorlesungen über die Entwicklung der Mathematik im 19. Jahrhundert*, vol. 1, Berlin: Springer.

Kodaira, Kunihiko, 1960, "On Compact Complex Analytic Surfaces, I", *Annals of Mathematics* 71(1), p. 111–52.

Kosharovsky, Yuli, 2017, *We Are Jews Again: Jewish Activism in the Soviet Union*, Syracuse: Syracuse University Press.

Krämer, Sybille / Totzke, Rainer, 2012, "Einleitung: Was bedeutet Schriftbildlichkeit? ", in: Krämer, Sybille / Cancik-Kirschbaum, Eva / Totzke, Rainer (eds.), *Schriftbildlichkeit: Wahrnehmbarkeit, Materialität und Operativität von Notationen*, Berlin: Akademie Verlag, p. 13–35.

Krämer, Sybille, 2014, "Trace, Writing, Diagram: Reflections on Spatiality, Intuition, Graphical Practices and Thinking". In: Benedek, András / Nyíri, Kristóf (eds.), *The Power of the Image. Emotion, Expression, Explanation*, Frankfurt am Main: Peter Lang, p. 3–22.

Kronecker, Leopold, 1881, "Ueber die Discriminante algebraischer Functionen einer Variabeln", *Journal für die reine und angewandte Mathematik* 91, p. 301–334.

Krömer, Ralf, 2007. *Tool and Object. A History and Philosophy of Category Theory.* Basel et al.: Birkhäuser.

Krömer, Ralf, 2013, "The set of paths in a space and its algebraic structure. A historical account", *Annales de la Faculté des sciences de Toulouse: Mathématiques*, Serie 6, 22 (5), p. 915–968.

Kulikov, Viktor, 1999, "On Chisini's Conjecture", *Izv. RAN. Ser. Mat.* 63, no. 6, p. 83-116.

Kulikov, Viktor, 2008, "On Chisini's Conjecture II", *Izv. RAN. Ser. Mat.* 72, no. 5, p. 63-76.

Lang, Serge, 1998, *Challenges*, New York: Springer.

Lakatos, Imre, 1976, *Proofs and Refutations: The Logic of Mathematical Discovery*, Cambridge: Cambridge University Press.

Langins, Janis, 1987, *La République avait besoin de savants: les débuts de l'Ecole polytechnique, l'Ecole centrale des travaux publics et les cours révolutionnaires de l'an III*, Paris: Belin.

Lanteri, Antonio, 1979, "Su un teorema di Chisini", *Atti della Accademia Nazionale dei Lincei. Classe di Scienze Fisiche, Matematiche e Naturali. Rendiconti* 66.6, p. 523-532.

Lanteri, Antonio, 1987, "Review of Catanese 'On a problem of Chisini' (1986)", MR0835794 (87g:14013), *Mathematical Review of the American Mathematical Society.*

Lê, François, 2015, *Vingt-sept droites sur une surface cubique: rencontres entre groupes, équations et géométrie dans la deuxième moitié du XIXe siècle*, Thèse de doctorat, Paris: Université Pierre et Marie Curie.

Lê, François, 2020, " 'Are the genre and the Geschlecht one and the same number?' An inquiry into Alfred Clebsch's Geschlecht", *Historia Mathematica* 53, p. 71-107.

Levich, Yevgeny, 1976, "Soviet dissidents (4): Trying to keep in touch", *Nature* 263, p. 366–367.

Lévy-Bruhl, Paulette, 1956–1957a, "Plans multiples et tresses algébriques", *Séminaire Dubreil. Algèbre et théorie des nombres* 10, exp. no. 3, p. 1-18.

Lévy-Bruhl, Paulette, 1956–1957b, "Tresses algébriques: quelques applications", *Séminaire Dubreil. Algèbre et théorie des nombres* 10, exp. no. 4, p. 1–18.

Lefschetz, Solomon, 1924, *L'analysis situs et la géométrie algébrique*, Paris: Gauthier-Villars.

Lemmermeyer, Franz / Roquette, Peter (eds.), 2006, *Helmut Hasse und Emmy Noether. Die Korrespondenz 1925-1935*, Göttingen: Universitätsverlag Göttingen.

Lenoir, Timothy (ed.), 1998, *Inscribing Science: Scientific Texts and the Materiality of Communication.* Stanford: Stanford University Press.

Linguerri, Sandra, 2012, "Cesarina Tibiletti Marchionna (1920-2005)", in: Linguerri, Sandra (ed.), *Dizionario biografico delle scienziate italiane (secoli XVIII-XX)*, vol. 2, Bologna: Pendragon, p. 165–170.

Liberman, Eran / Teicher, Mina, 2005, "The Hurwitz Equivalence Problem is Undecidable", preprint, in: arXiv:math.LO/0511153 (accessed on: 27.1.20)

Libgober, Anatoly, 2011 "Development of the theory of Alexander invariants in algebraic geometry", *Topology of Algebraic Varieties and Singularities*, Contemporary mathematics 538, Providence, RI: American Mathematical Society, p. 3–17.

Libgober, Anatoly, 2014, "Review of Alex Degtyarev's book: *Topology of algebraic curves. An approach via dessins d'enfants*", *Bulltin of the American Mathematical Society* 51(3), p. 479–489.

Livne, Ron, 1975, *Rigidity of the centre of $B_3$*, MA Thesis, department of mathematics, Tel Aviv: Tel Aviv University.

Livne, Ron, 1977, "Appendix II: A Theorem about the modular group", in: Moishezon, Boris, *Complex Surfaces and Connected Sums of Complex Projective Planes*, Berlin et al.: Springer, p. 223–230.

Livne, Ron, 1981, *On Certain Covers of the Universal Elliptic Curve*, Phd Thesis, department of mathematics, Harvard University, Cambridge, MA.

Lorenat, Jemma, 2014, "Figures real, imagined and missing in Poncelet, Plücker, and Gergonne", *Historia Mathematica* 42(2), p. 155-192.

Lorenat, Jemma, 2015, *Die Freude an der Gestalt: Methods, Figures, and Practices in Early Nineteenth Century Geometry*. Dissertation, Université Pierre et Marie Curie (Paris, France) and Simon Fraser University (Vancouver, Canada).

Loria, Gino, 1908, "Perspektive und Darstellende Geometrie", In: *Vorlesungen über Geschichte der Mathematik*, ed. Cantor, Moritz, vol. 4, Leipzig: Teubner, 579–637.

Luciano, Erika / Roero, Clara Silvia, 2012, "From Turin to Göttingen: dialogues and correspondence (1879-1923)", *Bollettino di Storia delle Scienze Matematiche* 32, p. 7–232.

Luciano, Erika, 2018, "From Emancipation to Persecution: Aspects and Moments of the Jewish Mathematical Milieu in Turin (1848-1938)", *Bollettino Di Storia Delle Scienze Matematiche* XXXVII, p. 127-166.

Lüroth, Jacob, 1871, "Note über Verzweigungsschnitte und Querschnitte in einer Riemann'schen Fläche", *Mathematische Annalen* 4–2, p. 181–184.

Manara, Carlo Felice, 1968, "Ricordo di Oscar Chisini", *Periodico di matematiche* (4) 46, p. 1–20.

Manara, Carlo Felice, 1987, "Ricordo di Oscar Chisini", *Rendiconti del Seminario Matematico e Fisico di Milano* 57(1), p. 11–29.

Manara, Carlo Felice, 1993, "In ricordo di un amico: Modesto Dedò", *L'insegnamento della matematica e delle scienze integrate* 16 (5-6), p. 415–423.

Manara, Carlo Felice, 1995, "Ermanno Marchionna. Commemorazione", *Rendiconti dell'Istituto Lombardo. Accademia di Scienze e lettere* V, n. 129, p. 167–172.

Mattheis, Martin, 2019, "Aspects of 'Anschauung' in the Work of Felix Klein". In: Weigand, Hans-Georg / McCallum, William / Menghini, Marta / Neubrand, Michael / Schubring, Gert (eds.), *The Legacy of Felix Klein*. ICME-13 Monographs. Springer, Cham, p. 93–106.

Marchionna, Ermanno, 1950, "Condizioni caratteristiche perchè una curva sia di diramazione per un piano multiplo", *Rendiconti Istituto Lombardo* 83, p. 655-664.

Marchionna, Ermanno, 1951, "Una nuova caratterizzazione delle curve di diramazione dei piani multipli", *Rendiconti Accademia Nazionale dei Lincei*, serie 8, 11, p. 170-177.

McLarty, Colin, 2007, "The Rising Sea: Grothendieck on Simplicity and Generality", in: Gray, Jeremy / Parshall, Karen (eds.), *Episodes in the History of Modern Algebra (1800–1950)*, Providence, RI: American Mathematical Society, p. 301–325.

McLarty, Colin, 2016, "How Grothendieck Simplified Algebraic Geometry", *Notices of the American Mathematical Society* 63(3), p. 256-265.

Mehrtens, Herbert, 1989, "Mathematics in the Third Reich: Resistance, Adaptation and collaboration of a Scientific Discipline", in: Visser, Robert Paul Willem et al. (eds.), *New Trends in the History of Science: Proceedings of a Conference Held at the Universtiy of Utrecht*, Amsterdam: Rodopi, p. 151-166.

Mehrtens, Herbert, 1990, *Moderne Sprache Mathematik. Eine Geschichte des Streits um Grundlagen der Disziplin und des Subjekts formaler Systeme*. Frankfurt am Main: Suhrkamp.

Mehrtens, Herbert, 1994, "Irresponsible purity: the political and moral structure of mathematical sciences in the National Socialist State", in: Renneberg, Monika / Walker, Mark (eds.), *Scientists, Engineers, and National Socialism*, Cambridge: Cambridge University Press, p. 324-338.

Mehrtens, Herbert, 1996, "Mathematics and war: Germany, 1900–1945", *Boston Studies in the Philosophy of Science* 180, p. 87-134.

Mehrtens, Herbert, 2004, "Mathematical Models", In: de Chadarevian, Soraya / Hopwood, Nick (eds.), *Models: The Third Dimension of Science*. Stanford: Stanford University Press, p. 276–306.

Miller, Haynes, 2000, "Leray in Oflag XVIIA: The origins of sheaf theory, sheaf cohomology, and spectral sequences", *Gazette des Mathematicens* 84, p. 17-34.

Milman, Vitali D., 2006, "Observations on the Movement of People and Ideas in Twentieth-Century Mathematics", trans.: R. Cooke, in: A. A. Bolibruch, Yu. S. Osipov, Ya. G. Sinai (eds.), *Mathematical Events of the Twentieth Century*, Berlin: Springer, p. 215–242.

Milnor, John W. / Stasheff, James D., 1974, *Characteristic Classes*, Princeton: Princeton university Press.

Miyaoka, Yoichi, 1983, "Algebraic surfaces with positive indices", *Classification of algebraic and analytic manifolds. Katata, 1982*, Boston: Birkhäuser, p. 281-301.

Moishezon, Boris, 1964, "A projectivity criterion of complete algebraic abstract varieties", *Izvestiya Akademii Nauk SSSR. Seriya Matematicheskaya* 28, p. 179–224.

Moishezon, Boris, 1966, "On *n*-dimensional compact varieties with n algebraically independent meromorphic functions, I, II and III", *Izvestiya Akademii Nauk SSSR. Seriya Matematicheskaya* 30, p. 133–174, 345–386, 621–656.

Moishezon, Boris, 1977, *Complex Surfaces and Connected Sums of Complex Projective Planes*, Berlin et al.: Springer.

Moishezon, Boris, 1981, "Stable branch curves and braid monodromy", *Lecture Notes in Mathematics* 862, Berlin: Springer, p. 107-192.

Moishezon, Boris, 1983, "Algebraic surfaces and the Arithmetic of braids I", *Progress in Mathematics* 36, Boston: Birkhäuser, p. 199-269.

Moishezon, Boris, 1985, "Algebraic surfaces and the Arithmetic of braids II", *Contemporary Mathematics* 44, p. 311-344.

Moishezon, Boris, 1990, *Сотворение мифа: заметки об идентификации "Юровского*, Jerusalem: Express.

Moishezon, Boris, 1992, "Opinion I. R. Shafarevich's Essay 'Russophobia' ", *The Mathematical Intelligencer* 14(1), p. 61–62.

Moishezon, Boris, 1993, "On cuspidal branch curves", *Journal of Algebraic Geometry 2*, p. 309-384.

Moishezon, Boris, 1994, "The arithmetic of braids and a statement of Chisini", *Contemporary Mathematics* 164, p. 151-175.

Moishezon, Boris, 2001, *Armenoids in Prehistory*, Lanham et al.: University Press of America.

Moishezon, Boris / N.N., 1972, "Personal Interview with Dr. Boris Moishezon", *Scientists Committee of the Israel Public Council for Soviet Jewry* 6, p. 2–3.

Moishezon, Boris / Teicher, Mina, 1987, "Simply-connected algebraic surfaces of positive index", *Inventiones Mathematicae* 89, p. 601–643.

Moishezon, Boris / Teicher, Mina, 1988, "Braid group technique in complex geometry I", *Contemporary Mathematics* 78, p. 425–555.

Moishezon, Boris / Teicher, Mina, 1990, "Braid group technique in complex geometry II: From arrangements of lines and conics to cuspidal curves", *Algebraic Geometry, Lecture Notes in Math.* 1479, p. 131-180.

Moishezon, Boris / Teicher, Mina, 1994, "Braid group technique in complex geometry IV", *Contemporary Mathematics* 162, p. 333–358.

Moishezon, Boris / Teicher, Mina, 1996, "Fundamental groups of complements of branch curves as solvable groups", *Israel Mathematical Conference Proceedings* 9, p. 329–346.

Monge, Gaspard, 1780 [1775], "Mémoire sur les propriétés de plusieurs genres de surfaces courbes et particulièrement sur celles des surfaces développables avec une application à la théorie générale des ombres et des pénombres", *Memoires de mathematique et de physique* 9, p. 382-440.

Monge, Gaspard, 1781, "Mémoire sur la théorie des déblais et de remblais", *Histoire de l'Academie Royale des Sciences*, Paris: Imprimerie Royale, p. 666–704.

Monge, Gaspard, 1801, *Feuilles d'analyse appliquée à la géométrie à l'usage de l'Ecole Polytechnique*, Paris.

Monge, Gaspard, 1807, *Application de l'analyse à la géométrie*, Paris: Librarie de I'Ecole Imperiale Polytechnique.

Monge, Gaspard, 1838, *Géométrie descriptive*, 6th ed., with appendices by M. Brisson, Paris: Bachelier.

Monge, Gaspard, 1847 [1785], "Des ombres", in: Olivier, Théodore (ed.), *Applications de la géométrie descriptive*, Paris: Carilian-Goeury et Dalmont, p. 26–35.

Mumford, David, 2009, "Zariski's Papers on the Foundations of Algebraic Geometry and on Linear Systems", in: Parikh, Carol, *The Unreal Life of Oscar Zariski*, Boston: Springer, p. 151–158.

Murre, Jacob P., 2014, "On Grothendieck's work on the fundamental group", in: Schneps, Leila (ed.): *Alexandre Grothendieck: A mathematical portrait*. Somerville, MA: International Press, p. 143-168.

Nastasi, Tina, 2010, *Federigo Enriques e la civetta di Atena*. Pisa: Plus.

Noether, Max, 1878, "Über die ein-zweideutigen Ebenentransformationen", *Sitzungsber der physikalisch-medizinischen. Societät zu Erlangen* 10, p. 81–86.

Noether, Max, 1888, "Anzahl der Moduln einer Klasse algebraischer Flächen", *Sitzungsber der Königlich Preussischen Akademie der Wissenschaften* 1888, p. 123–127.

Oka, Kiyoshi, 1951, "Sur les Fonctions Analytiques de Plusieurs Variables, VIII – Lemme Fondamental", *Journal of the Mathematical Society Japan* 3(1), p. 204–214.

Orevkov, Stepan Yu., 1998, "Realizability of a braid monodromy by an algebraic function in a disk", *Comptes rendus de l'Académie des Sciences. Paris Series I Mathematics* 326 (7), p. 867–871.

Palladino, Nicla / Palladino, Franco, 2009, "I modelli matematici costruiti per l'insegnamento delle matematiche superiori pure e applicate", *Ratio Mathematica* 19, p. 31–88.

Parikh, Carol, 2009, *The Unreal Life of Oscar Zariski*, Boston: Springer.

Parshall, Karen Hunger, 2022, *The New Era in American Mathematics, 1920–1950*, Princeton: Princeton University Press.

Parshin, Aleksei N., 2006, "Numbers as Functions: The Development of an Idea in the Moscow School of Algebraic Geometry"; trans. R. Cooke, in: A. A. Bolibruch, Yu. S. Osipov, Ya. G. Sinai (eds.), *Mathematical Events of the Twentieth Century*, Berlin: Springer, p. 297–330.

Peretz, Pauline, 2015. *Let My People Go. The Transnational Politics of Soviet Jewish Emigration During the Cold War*, New Brunswick, NJ: Transaction.

Persson, Ulf, 1981, "Chern invariants of surfaces of general type", *Compositio Mathematica* 43(1), p. 3-58.

Piatetski-Shapiro, Ilya, 1993, "Étude on life and automorphic forms in the Soviet Union", in: Zdravkovska, Smilka / Duren, Peter L. (eds.), *Golden years of Moscow mathematics*, Providence, RI: American Mathematical Society, p. 199-211.

Pinkus, Benjamin, 1989. *The Jews of the Soviet Union: The History of a National Minority*, Cambridge: Cambridge University Press.

Poincaré, Henri 1952 [1908], *Science and Method*, Dover, New York.

Poncelet, Jean-Victor, 1817-18, "Questions résolues. Solution du dernier des deux problèmes de géométrie proposés à la page 36 de ce volume; suivie d'une théorie des pôlaires réciproques, et de réflexions sur l'élimination", *Annales de mathématiques pures et appliquées* 8, p. 201–232.

Poncelet, Jean Victor, 1822, *Traité des propriétés projectives des figures. Ouvrage utile a ceux qui s'occupent des applications de la géométrie descriptive et d'opérations géométriques sur le terrain*. Paris: Bachelier.

Poncelet, Jean-Victor, 1828, "Sur la dualité de situation et sur la théorie des polaires réciproques, 2$^e$ article en réponse aux observations de M. Gergonne", *Bulletin des Sciences Mathématiques* 9, p. 292–302.

Pontryagin, Lev Semenovich, 1979, "Soviet anti-Semitism: reply by Pontryagin", in: *Science* 205, p. 1083–1084.

Popescu-Pampu, Patrick, 2016, *What is the Genus?*, Basel: Springer.

Popp, Herbert, 1970, *Fundamentalgruppen algebraischer Mannigfaltigkeiten*. Lecture Notes in Mathematics 176. Berlin, Heidelberg and New York: Springer.

Proctor, Robert N., 2008, "Agnotology: A Missing Term to Describe the Cultural Production of Ignorance (and Its Study)", in: Proctor, Robert N. / Schiebinger, Londa (eds.), *Agnotology: The Making and Unmaking of Ignorance*, Stanford: Stanford University Press, p. 1–33.

Puiseux, Victor, 1850, "Recherches sur les fonctions algébriques", *Journal de mathématiques pures et appliquées* 15, p. 365–480.

Raynaud, Michel, 1994, "Revêtements de la droite affine en caractéristique $p > 0$ et conjecture d'Abhyankar", *Inventiones mathematicae* 116, p. 425-462.

Reid, Miles, 1977, "Bogomolov's theorem $c_1^2 \leq 4c_2$," *Proceedings international symposium on algebraic geometry*, p. 623–642.

Remmert, Reinhold, 1995, "Complex analysis in 'Sturm und Drang'", *The Mathematical Intelligencer* 17, p. 4–11.

Remmert, Reinhold, 1998, "From Riemann surfaces to complex spaces", *Matériaux pour l'histoire des mathématiques au XXe siècle, Séminaires et Congrés 3*, Paris: Société Mathématique de France, p. 203–241.

Remmert, Volker R., 2017, "Kooperation zwischen deutschen und italienischen Mathematikern in den 1930er und 1940er Jahren", in: Albrecht, Andrea / Danneberg, Lutz / De Angelis, Simone (eds.), *Die akademische ›Achse Berlin-Rom‹?*, Oldenburg: de Gruyter, p. 305-321.

Rheinberger, Hans-Jörg, 1997, *Toward a history of epistemic things: synthesizing proteins in the test tube*, Stanford: Stanford University Press.

Riemann, Bernhard, 1851, "Grundlagen für eine allgemeine Theorie der Functionen einer veränderlichen complexen Grösse", in: Dedekind, Richard / Weber, Heinrich (eds.), *Gesammelte mathematische Werke und wissenschaftlicher Nachlass*, Leipzig: Teubner, 1892, p. 3–45.

Riemann, Bernhard, 1857, "Theorie der Abel'schen Functionen", in: Dedekind, Richard / Weber, Heinrich (eds.), *Gesammelte mathematische Werke und wissenschaftlicher Nachlass*, Leipzig: Teubner, 1892, p. 88–142.

Rowe, David E., 2013, "Mathematical Models as Artefacts for Research: Felix Klein and the Case of Kummer Surfaces", *Mathematische Semesterberichte* 60(1), p. 1–24.

Rowe, David E., 2017, "On Building and Interpreting Models: Four Historical Case Studies", *The Mathematical Intelligencer* 39(2), p. 6–14.

Sakarovitch, Joël, 1998: *Épures d'architecture; de la coupe des pierres à la géométrie descriptive, XVIe-XIXe siècles*. Basel: Birkhäuser.

Salmon, George, 1847, "On the degree of a surface reciprocal to a given one", *Cambridge and Dublin Mathematical Journal* 2, p. 65–73.

Salmon, George, 1849, "On the cone circumscribing a surface of the $m^{th}$ degree", *Cambridge and Dublin Mathematical Journal* 4, p. 187–190.

Salmon, George, 1852, *A Treatise on the Higher Plane Curves*. Dublin: Hodges and Smith.

Salmon, George, 1862, *A Treatise on Analytic Geometry of Three Dimensions*, Dublin: Hodges and Smith.

Salmon, George, 1865, *Analytische Geometrie des Raums*, second volume, trans. Fiedler, Wilhelm, Leipzig: Teubner.

Salmon, George, 1873, *Höheren ebenen Kurven*, 2$^{nd}$ edition, trans. Fiedler, Wilhelm, Leipzig: Teubner.

Salmon, George, 1874, *Analytische Geometrie des Raumes*, second volume, 2$^{nd}$ edition, trans. Fiedler, Wilhelm, Leipzig: Teubner.

Sarfatti, Michele, 2006, *The Jews in Mussolini's Italy*, Madison: The University of Wisconsin Press.

Sattelmacher, Anja, 2014, "Zwischen Ästhetisierung und Historisierung: Die Sammlung geometrischer Modelle des Göttinger mathematischen Instituts", *Mathematische Semesterberichte* 61(2), p. 131–143.

Sattelmacher, Anja, 2021, *Anschauen, Anfassen, Auffassen. Eine Wissensgeschichte Mathematischer Modelle*, Wiesbaden: Springer.

Saxon, Wolfgang, 1993, "Boris G. Moishezon, Columbia Professor Of Math, Dies at 55", *The New York Times*, August 27, 1993, Section B, p. 7.

Saxon, Wolfgang, 2004, "Alexander Lerner, Cybernetics Expert, Is Dead at 90", *The New York Times*, 6 July 2004, (accessed on: 29.12.2021).

Schappacher, Norbert, 2008, "How to describe the transition towards new mathematical practice: the example of Algebraic Geometry 1937 – 1954", in: *Oberwolfach Reports* 5(2), p. 1301–1304.

Schappacher, Norbert, 2011, "Rewriting Points", In: Bhatia, Rajendra / Pal, Arup / Rangarajan, Gita / Srinivas, Vasudevan / Vanninathan, Muthusamy (eds.), *Proceedings of the International Congress of Mathematicians* 2010 (ICM 2010), vol. 4, New Delhi: Hindustan Book Agency, p. 3258–3291

Schappacher, Norbert, 2015, "Remarks about Intuition in Italian Algebraic Geometry", Oberwolfach Reports, 47, p. 2805–2808.

Schneps, Leila (ed.), 2014, *Alexandre Grothendieck: A mathematical portrait*. Somerville, MA: International Press.

Scharlau, Winfried, 2017, *Das Glück, Mathematiker zu sein*, Wiesbaden: Springer.

Schiebinger, Londa, 2008, "West Indian Abortifacients and the Making of Ignorance", in: Proctor, Robert N. / Schiebinger, Londa (eds.), *Agnotology: The Making and Unmaking of Ignorance*, Stanford: Stanford University Press, p. 149–162.

Scholz, Erhard, 1980, *Geschichte des Mannigfaltigkeitsbegriffs von Riemann bis Poincaré*, Boston: Birkhäuser.

Scholz, Erhard, 1999, "The Concept of Manifold, 1850–1950", in: Ioan, James Mackenzie (ed.), *History of Topology*, Amsterdam: North-Holland, p. 25–64.

Schubring, Gert, 2005, *Conflicts Between Generalization, Rigor, and Intuition: Number Concepts Underlying the Development of Analysis in 17th-19th Century France and Germany*. New York: Springer.

Seidl, Ernst / Loose, Frank / Bierende, Edgar (eds.), 2018, *Mathematik mit Modellen. Alexander von Brill und die Tübinger Modellsammlung*, Tübingen: Museum der Universität Tübingen.

Segre, Beniamino, 1929, "Esistenza e dimensioni di sistemi continui di curve piane algebriche con dati caratteri", *Rendiconti dell'accademia nazionale dei Lincei*, ser. 6, 10, p. 31–38.

Segre, Beniamino, 1930, "Sulla Caratterizzazione delle curve di diramazione dei piani multipli generali", *Memorie della Reale accademia d'Italia, Classe di scienze fisiche, matematiche e naturali* 4, p. 5–31.

Segre, Beniamino, 1987, *Opere scelte*, vol. 1, Bologna: Cremonese.

Segre, Corrado, 1891, "Su alcuni indirizzi nelle investigazioni geometriche. Osservazioni dirette ai miei studenti", *Rivista di Matematica* 1, p. 42-66.

Sernesi, Edoardo, 2012, "The Work of Beniamino Segre on Curves and Their Moduli", in: Coen, Salvatore (ed.), *Mathematicians in Bologna 1861–1960*, Basel: Springer, p. 439–450.

Serre, Jean-Pierre (1960), "Revêtements ramifiés du plan projectif", in: *Séminaire Bourbaki: années 1958/59 - 1959/60*, exposés 169-204, p. 483-489.

Severi, Francesco, 1921, *Vorlesungen über algebraische Geometrie*, Leipzig: Teubner.

Simili, Raffaella, 1989, *Federigo Enriques, Filosofo e Scienziato*, Bologna: Cappelli.

Sinaceur, Hourya, 1991, *Corps et modèles*, Paris: Vrin.

Shafarevich, Igor R., 1967, *Algebraic Surfaces*, Providence, Rhode Island: American Mathematical Society.

Shifman, Mikhail A. (ed.), 2005, *You failed your math test, Comrade Einstein*, Singapore: World Scientific Publishing.

Slembek, Silke, 2002, *Weiterentwicklung oder Umbruch? Zu Oscar Zariskis Arithmetisierung der algebraischen Geometrie*, Ph.D. thesis, University of Strasbourg and University of Mainz.

Stillwell, John, 2012, "Poincaré and the early history of 3-manifolds", *Bulletin of the American Mathematical Society* 49, p. 555-576.

Svetlikova, Ilona, 2013, *The Moscow Pythagoreans: Mathematics, Mysticism, and Anti- Semitism in Russian Symbolism*, New York: Palgrave.

Sylvester, James Joseph, 1861, "Note sur les 27 droites d'une surface du 3e degré", *Comptes rendus des séances de l'Académie des sciences* 52, p. 977–980.

Teicher, Mina, 1975, *Riemann-Roch theorem for algebraic varieties of dimension 3*, MA thesis in mathematics, Tel Aviv: Tel Aviv University.

Tel Aviv University Curriculum, 1973, *Curriculum for students of the faculty of exact sciences for the year 1973–74*, Tel Aviv: Tel Aviv University.

Tel Aviv University Curriculum, 1974, *Curriculum for students of the faculty of exact sciences for the year 1974–75*, Tel Aviv: Tel Aviv University.

Tel Aviv University Curriculum, 1975, *Curriculum for students of the faculty of exact sciences for the year 1975–76*, Tel Aviv: Tel Aviv University.

Tel Aviv University Curriculum, 1976, *Curriculum for students of the faculty of exact sciences for the year 1976–77*, Tel Aviv: Tel Aviv University.

Tel Aviv University Curriculum, 1977, *Curriculum for students of the faculty of exact sciences for the year 1977–78*, Tel Aviv: Tel Aviv University.

Terracini, Alessandro, 1968, *Ricordi di un matematico. Un sessantennio di vita universitaria*, Rome: Ed. Cremonese.

Tibiletti, Cesarina Marchionna, 1952a, "Costruzioni a priori della sestica con nove cuspidi", *Rendiconti del R. Istituto Lombardo Accademia di Scienze e Lettere* 85, p. 207–220.

Tibiletti, Cesarina Marchionna, 1952b, "Piani tripli e piani quadrupli con la stessa curva di diramazione", *Rendiconti Accademia Nazionale dei Lincei* (8) 12, p. 537–543.

Tibiletti, Cesarina Marchionna, 1955, "Trecce algebriche di curve di diramazione: costruzioni ed applicazioni", *Rendiconti del Seminario Matematico della Università di Padova* 24, p. 183–214.

Tibiletti, Cesarina Marchionna, 1999, "Uno sguardo su Matematica e Matematici nell 'Universita' degli Studi di Milano dal 1924 al 1974", *Rendiconti del Seminario Matematico e Fisico di Milan* 69(1), p. 193–236.

Tobies, Renate, 2021, *Felix Klein. Visions for Mathematics, Applications, and Education*, Birkhäuser: Cham.

Tournès, Dominique, 2012, "Diagrams in the theory of differential equations (eighteenth to nine-teenth centuries)", *Synthese* 186, p. 257–288.

Turri, Lucia, 2011, "History And Becoming Of Science In Jean Cavaillès", *Rivista Italiana di Filosofia Analitica Junior* 2(2), p. 60-79.

Tuana, Nancy, 2008, "Coming to Understand: Orgasm and the Epistemology of Ignorance", in: Proctor, Robert N. / Schiebinger, Londa (eds.), *Agnotology: The Making and Unmaking of Ignorance*, Stanford: Stanford University Press, p. 108–145.

Vakil, Ravi, 2017, *The Rising Sea: Foundations of Algebraic Geometry*, online at: http://math.stanford.edu/~vakil/216blog/FOAGnov1817public.pdf, (accessed on: 31.1.20).

Vallès, Francois, 1826, "Géométrie des surfaces courbes. Démonstration d'une propriété générale des lignes de contact des surfaces courbes avec les surfaces coniques circonscrites", *Annales de mathématiques pures et appliquées* 16, p. 315-322.

Van de Ven, Antonius, 1978, "Some recent results on surfaces of general type", *Séminaire Bourbaki* 19, vol. 1976/77, exposés 489-506, p. 155-166.

van Kampen, Egbert R., 1933, "On the fundamental group of an algebraic curve", *American Journal of Mathematics* 55, p. 255–260.

Vanden Eynde, Ria, 1999, "Development of the Concept of Homotopy", in: James, I. M. (ed.), *History of Topology*, p. 65-102.

Vershik, Anatoly, 1994, "Admission to the mathematics faculty in Russia in the 1970s and 1980s", *Mathematical Intelligencer* 16(4), p. 4–5.

Vesentini, Edoardo, 2005, "Beniamino Segre and Italian geometry", *Rendiconti di matematica,* ser. 7, 25, p. 185–193.

Volkert, Klaus, 1986, *Die Krise der Anschauung: eine Studie zu formalen und heuristischen Verfahren in der Mathematik seit 1850*, Göttingen:Vandenhoeck & Ruprecht.

Volkert, Klaus, 2018, "Mathematische Modelle und die polytechnische Tradition", *Siegener Beiträge zur Geschichte und Philosophie der Mathematik* 10, p. 161–202.

Volkert, Klaus, 2019, "Otto Wilhelm Fiedler and the Synthesis of Projective and Descriptive Geometry", in: Barbin, Évelyne / Menghini, Marta / Volkert, Klaus (eds.) *Descriptive Geometry, The Spread of a Polytechnic Art. International Studies in the History of Mathematics and its Teaching*. Cham: Springer, p. 167–180.

von Staudt, Georg Karl Christian, 1889, *La geometria di posizione*, trans. Pieri, Mario, Torino: Fratelli Bocca.

Voronel, Alexander, 1991, "Jewish Samizdat", in: Ro'i, Yaacov / Beker, Avi (eds.), *Jewish Culture and Identity in the Soviet Union*, New York: New York University Press, p. 255-261.

Yau, Shing Tung, 1977, "Calabi's conjecture and some new results in algebraic geometry", in: *Proceedings of the National Academy of Sciences of the United States of America, National Academy of Sciences* 74 (5), p. 1798–1799.

Wahl, Jonathan M., 1974, "Deformations of Plane Curves with Nodes and Cusps", *American Journal of Mathematics* 96(4), p. 529-577.

Weil, André, 1946, *Foundations of Algebraic Geometry*, New York: American Mathematical Society.

Winckel, L., 1874, "Ueber die Verschiebung der Axe derjenigen Tunnels welche in Curven liegen", *Zeitschrift für Vermessungswesen* 3, p. 125-128.

Zalamea, Fernando, 2012, *Synthetic Philosophy of Contemporary Mathematics*, trans. by Zachary Luke Fraser, New York: Sequence Press.

Zalamea, Fernando, 2019, *Grothendieck. Una Guía A La Obra Matemática Y Filósofica*. Edtorial Nomos.

Zalamea, Fernando, 2020, "Grothendieck: A Short Guide to His Mathematical and Philosophical Work (1949–1991)", in: Sriraman, Bharath (ed.), *Handbook of the History and Philosophy of Mathematical Practice*, Springer, Cham. https://doi.org/10.1007/978-3-030-19071-2_27-1

Zappa, Guido, 1942, "Caratterizzazione delle curve di diramazione delle rigate e spezzamento di queste in sistemi di piani", *Rendiconti del Seminario Matematico della Universià di Padova* 13, p. 41–56.

Zappa, Guido, 1943, "Sulla degenerazione delle superficie algebriche in sistemi di piani distinti, con applicazioni allo studio delle rigate", *Atti R. Accad. d'Italia, Mem. Cl. Sci. FF., MM. e NN* 13 (2), p. 989–1021.

Zappa, Guido, 1945, "Sul gruppo fondamentale delle curve di diramazione delle superficie algebriche suscettibili di spezzarsi in sistemi di piani", *Annali di Matematica Pura ed Applicata*, 24(1), p. 139–151.

Zappa, Guido, 1997, "Matematici al tempo del fascismo. Ricordi di un vecchio docente", *Bollettino dell'Unione Matematica Italiana* 2-A.1, p. 37-40.

Zariski, Oscar / Samuel, Pierre, 1958, *Commutative Algebra*, vol. 1, Princeton: D. van Nostrand.

Zariski, Oscar, 1928 "Sopra il teorema d'esistenza per le funzioni algebriche di due variabili", in: *Atti del Congresso Internazionale dei Matematici*, Bologna 3–10 Settembre 1928, vol. 4, Bologna: Zanichelli, p. 133–138.

Zariski, Oscar, 1929a, "On the Problem of Existence of Algebraic Functions of Two Variables Possessing a Given Branch Curve", *American Journal of Mathematics* 51, p. 305–328.

Zariski, Oscar, 1929b, "On the linear connection index of the algebraic surfaces $z^n = f(x, y)$", *Proceedings of the National Academy of Sciences* 15, p. 494-501.

Zariski, Oscar, 1931a, "On the irregularity of cyclic multiple planes", *Annals of Mathematics* (2) 32, p. 485-511.

Zariski, Oscar, 1931b, "On the Non-Existence of Curves of Order 8 with 16 Cusps", *American Journal of Mathematics* 53(2), p. 309-318.

Zariski, Oscar, 1932, "On the Topology of Algebroid Singularities", *American Journal of Mathematics* 54(3), p. 453-465.

Zariski, Oscar, 1935, *Algebraic Surfaces*, Berlin: Springer.

Zariski, Oscar, 1936, "On the Poincaré group of rational plane curves", *American Journal of Mathematics* 58, p. 607-619.

Zariski, Oscar, 1937, "On the topological discriminant group of a Riemann surface of genus $p$", *American Journal of Mathematics* 59, p. 335–358.

Zariski, Oscar, 1948, "Review: André Weil, Foundations of algebraic geometry", *Bulletin of the American Mathematical Society* 54 (7), p. 671–675.

Zariski, Oscar, 1950, "The fundamental ideas of abstract algebraic geometry", *Proceedings of the International Congress of Mathematicians* 2, p. 77–89.

Zariski, Oscar, 1958, "The purity of the branch locus of algebraic functions", *Proceedings of National Academy of Sciences*, U. S. A. 44, pp. 791–796.

Zariski, Oscar, 1969, *An Introduction to the Theory of Algebraic Surfaces*, Berlin/Heidelberg: Springer.

Zariski, Oscar, 1971, *Algebraic Surfaces. With appendices by S.S. Abhyankar, J. Lipman and D. Mumford,* 2nd ed., Berlin: Springer.

Zaslavsky, Victor / Brym, Robert, 1983, *Soviet-Jewish Emigration and Soviet Nationality Policy*, London and Basingstoke: The Macmillan Press.

Zimmerman, Joshua D. (ed.), 2005, *Jews in Italy under Fascist and Nazi Rule, 1922–1945*, Cambridge: Cambridge University Press.

Zdravkovska, Smilka / Shafarevich, Igor R., 1989, "Listening to Igor Rostislavovich Shafarevich", *The Mathematical Intelligencer* 11, p. 16–28.

# Index

Printed in the United States
by Baker & Taylor Publisher Services